T0136243

SUSTAINING WILDLANDS

EDITED BY
AARON J. POE
AND
RANDY GIMBLETT

SUSTAINING WILDLANDS

Integrating Science and Community in Prince William Sound

THE UNIVERSITY OF ARIZONA PRESS
TUCSON

Additional content related to this book can be found at www.sustainingwildlands.com.

The University of Arizona Press
www.uapress.arizona.edu

Printed in the United States of America
22 21 20 19 18 17 6 5 4 3 2 1

ISBN-13: 978-0-8165-3564-4 (cloth)

Cover design by Leigh McDonald
Cover art by Anchorage-based artist Terry Josey, composed as a vision of the ecological system of Prince William Sound

Publication of this volume was made possible in part by funding from the U.S. Forest Service and Pacific Northwest Research Station, and with a grant from the Provost's Author Support Fund at the University of Arizona.

Library of Congress Cataloging-in-Publication Data
Names: Poe, Aaron J., editor. | Gimblett, H. Randal (Howard Randal), editor.
Title: Sustaining wildlands : integrating science and community in Prince William Sound / edited by Aaron J. Poe and Randy Gimblett.
Description: Tucson : The University of Arizona Press, 2017. | Includes bibliographical references and index.
Identifiers: LCCN 2017002334 | ISBN 9780816535644 (cloth : alk. paper)
Subjects: LCSH: Ecosystem management—Alaska—Prince William Sound. | Sustainable tourism—Alaska—Prince William Sound. | Ecotourism—Alaska—Prince William Sound. | Nature—Effect of human beings on—Alaska—Prince William Sound.
Classification: LCC QH76.5.A4 S87 2017 | DDC 333.7209798/3—dc23 LC record available at https://lccn.loc.gov/2017002334

∞ This paper meets the requirements of ANSI/NISO Z39.48-1992 (Permanence of Paper).

We dedicate this book to the residents of the six key communities of Prince William Sound: Anchorage, Chenega Bay, Cordova, Tatitlek, Whittier, and Valdez. Your dependence on the region ranges from direct sustenance and economic gain to recreation, renewal, personal inspiration, and cultural vitality. You are an inextricable part of the Sound and have repeatedly demonstrated that to us through our work to understand human use of this landscape. It's our hope that in return, this volume helps show pathways by which managers and stakeholders can work together to develop the more rigorous and adaptive planning approaches needed to ensure a sustainable future for the Sound.

CONTENTS

ACKNOWLEDGMENTS

NUMEROUS INDIVIDUALS AND organizations have made substantial contributions to this volume. Specifically, we wish to thank the *Exxon Valdez* Oil Spill Federal Trustees for their generous support of this work as well as many of the research endeavors described herein. In the years since the 1989 *Exxon Valdez* oil spill, substantial efforts have been made by the trustees to promote the recovery of Prince William Sound. Their support of this volume as further progress in that direction is greatly appreciated. The Chugach National Forest provided significant contributions of in-kind support in terms of staff time as well as leadership support from former Forest Supervisor Joseph L. Meade, who helped launch this work. The University of Arizona School of Renewable Natural Resources and Environment similarly invested substantial staff time to develop what became a great collaboration between the university and the Chugach National Forest.

We are indebted to the 28 scientists and managers who invested their time to develop and submit individual chapters. Their perspectives are joined by those of 13 Prince William Sound community residents—several of whom began exploring the region years before the oil spill that brought so much research and management interest to the Sound. We greatly appreciate the willingness of all these individuals to stick with us through revisions and to provide suggestions that helped shape the direction of this volume.

Finally, we offer heartfelt thanks to our esteemed colleague Andromeda Romano-Lax, who provided invaluable assistance in copyediting and proofing the manuscript. She also helped us bring in the voices of Prince William Sound residents for the "Stakeholder Essay" pieces and integrate them into the larger volume. As a local Alaskan author of several fiction and nonfiction works, her voice helped soften some of the technical language that too often makes scientific books less accessible to broader audiences.

INTRODUCTION

LEARNING FROM THE SOUND

Introducing This Place and This Volume

AARON J. POE, RANDY GIMBLETT, AND DALE J. BLAHNA

WHEN THE *EXXON VALDEZ* oil tanker ran aground on Bligh Reef on March 24, 1989, spilling 11 million gallons of oil, it changed the future of Prince William Sound forever. This catastrophic event disrupted the region's biological system, directly killing countless individual birds, marine mammals, and fish, and poisoning habitats that to this day no longer support some of the species they did prior to the spill (EVOS Trustee Council 2010). Less well documented and poorly understood are the impacts that were felt by the region's human communities—impacts that have profoundly altered the way people use this region. Two decades later, changes in recreation use have run counter to what was expected in the months and early years following the spill.

People haven't avoided Prince William Sound as a legacy of this ecological disaster. In fact, the opposite has been true, particularly in the last two decades (Fay 2008). The terrible damage done to such a picturesque and seemingly pristine place captivated people from around Alaska, the United States, and the world. A region formerly known to few became a magnet for tourists and visitors (Twardock 2000), including some of the many thousands of folks who spent years attempting to clean oil from beaches following the spill. Subsequent efforts to boost the region's economic vitality resulted in increased access and the development of commercial enterprises focused on providing high quality nature tourism experiences in the area's remote bays and fjords (Colt et al. 2002). This increase in recreation caused many to worry that the very ability to experience wilderness or see plentiful wildlife in pristine settings may again be threatened—not by oil this time, but by the very humans seeking those wilderness experiences (Murphy et al. 2004). Managers and communities in the region wrestle with these concerns, recognizing that increased use allows for more diverse community economies and grows a bond between this precious place and visitors who can become its future advocates.

INTRODUCTION

An enduring paradox for the management of public lands is the need to accommodate increasing levels of human uses while reducing the environmental impacts of those uses. This is the crux of managing sustainable human uses of natural environments (Keough and Blahna 2006). Managers must continue to protect these areas, simultaneously maximizing their use in order to promote their relevance and connections to broader societal goals. In so doing they can show how wildlands are linked to robust economies and promote health and well-being for sustainable communities (Meffe et al. 2002; Gunderson and Holling 2002). This is vital when land and resource managers are increasingly asked to prove the value of a system of largely undeveloped public lands and the agencies entrusted with their stewardship.

But how do we manage to accommodate human uses and protect environmental values at the same time? Historically, land management agencies used standard practices and principles of outdoor recreation management like recreation carrying capacity, limits of acceptable change, recreation opportunity spectrum, and scenery management system (Hendee et al. 1990; Manning 2007). But most of these principles are dated and focus on traditional forms of recreation like hunting, fishing, camping, and hiking (Stankey 1999; Cerveny et al. 2011). Current demands on natural environments are more diverse and often conflicting, and agency goals for managing human uses are far more complicated. Today's land managers are juggling requirements for cultural and historical preservation, Native American treaty rights, gathering of nontimber and other forest products, spiritual and symbolic meanings of special places, subsistence and traditional harvesting activities, and a history of roads, trails, and facilities systems that are no longer economically sustainable (Williams et al. 1992). Large-scale ecosystem stressors like climate change, invasive species, and widespread habitat loss and fragmentation further complicate the picture for public land managers (Cole and Yung 2010).

This book provides a case study of an integrated approach to understanding and planning for human uses in a large, natural ecosystem: Prince William Sound in Southcentral Alaska. It provides an ideal case study because it is a relatively intact wildland setting with few managing agencies, communities, and private companies. Due to the grounding of the *Exxon Valdez* and the ensuing environmental and socioeconomic disaster, a multitude of studies were conducted to help track environmental changes, human travel and use patterns, and the status of "injured" resources in the

Sound. This volume presents summaries of many of these studies and shows how the results were used in an issue-based planning process to develop recommendations for managing human uses in the Sound in an integrated and sustainable way. While this has been an evolving and exploratory effort, we believe it meets many objectives of sustainable human use planning, and offers lessons for public land management agencies and stakeholders to use in future planning efforts.

A STARTING POINT: SUSTAINABLE RECREATION

A new initiative of the U.S. Forest Service called the Framework for Sustainable Recreation offers a unique policy solution to current recreation challenges by "providing desired benefits through the science of recreation management for present and future generations" (USDA Forest Service 2010). The goals of the strategy are to unite diverse interests, create and strengthen partnerships, focus scarce resources on mission-driven priorities, connect recreation benefits to communities, provide for changing urban populations, and most importantly, sustain and expand the benefits to America that quality recreation opportunities provide. Sustainable recreation planning recognizes that management should be driven by the landscape and its people—not administrative boundaries (internal or external)—and seeks to advance integrated and collaborative planning, and management through innovative approaches (USDA Forest Service 2010).

A critical requirement of the Sustainable Recreation Framework is the need to develop management plans at multiple scales of analysis. This is a major departure from traditional management that tends to focus on individual sites, trails, campgrounds, or other recreation settings. At the site level, for example, it may seem reasonable that restricting human use is the best way to protect environmental or social integrity. But looking at the landscape or regional implications of restricting use often results in the *opposite* conclusion; restricting access may exacerbate both physical and social impacts as use is displaced to more pristine or low use areas (Keough and Blahna 2006; Blahna 2007). Restricting users, or certain types of use, comes at a cost—not only in diversity of constituent support but also in unintended resource or social impacts on different parts of the managed landscape as use is shifted. It is vital to step back and take a broader view of both potential impacts and longer-term conservation benefits of allowing for sustainable use.

To accomplish sustainable human use planning, agencies and partners need rigorous characterizations of human use, including an understanding of spatial and temporal distribution of use, as well as a characterization of the motivations of users (McCool and Cole 2001). This type of information requires deliberate investments in social science investigations, which are often a lower priority for land managers and scientists alike (Endter-Wada et al. 1998; Sievanen et al. 2012). Furthermore, there is a critical need for investigations of potential harm to biological and cultural resources that go beyond simple assumptions like *use-equals-impact* (McCool and Stankey 2004). An understanding of the timing, location, and specific effects of human interactions with species or sites of concern, as well as the specific mechanisms of those effects, should be of vital importance to managers attempting to balance use with ecosystem conservation (Keough and Blahna 2006).

Sustainable recreation planning also emphasizes that renewed efforts to manage increasing human use must be combined with greater efforts to educate and truly engage the public (USDA Forest Service 2010). Rapid changes in the Sound and its human use dynamics since the oil spill have not been accompanied by equivalent efforts to engage the public to reflect upon the future of the region. An effort to understand this region, the future sustainability of its communities, and the needs of its many users must involve data and scientific inquiry as well as opportunities for sharing individual perspectives.

In this volume we explore human use in the context of an *ecosystem service*—or the benefit that humans receive from naturally functioning ecosystems (Millenium Ecosystem Assessment 2005). Challenges for this approach abound, including what can appear to be a focus on commodification, or counting users instead of focusing on the benefits they derive from wildland settings (e.g., Chan et al. 2012). A few of our chapters (1, 16, and 18), and certainly our stakeholder essays, aim to go beyond counting users and look into the motivations, sense of place, and values they derive from the Sound. Ultimately, when it comes to decisions like those that managers must make about allocating use, there is very much a need for a rigorous description of the distribution and intensity of use *as well as* an understanding of the values derived from that use.

This book illustrates one approach for pairing rigorous and innovative research and empirical data with local ecological knowledge and community perspectives shared through stakeholder essays. Its hybridized content exemplifies the complex, multifaceted, and often value-laden challenges faced by managers seeking to make decisions in an absence of information and with broad diversity in public perspectives. It highlights how aspects of these decisions may be *informed* by data, but many will nonetheless revolve around human impressions and perceptions difficult to influence even with the best research and communication efforts.

WHAT IS THIS PLACE WE CALL "THE SOUND"?

In chapter 1, Alaskan author Marybeth Holleman describes Prince William Sound as "a place of extremes and convergences—of weather, geography, wildlife, and experiences. One of the most active seismic regions in the world, it anchors the eastern edge of the Ring of Fire; it's the northernmost reach for the temperate rain forest; it's rest stop and destination for tens of thousands of migrating birds and marine mammals. Arctic terns fly in from South America; humpbacks swim from Baja California and Hawai'i. It's a biological and geographic center for the Pacific, where temperate and subarctic environmental conditions overlap. Like the center of a wheel, the Sound is the apex of Alaska's Pacific shores that curve from southeast Alaska to the Aleutian Islands. A 3,500-mile-long undulation of capes and fjords and islands and islets and sea stacks, Prince William Sound is encircled by three mountain ranges draped in ice fields from which more than 150 glaciers pour, dozens of them tidewater. It is an expansive, meandering maze."

In chapter 3, Paul Twardock, chair of the Outdoor Studies Department at Alaska Pacific University, explores the history of human use and recreation in the region. He concludes that there has been a long tradition of human use in the Sound, a place inhabited for some 5,000 years, with the label of wilderness applied only recently. During the late 1960s, recreationists seeking wilderness explored the Sound, leading to resurgence in human use during the 1970s and 1980s. The *Exxon Valdez* oil spill of 1989 had major short- and long-term impacts on tourism and recreation.

Human use and recreation tourism in the Sound have increased and changed over time—as, for example, cruise ships altered their routes and glacier tour operators added trips. Recreational boating, kayaking, and sport fishing have increased dramatically in the last decade and are expected to continue increasing (Bowker 2001). With continuing importance of commercial and subsistence fishing, there is growing concern among management agencies that increased competition and conflict may threaten the ability of the region to sustain increasing uses.

WHAT PROCESSES ARE AT WORK IN THE SOUND AND HOW DO THEY AFFECT HUMAN USE?

In chapter 1, Marybeth Holleman describes how water shapes Prince William Sound: "water sustains it, creates and contains it. Glaciers pouring to sea, pushing rock before them and carving U-shaped valleys from solid rock; streams rocketing down from the mountains to the sea, carving and carrying silt; rains stippling rock and nurturing a dense rain forest growing on the thinnest of soils, all weathering and shaping the Sound. Water links land and sea, where Sitka black-tailed deer, brown and black bears, and river otters come to the shore to feed, and seabirds come to the land to nest. And water is how people get here, how they navigate this place. There are no roads and very few trails; it's by boat that people enjoy what water has created."

In chapter 2, oceanographer Ted Cooney describes the relationships between the living and nonliving parts of the system and concludes that they "portray a complicated set of dependent populations and processes that result in the annually pulsed production of organic matter, which is then moved through an immense food web, with consequences for both marine and freshwater environments." This rich food web provides a unique resource that supports many species of marine mammals and their freshwater counterparts in this ecosystem that is still recovering from the *Exxon Valdez* oil spill.

The Sound is a place of many extremes; it is not a place for the weak of heart. In chapter 5, social scientist Greg Brown suggests that many who are fortunate enough to have visited the Sound recognize it as a very special place and even become its advocates. Scenic beauty, biological richness, and seemingly limitless nature-based recreation opportunities combine to make the Sound a Mecca for outdoor enthusiasts, offering a valuable economic service to the region's communities. Alaskans who live in or near the Sound experience a relationship characterized by multiple meanings embedded within social, ecological, historical, and cultural values. The intersection of these values with the need to derive an economic benefit from sharing this special place with "outsiders" remains a constant source of dynamic tension for residents of local communities.

WHY IS THIS WILDLAND SETTING SO VALUABLE?

The voices from a dozen essays—contributed by stakeholders who live, work, and play in the Sound—help to answer this important question. Key among them is Marybeth Holleman, who opens the book in chapter 1 by reminding us how significant and unique each and every experience can be when visiting the Sound. She describes a vast, open, and unique landscape where one can get lost, out of sight and out of contact with others. She concludes that everyone who chooses to visit the Sound does so because of "the desire to be in a place where the human footprint is negligible, where wildlife abounds unfettered, where wilderness still reigns." Subsistence harvest is also a critical part of the rural Alaskan lifestyle for Prince William Sound residents and is specifically protected under U.S. law. It is widely recognized in Alaskan land and wildlife management that subsistence harvest provides irreplaceable cultural, spiritual, personal, and sustenance value—a relationship described eloquently by an Eyak tribal elder, Patience Andersen-Faulkner (chapter 9).

Confirming what Holleman intuitively describes, Randy Gimblett (chapter 7), social scientist, finds that many Prince William Sound visitors seek solitude. But more specific choices about destinations were made based on good fishing, glacier viewing, and wildlife viewing. The ability to view wildlife was the only activity identified as "very important" to all users. Similarly, solitude was a strong motivator for survey respondents, but only 10 percent identified it as a prominent reason for choosing their destination.

The Sound is more than a beautiful and scenic landscape. It is the ideal mirror to reflect the complexity of Alaskan identity that has been forged while living in a challenging and inspirational northern environment. Alaskans perceive themselves as more resourceful, risk taking, independent, and wilderness seeking than "outsiders," aka non-Alaskans. Greg Brown (chapter 5) reminds us that there are many values associated with the Sound, but not surprisingly, the values of the greatest importance are its scenic beauty, recreational opportunities, and biological richness. Brown's mapping work also reveals economic opportunities are dominant in parts of the Sound, and he identified "hot spots" that are associated with these values and revealed in clusters near prominent natural landscape features, such as Columbia Bay and Montague Island, and the communities of Cordova, Valdez, Whittier, Tatitlek, and Chenega Bay.

WHAT IMPACT DID THE *EXXON VALDEZ* OIL SPILL HAVE ON HUMAN USE IN THE SOUND?

In the years since the *Exxon Valdez* oil spill (EVOS), the Sound has experienced numerous changes. The immediate effects on the environment are well documented: thousands of animals and birds perished, and thousands of miles of shoreline were oiled. In 2014, 10 of 28 injured resources were considered to have recovered from the effects of the spill, while 13 resources were believed to still be recovering. Though much has been documented with respect to the effects on biological communities and resources impacted by the spill, impacts to the culture and economies as well as the sociological well-being of Prince William Sound communities are less well understood.

The Sound also supports significant subsistence and private sport harvest activities as well as a thriving commercial industry for fish and game. There are perceptions from subsistence users and other harvesters that the effects of the oil spill have reduced wildlife productivity in the Sound. Wildlife biologist Aaron Poe and colleagues undertook a study (chapter 9) to characterize harvest patterns to provide managers with contemporary insights into the recovery of the subsistence human service that was injured by the oil spill.

The cleanup also brought social impacts. Over 10,000 people worked on the cleanup as spill responders, resulting in a long-term increase in the population of Valdez (Wooley and Haggerty 1993). Community impacts are described for the small remote community of Chenega Bay by resident and researcher Kate McLaughlin (chapter 4). She describes a new infusion of cash into the community as they assisted with the spill cleanup, combined with a fear of depending on the harvest of species contaminated by the spill that had sustained this community for generations.

A novel type of social impact is described by Twardock (chapter 3), who documents how many cleanup workers experienced the Sound for the first time and later returned to experience the Sound as a destination for recreation. Exxon settled with the state and federal governments, paying $100 million in criminal charges and $900 million in a civil settlement. The money funded extensive biological studies, which brought more people to the region to study the effects of the spill. It also funded infrastructure improvements that allowed greater access, including state park cabins. The increased publicity, easier access, and better facilities contributed to a steady increase in recreation use since the spill.

In chapter 4, Poe and Gimblett document a transition of looking at direct effects from the oil spill to the indirect effects of increased human use following the spill—and the potential concern for cumulative effects (i.e., the oil spill plus new human use) on the Sound's ecosystem. They suggest that successful recovery of recreation in the Sound is likely dependent on recognizing and facilitating key recreation opportunities sought by users in the region while maintaining a spectrum of available other experiences appropriate for wildland settings. They conclude that by allowing use from diverse groups of stakeholders, a broader section of society will come to support the region's resources—ultimately improving their odds for recovery and persistence into the future.

WHAT ARE THE CURRENT HUMAN USE DYNAMICS IN THE SOUND?

The Prince William Sound region includes the towns of Whittier, Valdez, and Cordova, as well as the Alaska Native villages of Tatitlek and Chenega Bay. As Twardock describes in chapter 3, these communities combined have fewer than 7,000 residents—though Anchorage, which has about 40 percent of Alaska's population, is within close proximity.

Over the past decade, the Sound has continued to experience increased human use. The growth of the recreation and tourism sector in Alaska has been accompanied by improved access to the region. In the western Sound in particular, the opening of the Whittier access road in 2000 has led to both increased personal and commercial recreation/tourism use. In 2001, there were 180,000 vehicles going through the tunnel to Whittier, and that increased to over 240,000 by 2007. More dramatic is the number of visitors entering the Sound, which has increased from 250,000 in 2001 to over 475,000 in 2007 (Fay 2008).

The geographic pattern of recreation use is not evenly dispersed across the Sound, but concentrated near destination landscape features like the tidewater glaciers of Columbia Bay, Harriman Fjord, and Blackstone Bay, as well as at key seasonal fishing locations. Use is also concentrated near the access communities of Whittier, Valdez, and Cordova. Some of the earliest use mapping efforts in the region by Brown

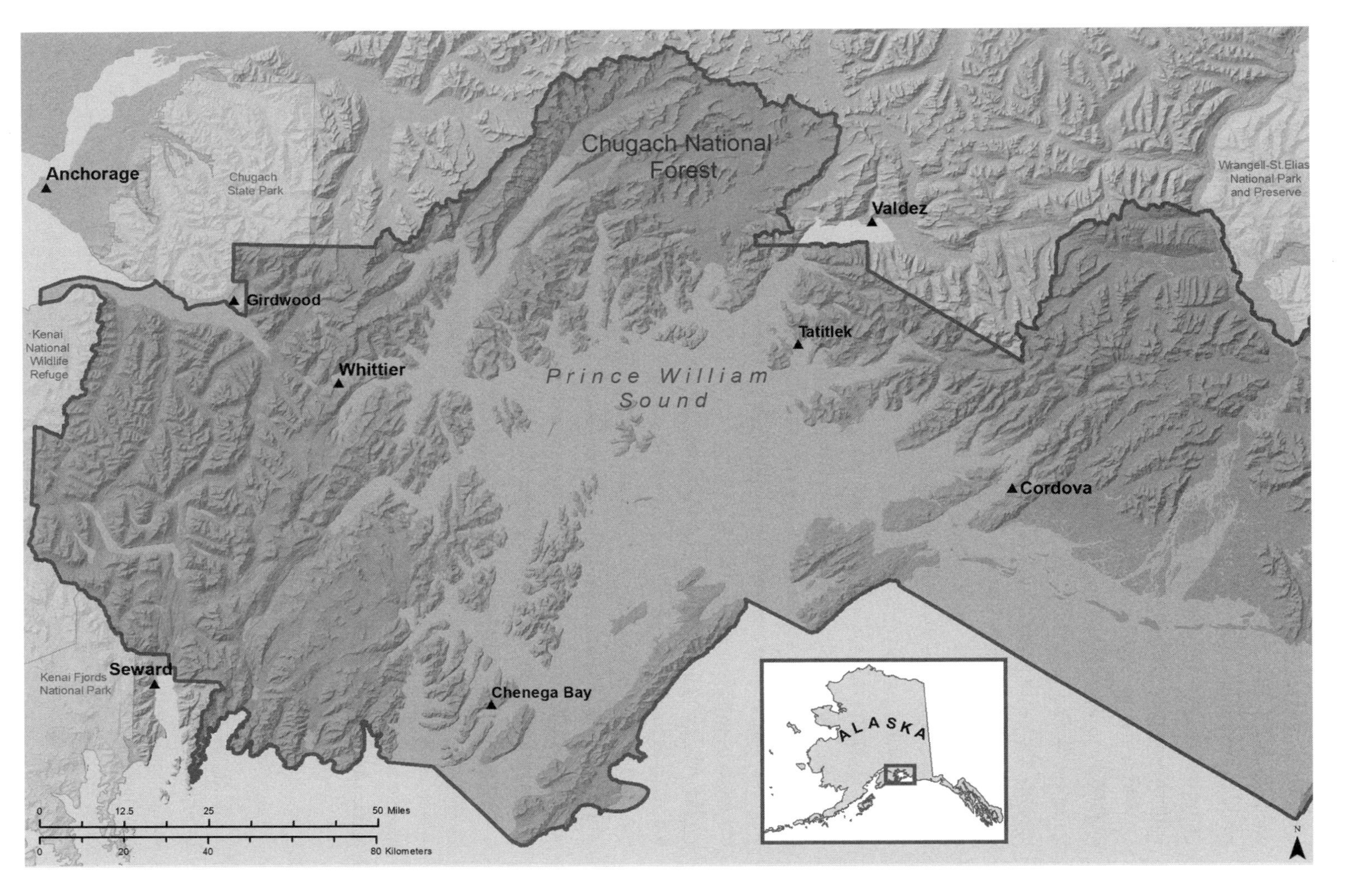

FIGURE I.1. The communities of Prince William Sound.

(chapter 5) provide this understanding with similar conclusions about human use concentrations documented by spatial analysts Chandra Poe and Samantha Greenwood in their human use hot spots study (chapter 6). They found a strong seasonal pattern of use, with the vast majority of recreation happening during summer months, with an exception for hunting in the fall and spring.

With a 100 percent increase in black bear sport harvest between 1995 and 2001 (ADF&G 2008), bear hunting is a significant example of increasing recreation activity in the Sound with the potential for user conflict. Spatial analyst Spencer Lace explores the use of geographic information systems (or GIS) with bear harvest data sets and a contemporary spatial summary of other recreation use in the Sound. He concludes that interaction between bear hunting activities and kayaking is not a widespread problem across the Sound, but that there are some key areas where bear hunter/kayaker interactions are likely to occur. Hunting is described by Poe and colleagues (chapter 19) as a "key experience" sought by individuals using the uplands of the Sound. They fear that U.S. Forest Service managers' lack of knowledge associated with hunters and the sensitivity of this group to changes in use levels, competition, and other recreation groups make them especially vulnerable to experiencing declines in quality of experience in the Sound.

The Alaska National Interest Lands Conservation Act (ANILCA) requires that federal land managers consider the effects of management on subsistence activities. Subsistence-eligible residents in the Sound include people of Alaska Native heritage and other rural residents from the communities of Chenega Bay, Cordova, Tatitlek, and Whittier. Fish comprise the majority of resources taken, but there is also significant use of other species, including Sitka black-tailed deer, black bear, marine mammals, mountain goats, waterfowl, seabirds, river otters, and mink. Research conducted by Poe and colleagues (chapter 9) summarizes the spatial and seasonal distribution as well as the intensity of subsistence harvest. They found the majority of subsistence harvest happens in relatively close proximity to communities in the region and that participation in subsistence harvest appeared to be on the decline.

ARE THERE IMPACTS RESULTING FROM INCREASED VISITOR USE OF THE SOUND?

Increasingly, some managers want to demonstrate the relevance of the wildlands by sustaining robust and diverse stakeholder constituencies. This can present potential challenges relative to environmental impacts as well as social impacts stemming from crowding at key landscape destinations, or conflicts from overlapping uses by incompatible user groups (i.e., classic disputes between motorized and nonmotorized users).

A scientific survey conducted by Gimblett and colleagues (chapter 7) found that despite management concerns about the negative effects of increasing use, crowding was not a problem for most visitors. No respondents reported displacement as a result of interactions with others, and an overwhelming proportion (95 percent) of visitors reported experiences that were as good, or better, than expected. However, in focus groups of longtime Sound users, Bob Itami and colleagues (chapter 8) found that hunters and some kayakers were negatively impacted by increasing numbers of specific user types—but those impacts had more to do with specific behaviors than use density per se. An essay from Anchorage resident Lynn Highland (chapter 8) is a window into some of these perspectives held by longer-term Sound users that were also documented by Itami.

Poe and colleagues (chapter 9) evaluate the overlap of subsistence harvest with recreation activity in the Sound in order to evaluate potential competition or displacement resulting from other human uses. The study's authors found a low level of displacement was occurring mostly in immediate proximity to communities, and this displacement is caused primarily by other local harvesters. The study concludes that reduced household subsistence harvest effort in recent years is caused more by individual lifestyle changes and perceptions about lack of resources than by concerns about competition or user conflict.

Managers find campsite assessment studies useful for looking at resource impacts from increasing use and have studied site degradation in an attempt to understand potential decreases in quality of recreation user experience. In chapter 17, social scientists Paul Twardock and Christopher Monz describe extensive work studying recreation impacts in backcountry situations. Their pioneering work in measuring campsite impacts in the Sound since the 1990s suggests that there have been significant changes in conditions at campsites. With this study documenting an increase in the number and size of sites, the management practice of "hardening" primitive campsites is explored by Maryann Fidel (chapter 16). Her study points to the tradeoffs between developing campsites with the intent of concentrating use or preventing resource impacts with the cost of changing the character of a wildland setting.

ARE THERE HUMAN-CAUSED WILDLIFE DISTURBANCES IN THE SOUND?

Since many species injured by the *Exxon Valdez* oil spill have been slow to recover, some scientists are also concerned that human use could be a cumulative impact on species affected by the spill. These concerns helped launch an early effort by wildlife ecologist Lowell Suring and colleagues (chapter 11) focused on evaluating potential future levels of human use relative to two species also harmed by the spill (harbor seals and pigeon guillemots). Suring and Poe (chapter 12) conducted a contemporary spatial analysis and report that there are relatively few areas—19 primary hot spots representing less than 1 percent of the Sound—where high-intensity recreational boating activity overlaps with species assemblages that were also affected by the oil spill. They also identified 182 secondary hot spots that may warrant future monitoring or study of potential recreation impacts. When making disturbance risk assessments, they advocate for the need for spatially explicit characterizations of human use intensity—equivalent in rigor to those traditionally used by ecologists to characterize species distribution and habitat use.

Graduate student Jessica Fraver (chapter 13) investigated the influence of a suite of human activities common along protected Alaskan shorelines—such as regulated harvest, commercial fishing, and recreation—on river otter abundance and behavior. Though field observations suggested that otters use beaches that were also used by campers, results of more rigorous spatial analyses indicated limited potential for interactions between river otters and campers based on shoreline feature preferences. Similarly, Poe and colleagues (chapter 15), concerned about beach campsites near nesting black oystercatchers, found that though such an association was true at the landscape scale, substantial separation occurred at the site level—with separation distances averaging more than one mile. In contrast, graduate student Laura Kennedy (chapter 14) found that human use overlapped with shoreline features associated with the presence of harlequin ducks (a species still recovering from the oil spill) and common mergansers. Her spatial analysis was paired with empirical observation of behavioral disturbance at those sites, which indicated the potential for disturbance to these sea duck species.

These three studies exemplify the type of research needed by wildland managers facing questions about the sustainability of human use and wildlife impacts. Spatially explicit analyses and monitoring of specific human-wildlife interactions are needed to corroborate managers' concerns that tend to be based on anecdotal evidence or individual field observations. The nuances associated with place-specific variables and species-specific responses, including the potential for habituation, are further complicated when considered alongside impacts from other environmental stressors which *could* be far greater. It is important to point out that very little is known about many of these other potential stressors on wildlife in the Sound—making conclusions about the ultimate effects of human interactions with wildlife very challenging.

HOW CAN SUSTAINABLE HUMAN USE MANAGEMENT BE ACHIEVED IN THE SOUND?

Although the U.S. Forest Service (specifically the Chugach National Forest) manages approximately 80 percent of the land in the Sound, land and human use management is shared by local, state, and private landowners, among them the Eyak, Chenega, Tatitlek, and Chugach Alaska Native Corporations. Consequently, there is no single unified policy direction nor is there one overarching approach to management. The Prince William Sound Framework or "the Framework" initiated by the Chugach National Forest is the largest interdisciplinary planning effort attempted in the Sound since the oil spill restoration plan was completed in the 1990s. Poe and Gimblett (chapter 4) describe how this effort grew out of the Chugach's need to improve its management of recreation and tourism in the Sound based on perceptions of increased use and potential for user conflict. It evolved into an approach based on ecosystem management principles that parallels the Forest Service's 2010 Recreation Sustainability Framework (USDA Forest Service 2010).

Poe and colleagues (chapter 19) demonstrate how these broad principles of sustainable recreation can be put into practice through integration of social and ecological information, analysis at multiple spatial scales (ranging from regional to landscape to site-specific levels), issue-based planning, and collaboration with stakeholders. They recommend changes in operational approaches that would encourage the Forest Service to focus on managing user expectations relative to issues like crowding or user conflict. They suggest that if users have realistic expectations about the experience they should expect in wildland settings, as well as good information about the other types of users they are likely to encounter, there is less potential for low-quality user experiences. Furthermore, they suggest an approach of concentrating human use rather than dispersing it across the Sound in order to minimize impacts and maximize opportunities for a greater diversity of individuals to use and enjoy wildland settings. Finally, they recommend establishing management partnerships that can leverage the assets of stakeholders *and* increase opportunities for citizen involvement. Such partnerships help managers meet targeted needs relative to resource monitoring or environmental education while, simultaneously, truly engaging stakeholders in helping to sustain wildlands.

USING THIS VOLUME

This volume is grounded in the science foundations of its technical chapters. In total, 28 scientists and managers made contributions. Their work is joined by perspectives from a dozen residents of the region who share local knowledge reflecting the values of the region through a series of invited essays and research notes. These are the perspectives of the individuals who live, work, and play in the Sound and who are on the receiving end of the science and management endeavors described in this book's technical chapters. Their perspectives complement those chapters by underscoring the difficulty of managing such a region sustainably, and at the same time help make the compelling case for doing so.

Another reason for integrating research results, management tools, and individual public perspectives is to help remove communication barriers between managers and stakeholders—a vital step toward sustainable recreation management. Stakeholder reflections serve to expand the perspectives of managers and researchers looking for new tools and approaches among the book's pages—reminding them of the societal complexities associated with the wildland systems they are attempting to understand and manage. Similarly, by providing short, descriptive essays that intersect with research and management chapters, a broader public audience can be invited into the volume's content and the issues facing the Sound. Ultimately it is our hope that this approach, combined with concise introductions and conclusions for the book's technical chapters ("Sound Bites" and "Lessons Learned") will broaden the audience beyond managers, researchers, academics, students, and conservation organizations—to include public stakeholders for wildlands.

Chapter 1 describes the study area using a first-person look at what it means to experience the Sound. This individual narrative shows how individual experiences interact to shape complex perceptions about the value of wilderness. It is followed by chapter 2, a thorough characterization of the geophysical setting of the region and its ecological processes. Chapter 3 provides a brief history of human use in the region and concludes with the disaster of the *Exxon Valdez* oil spill; chapter 4 explores the impacts that the spill had on social and economic services in the region and the resulting management responses. Chapters 5 through 19 are technical chapters contributed by researchers and managers working in the region and are presented as complete works, including orientation to the region. We understand that this comes with some potential price in content redundancy as individual authors introduce the Sound within the context of the specific management issues they are attempting to address. Given that this volume is also available as an e-book through which readers are encouraged to download individual technical chapters for their electronic libraries and share essays by e-mail and social media, this self-contained approach to content pieces is necessary.

Some readers may prefer to focus on individual research and management chapters or only skim those in favor of stakeholder reflections shared in essays. We hope to lure those casual readers into technical chapter content through the Sound Bites and Lessons Learned that share the key results and broad themes resulting from the work described. Likewise, we hope scientists and managers may gain perspective from individual stakeholder essays which allow them to better understand the diversity of stakeholder needs and motivations that must be considered for a truly integrated approach to managing landscapes like Prince William Sound.

This volume's goal is to help broaden the conversation about the science that informs management of wildlands like the Sound. We hope that the examples shared elevate the level of sophistication in which human use can be studied as a way to improve future recreation management efforts while engaging stakeholders. We hope these techniques will ultimately allow managers to move beyond simple number-based recreation capacity schemes and address these complex systems using social science with the same level of rigor currently applied to other ecosystem components.

We also hope to convey an appreciation for the challenges of applying science to the management of human use in wildlands. Even when management can be informed by rigorous social and ecological science, managers must recognize that these settings are depended upon by stakeholders who may operate from a different base of knowledge. In order to be sustainable, the policies that managers put in place must deftly navigate the complexities of science and individual human experience—truly honoring the contributions and understanding the limitations of each. Success in this endeavor requires regular and meaningful engagement with stakeholders of public lands. The complexities associated with sustainable management of wildland settings necessitate true, two-way exchanges of knowledge that go beyond public testimony proceedings and 90-day comment periods.

Finally, we hope to convey the need for a profound change in perspective as managers approach questions of human use of wildlands. Too often, human use is only thought of in terms of the disruptive impact that it can have on "pristine" settings or sensitive species. While certainly there are legitimate concerns about environmental impacts, this default approach can result in managers having the mindset of consistently looking only for ways to reduce human use. This can result in artificial caps on the numbers of users or types of activities allowed on public lands, which in turn decrease the overall societal relevance of those lands and the agencies tasked with their management. In an era of increased scrutiny of public investments and overall shrinking budgets, managers can't afford decreased public support. A shift in perspective that recognizes restrictions in use as having long-term potential harm toward the societal value of wildlands must be weighed against concerns about potential impacts from human use.

SUSTAINING WILDLANDS

CHAPTER 1

WHAT WILDERNESS GIVES

MARYBETH HOLLEMAN

MY PADDLE IS USELESS. The ice is so thick, the bergs as tightly packed as a thousand-piece jigsaw puzzle, that there's simply no water to slip my paddle into. So I push on ice, shove it so hard that if I were in water, I'd roll my kayak. But nothing happens except the ice pushing back, and bergs all around crunching and rumbling like a crowd of hecklers.

In every direction is ice, spreading across the fjord to steep rock faces and towering, snow-clad peaks. Two large bergs have me stuck. I can't turn, can't go forward, all I can do is go backwards through the sliver of a lead that brought me here. But I can see the shore where I believe there's still enough open water to carry us out of the fjord.

My kayaking partner, Carol, who has been following close behind me, now strikes out on another wedge of water, and together we slowly grind and shove and push our way. Around us on larger icebergs are the dark crescent shapes of harbor seals, all of them turned to look at us, no doubt wondering what these two thin blue boats are doing in their icy world.

At first imperceptibly and then more gradually, the floe becomes thinner, and then, finally, we are out of it, into a sliver of blue water along the opposite shore. After all that ice, the water feels smooth as silk. We paddle fast, urged on by tide and wind, out of Nassau Fjord and into Icy Bay. We stop on a wide sand beach at Gaamaak Cove and sit, surrounded by beached icebergs, laughing with the relief of those who have overcome great obstacles. Prince William Sound has surprised me again.

This is day four of an early July ten-day trip to southwestern Prince William Sound. A water taxi carried us, our gear, and our boats to Ewan Bay. We planned to paddle to Chenega Bay on Evans Island to take the Alaska state ferry M/V *Tustumena* back to Whittier.

Yesterday afternoon, we paddled into Nassau Fjord and spent a lovely full-moon evening there, listening to the crack and thunder of the calving glacier, filtering water from a nearby stream, watching a black bear forage along another beach, and seeing the moon rise as the snowy peaks all around us turned pink with the setting sun.

We reveled in having all this to ourselves. Now we know why: the ice floe must have gathered behind us, so that no others—motorboats or kayaks—dared enter. No other boats visited Chenega Glacier that day; no other people were in sight.

Though I've kayaked in Prince William Sound for over 20 years, I've never been caught in an ice floe like that before. So I didn't know just how dire was our situation. As long as we stayed in our kayaks, we were likely fine, though I didn't know how long we'd be caught in that ice floe. Several tides back and forth, pushing the ice even tighter together, could have kept us there a long time. Or worse, the ice could have shoved at us to a point of capsize. Hundreds of large ships throughout history have been crushed and sunk by sea ice. We were surrounded by bergs the size of houses, and there were times when bergs began to lift my boat and I felt my balance slipping. Once in the water, what would we have done? Climbed out on an iceberg, laid there like a seal with no blubber, hoped for rescue?

But the danger is now mere conjecture. We pull our lunch out of a dry bag and sit in sunlight, beached icebergs towering around us.

"That's the crunchiest kayaking I've ever done!" says Carol, and all I can do is laugh.

Prince William Sound is a place of extremes and convergences—of weather, geography, wildlife, and experiences. One of the most active seismic regions in the world, it anchors the eastern edge of the Ring of Fire; it's the northernmost reach for the temperate rain forest; it's rest stop and destination for tens of thousands of migrating birds and marine mammals. Arctic terns fly in from South America; humpbacks swim from Baja California and Hawai'i. It's a biological and geographic center for the Pacific, where temperate and subarctic environmental conditions overlap.

Like the center of a wheel, the Sound is the apex of Alaska's Pacific shores that curve from Southeast Alaska to the Aleutian Islands. A 3,500-mile-long undulation of capes and fjords and islands and islets and sea stacks, Prince William Sound is encircled by three mountain ranges draped in ice fields from which more than 150 glaciers pour, dozens of them tidewater. It is an expansive, meandering maze.

I like that. I like reading the map of the Sound and tracing the fingers of water yet to explore, the beaches and ridges yet to walk, the headlands and islands yet to stand upon.

My first summer in Alaska, I sold tickets on the train that ran between Portage, on the road system, and Whittier, on the shores of Prince William Sound. This was before the road, built in 2000, connected Whittier to Anchorage; the train was the only way to get there.

Selling tickets on that train, I met the great cross section of those who visit and those who live in Alaska: Anchorage residents with coolers and dogs heading out on their pleasure boats; college students with backpacks ready for a summer's adventure; neatly dressed tourists just embarked from their cruise ships; fishermen just back from an opener smelling of fish guts.

Once I met a man who had fished the waters of the Sound for 23 years. Tall and lean, with a long black beard and a black cap, he'd told me story after story in a deep Russian accent about his adventures in the Sound.

"I could spend many lifetimes exploring Prince William Sound," he had said, "and still not see it all."

Now I've spent time out here for 25 years, kayaking, hiking, riding on ferries and sailboats, fishing boats and pleasure boats. I've camped on beaches and stayed in Forest Service cabins; I've come out so early that one spring we had to dig snow off the beach to create a campsite, and so late in the fall that winter storms stranded us.

On this trip with Carol, I am once more amazed at how this place continues to surprise me, at how much of it I've yet to experience. I'm also flooded, at every corner, with memories of all the other times I've had out here.

Emerging from the tunnel between Bear Valley and Whittier, I remembered once more my first view of Prince William Sound. Growing up in the Appalachian Mountains and spending time on North Carolina's Outer Banks, I'd always wished to live where mountains and the sea came together. When I saw Prince William Sound, with its ice-clad mountains dropping sharply into the crystalline blue sea, I knew I'd found it.

Boarding the water taxi at the harbor, I recalled that first summer when I'd come upon two kayakers sitting at the dock, wet sleeping bags, tent, raingear, clothes, and maps spread around them to dry in the first sunshine in 12 days. When I asked them where they'd paddled, they picked up a soggy topographic map and showed me their route. They'd spent two weeks of rainy weather paddling the entire western side, from Chenega Bay to Whittier, at least 75 miles around islands, across straits, past glaciers, and through passages. I was astonished that such an adventure was even possible and yearned to do it myself. They gave me their wet maps to keep, to harbor my own dreams for such a trip.

After being dropped off on a small beach in Ewan Bay, Carol and I paddled to a stream to collect water. I'd hiked up this stream before, with my two-year-old son, and we'd bathed him in a still-water pond, while at the far end a common loon had let loose its sonorous cry. We passed another small knoll, where once I'd climbed through a series of meadows, following a slick track made from one muskeg pond to another, and there at the ridgeline saw three river otters playing in a larger pond, tossing a pink salmon from adult to pup. At another stream, I recalled staying at the Culross Passage cabin in midsummer, standing on the porch as a half dozen black bears swiped salmon from the stream just yards away.

And there are memories, too, of the biggest agent of change in Prince William Sound's recent history: the 1989 *Exxon Valdez* oil spill. The year of the spill, I'd volunteered at a camp on Perry Island, my eyes burning and head aching from oil on the beach. I'd walked to an eagle's nest with a researcher, where he'd found oil on the egg, brought back on the legs of the parents. I'd observed a fishing boat hauling oiled kelp into its hold while a sea otter floated nearby. Walking a beach in West Twin Bay two years after the spill, I'd found tar balls bigger than my fist. Visiting Sleepy Bay Beach eight years after the spill, I'd watched technicians and residents of Chenega Bay try yet another method to wrest oil from the rocky substrate.

Memories of that time, and the damage it still wreaks, weave tightly into my experiences of Prince William Sound. Even more so this trip: I'm kayaking in the Sound while in the Gulf of Mexico, oil continues to gush from BP's deep water well. The BP spill has been like a 21-year-long echo for those of us who experienced firsthand the oil spill in Prince William Sound. It sits heavy on my heart, a weight I carry this entire trip.

Yet this weight is somehow balanced by all the joyful memories, and more so by how this place can't help but draw me in to the present moment. That's the lesson of the ice floe: be here, now, in this moment, in this place. Prince William Sound never ceases to amaze me. It is still, as naturalist John Burroughs wrote in 1899, an enchanted circle.

After navigating the ice floe, Carol and I cross Icy Bay to camp at a favorite beach, one with a small cave where I've sat out many a rainy day. As if on cue, the next morning we awaken to rain pattering on our tents. It's not a driving rain, and we certainly could paddle in it, but decide to make use of the cave and hunker down for a day of reading, sleeping, eating, and walking the beach.

Prince William Sound is a stunning place, but it's not an easy place. To be here is to expect rain and wind; to expect to need warm layers and hats and gloves and raingear; to have a marine radio and extra food for being stranded by weather; to be uncomfortable. To be here is to live water.

Water shapes Prince William Sound; water sustains it, creates and contains it. Glaciers pouring to sea, pushing rock before them and carving U-shaped valleys from solid rock; streams rocketing down from the mountains to the sea, carving and carrying silt; rains stippling rock and nurturing a dense rain forest growing on the thinnest of soils, all weathering and shaping the Sound. Water links land and sea, where Sitka black-tailed deer, brown and black bears, and river otters come to the shore to feed, and seabirds come to the land to nest. And water is how people get here, how they navigate this place. There are no roads and very few trails; it's by boat that people enjoy what water has created.

Water, the weather it creates and remoteness it engenders, tempers visitation so those who do come can find the remove that wilderness alone provides. This is what draws people here, as much as the beauty of the place and diversity of wildlife: the solitude.

It has, in the past two decades, become more popular. The oil spill put this place before the world's eye, and the 2000 road between Portage and Whittier made it more accessible. A wholly solitary experience is increasingly rare.

Our second day out, after lunch and a walk down a long sand beach along Icy Bay, we rounded a headland and were astonished by a group of 14 kayakers. They were all from Israel, and this was their first time in kayaks, in Prince William Sound, and Alaska. The decks of their kayaks were loaded with gear, and they bumped into each other as they gathered around us to tell us all they'd seen: humpback whales, harbor seals, icebergs, salmon. While at first Carol and I had been irritated by their noisy presence marring our wilderness experience, we came to appreciate their wonder and amazement; it reminded us how it was for us the first time we'd set eyes on this place.

Still, though, the vastness of the Sound makes the experience of having a beach, or a cove, all to yourself, possible. So when, on our rainy day at Cave Camp, a small skiff carrying a man and a dog comes ashore, we're taken aback at having to share "our beach."

He strikes up a conversation and invites us back to his boat for dinner. Wet and cold and curious, we agree. He and his wife serve us a bowl of shrimp from the waters on which we float and tell us about their adventures in the Sound. A firefighter and a computer technician, they spend much of the summer months on this boat, exploring the Sound. They catch salmon and shrimp, walk beaches, and watch whales. Here in Knight Island Passage in summer, humpbacks abound, feeding on krill. He tells us of one humpback that, he says, must have been enamored of the sound of his engine. The whale circled the boat several times, spy-hopping so close they thought he'd land on the deck, and even rubbing against the sides of the boat, but so carefully there was no damage.

After returning to our camp, Carol and I talk about why people come to Prince William Sound, what we all—the Israeli kayakers, the pleasure boat couple, and us—have in common. It's the desire to be in a place where the human footprint is negligible, where wildlife abounds unfettered, where wilderness still reigns. No matter who we are back in our towns and cities, our jobs and careers, there's a common denominator out here: the way being in wilderness makes us more fully human. It's a paradox, that we must leave behind our daily lives in order to become more ourselves. This is what wilderness gives. And as the wilderness of Prince William Sound brings us home to ourselves, those of us who return again and again show a strong and enduring allegiance to it.

Over bites of shrimp, we'd told the couple that we'd seen a black bear along the shoreline on our way to camp. They'd told us how rare it is now to see a black bear, how sport hunting has increased exponentially. Hunters shoot bears from boats, though it's illegal; enforcement in such a large and remote place is difficult. So this couple does what they can to keep hunters from such illegal activity.

Prince William Sound seems to elicit this kind of desire to protect the place by direct means. On our ferry ride back to Whittier from Chenega Bay, Carol and I speak with three other kayakers. They bring out their maps and show us their route, how they paddled the outer shores and hiked up beside icefields for views of calving glaciers. A retired glaciologist, he is all too aware of how climate change is affecting the glaciers, the waters, wildlife and forests of the Sound.

Later in the summer, I will spend a weekend on a volunteer beach cleanup with others who have been doing this beach cleanup work for decades. I've been on a few of them already—the first time, with my eight-year-old son, we found as much to keep as to throw away: a weathered paddle, a boat seat, buoys, and nets. These beach cleanups used to find mostly derelict fishing gear, but now, increasingly, the haul is overwhelmingly dominated by one object: the ever-pervasive plastic bottle. Walking beaches in rain, pulling trash from between driftwood, I'm reminded that, as much as people come to the Sound to recreate and experience an untrammeled natural place, they often come to defend it, to look for ways to give back to this wild place that has given so much to them.

Carol and I spend our days paddling shorelines where salmon jump all around us, where sea lions patrol embayments, where jellyfish float languidly beneath our paddles. We're enamored of the fecundity of this place; even as we're aware of how it's changed, we revel in what remains. We spend a night on a beach facing a group of islands called the Pleiades. The next morning, we both sit on the beach in morning sunlight, with our notebooks and our coffee, writing. At a now-familiar sound, we look up. It's whale breath, and there, just yards from where we sit, surfaces yet another humpback whale.

On our eighth day, the winds change. Yesterday, we crossed Port Bainbridge in seas so calm they were glassy. But today, as we turn the corner from our campsite to paddle down the wide passage, we're hit with a wind that grows the farther down the passage we go. So we slip into a cove, paddling close to the rocky shoreline, and find a narrow slit between rock faces just wide enough for a kayak.

There we enter another world. It is quiet, calm. The waters are so still we can see as well into them as we can into the forest edging the shore. Gazing into the forest of bright green eelgrass and sea lettuce, my eye catches some glimmer of blue. At first I think it's a reflection of my blue kayak, but it moves, wriggles, flashes blue then yellow. It's a small swimming creature, about the size and shape of my pinky finger, snaking through the water. Its head is yellow, its body blue and green, and all of it iridescent.

I've never seen this animal before. I've never been to this quiet lagoon before. One could pass by this shore every day and never find this little lagoon. And that's how it is here in Prince William Sound. That's why I keep coming back.

STAKEHOLDER ESSAY

MUSINGS ON WILDNESS AND PRINCE WILLIAM SOUND

BILL SHERWONIT

HERE'S A CURIOUS THING, something I hadn't thought about for years until this August morning, after savoring Marybeth Holleman's essay, "What Wilderness Gives": the first wild place I visited in Alaska was Prince William Sound.

In June 1974, I spent a few days on Latouche Island, at the Sound's southwestern fringes. I didn't go there for pleasure or adventure, but to work. (Although I suppose that job was its own sort of adventure.) A geologist at the time, brand new to the state, I was assigned to a claim-staking crew. Relying on that trickster, memory, what I recall now is the cold, wet weather, the thick gray, oppressive clouds, an abundance of snowfields (at least at higher elevations), and a forested island of large evergreen trees. To be honest, I don't recollect the target of our staking effort, only that the days were long and exhausting and not much fun.

A quick Google search informs me that several copper prospects were discovered and staked on Latouche in the late 1890s, so copper was likely our target too. Developed into a mine in the early 1900s, one of the original prospects eventually produced substantial copper ore and sparked the birth of a town; nearly 4,000 people reportedly once inhabited the island. Eventually both the mine and town were abandoned, but some ruins remain. So Latouche Island isn't exactly wilderness, at least as we usually imagine it. But it has stayed with me as a place of raw wildness.

After returning to Anchorage, I joined a crew bound for the Central Brooks Range. In a matter of days—maybe even my first day in the range—I fell in love with that mountainous Arctic landscape. Latouche, meanwhile, became something of an afterthought, an interesting but largely forgettable sidenote to the start of my love affair with Alaska, and especially its wildlands and waters. Despite the remoteness and challenges of getting there, I've repeatedly returned to the Brooks Range and journeyed deep into its wilderness. Yet I've barely touched the wildness of Prince William Sound's "3,500-mile-long undulation of capes and fjords and islands and islets and sea stacks."

I can't help but wonder: if the days had been sunny and mild and I'd had more free time to explore Latouche Island and its surrounding waters, or maybe had some memorable wildlife encounter, would I have become enchanted enough with the place to further explore Prince William Sound? First impressions are indeed sometimes crucial in determining what shape a relationship takes.

So, unlike Marybeth Holleman and many other Alaskan friends, I have not become intimate with Prince William Sound. Instead, through their stories and images and enthusiastic advocacy, I've been something of a distant admirer. And there is clearly much to admire and love and protect in the Sound, wilderness and wildness galore in the form of glaciers, ice-choked bays, sheltering forests, rugged mountains, bears and berries, pods of whales and teeming birds and salmon, shrimp and secluded ponds, on and on it goes, with the additional blessings of solitude and natural quiet, and the opportunity to pay close attention to wild landscapes and seascapes, to embrace the wild neighbors that most of us too often and too easily ignore in our overly hectic and self-centered urban lives.

As I've written elsewhere, I didn't clearly understand the importance of wild nature to my life when I came north to Alaska at the age of 24. Or perhaps better put, I'd gradually forgotten its importance, something I'd intuitively understood as a young boy. In Alaska (both while working as a geologist in the mid-1970s and after resettling here in 1982, reincarnated as a writer), I gradually rediscovered my early passions and reformed the primal bonds that connect me to the more-than-human world. Along the way, I have redefined—and continue to explore—what wildness means to me, and its relevance to my life and the larger American culture.

Even in our high-tech, polluted, and increasingly urban world of the early twenty-first century, wild nature is all around us, all the time. Here in Anchorage, it's easier to notice the city's wilds, what with moose and bears roaming our streets and salmon swimming our creeks. But I also think of the trees and mushrooms in our yards, the mosquitoes and aphids that pester us and our plants in summer, the birds that sing and call throughout the year, the spiders and mice that sometimes make their way into our houses, the rain and snow and sunshine, the germs and bacteria we carry around.

The wildness also lurks within us. Our bodies, our imaginations, our dreams and emotions and ideas are wild. But in going about our busy, modern lives, we consciously or unconsciously suppress, deny, ignore, or forget our wildness.

Still, the wild animal remains, waiting for release. And—naturally—it's most easily set free in wild surroundings free of artifice and development. Free, largely, of the human touch. That's why the *feeling* of wildness most deeply resonates within us when we enter wilderness, whether in the high Arctic or Prince William Sound. And for many of us, the longer we stay in the "wilds," the more connected, refreshed, invigorated, and even healed we feel. There's a sense of being at ease, and sometimes even of being one with nature (which of course we are, whether or not we sense it). Something shifts *inside*.

That's what happened, I think, when I came to Alaska in the midseventies and spent several summers working deep in the wilderness, mostly within the Brooks Range. And it's clearly what has happened to Marybeth and several others I know in their explorations of Prince William Sound.

This is one of the reasons we humans need wilderness: for the deepened connection it provides to the larger, wilder world we inhabit. But there's a larger need beyond any gifts the wilderness gives to us humans: for the wildlands and waters themselves and the multitude of wild beings that inhabit those places. These other life-forms and landscapes have value in and of themselves, regardless of what they give to our species.

Besides an embrace of wilderness and wildness, we have a deep need for stories, such as the ones shared in this book. Now in my sixties, I recall the many wild places (both in Alaska and beyond) I've hoped or planned to explore but haven't. Most I will never visit. So, I—we—depend on stories to know the nature of places like Prince William Sound, stories about whales and salmon and bears, yes, but also stories about the ocean and weather's many moods; stories about jellyfish and iridescent, pinky-sized creatures that wriggle through the shallows; stories about the people who live and work and recreate here, who depend on the Sound's wild resources, its natural treasures. We need stories, too, that caution us about the harm that can be done to such places, the piles of plastic trash, illegal and wasteful harvests, many tons of oil ravaging a land- and seascape. And of course we need stories of restoration and hope, stories that show how we can more reverently and lovingly live on this wild, magical planet.

CHAPTER 2

THE GEOPHYSICAL SETTING AND MARINE ECOLOGY OF PRINCE WILLIAM SOUND

TED COONEY

INTRODUCTION

LOCATED AT THE NORTHERN APEX of the Gulf of Alaska, Prince William Sound is a large, semienclosed oceanic embayment of nearly 10,000 km^2. The Sound is backed to the north and west by the high, rugged Chugach Mountains, and fronted on the open ocean side by Hinchinbrook and Montague Islands. Several other small and large islands are scattered throughout the region. The shoreline extends for approximately 3,500 km, and there are locations where depths exceed 600 m (Hood 1986).

The northern Gulf of Alaska, including Prince William Sound, receives extraordinary amounts of rainfall over the course of a year: up to 4.0 m have been gauged at some locations. Either running off immediately, or stored as snow in the late fall, winter, and early spring, this precipitation eventually enters the coastal ocean, where it contributes to sound-wide and along-shelf currents. Significant amounts of freshwater also enter the Sound from tidewater and outwash glaciers, and from nearby rivers. The largest, the Copper River, drains a watershed of nearly 25,000 mi^2 in Southcentral Alaska (Weingartner 2007; Mundy and Olsson 2005).

The marine climate of the entire northern Gulf of Alaska is reflective of seasonality in the strength and location of the Aleutian Low Pressure system, a massive marine and atmospheric weather phenomenon that is well developed each year in the late fall, winter, and early spring (Wilson and Overland 1986). This condition produces a progression of marine cyclones (low pressure storms) that, after crossing the gulf from west to east, eventually impact the coastal and inland weather of Alaska, British Columbia, and the continental United States. The Aleutian Low Pressure system is replaced during the summer by the North Pacific High. This quite different condition brings a relaxation in storm frequency and magnitude. The seasonal cycle of vigorous onshore weather in the "winter" and more complacent "summer" weather plays an important modifying role for the ecology of the region.

In addition to these annually repeating seasonal weather patterns, the oceanography of the Gulf and the Sound is also impacted by other climate cycles of decadal scale and longer; the Pacific Decadal Oscillation and the El Niño Southern Oscillation are examples (Mantua et al. 1997).

The sections that follow present a general description of the biogeophysical setting of the northern Gulf of Alaska and what is known about the ecology of Prince William Sound. For a reader desiring more detail, there are other reviews available. The author recommends Hood and Zimmerman 1986, Shaw and Hameedi 1988, Mundy and Hollowed 2005, and Spies 2007.

THE PHYSICAL ENVIRONMENT

Understanding the interactive roles among the nonbiological components (substrates, winds, freshwater input, water density, temperature, and salinity) of a marine ecosystem creates a useful knowledge base for describing immediate, seasonal, and longer-term responses of living populations whose life history strategies have evolved to assure reproductive success over time. This section reviews what is generally understood about the physical environment of Prince William Sound and the nearby bordering shelf and open ocean.

GEOMORPHOLOGY AND TECTONICS

The dynamic continental margin of the northern Gulf of Alaska has been formed and reworked over geological time by forces of ice, ocean currents, and the tectonic shifts of large basin-scale crustal plates. The resulting terrestrial, shelf, and coastal habitats exhibit a diverse array of geological features, including a steep and relatively young coastal mountain range and associated glaciers, numerous estuaries and fjords, cross-shelf drowned river valleys, and a patchwork of sediment types ranging from very fine to coarse and rocky. Delivered from the nearby eroding mountains by ice flows, rivers, and streams, these different sediment environments provide a variety of habitats for contemporary seabed communities (Hampton et al. 1986).

Prince William Sound and its many peripheral side fjords and estuaries continue to be shaped by geological forces. The deep waters (more than 300 m) annually receive large amounts of fine-grain glacial sediments—blue muds low in organic matter content—that create a relatively stable but unproductive deep seabed (benthic) habitat. Retreating tidewater glaciers, like the Columbia Glacier, not only supply suspended sediments annually, but also expose entirely new fjord environments. The nearby Copper River is a source of significant sediment input, and the Hinchinbrook sea-valley provides a cross-shelf conduit through which dense shelf water can flow into the deeper regions of the Sound each year.

Depths exceeding 400 m formed by past ice scouring and tectonic activity in the Sound provide a rare inshore environment that occupies about 10 percent of the region. These deep waters and seabed support a sprinkling of species (squids, meso-pelagic fishes, and deepwater shrimps, among others) that otherwise would not occur in the Sound. These rare coastal depths also afford local overwintering habitat for oceanic interzonal species (like the large body copepod, *Neocalanus*) that mature in the surface layers but require deep water for reproductive purposes. Some believe that this unique "black hole region" serves to trap fauna swept into the Sound from the bordering shelf each year.

The great Alaska earthquake of 1964, with its origin in central Prince William Sound, is a contemporary example of tectonic forces substantially altering the marine environment. Regional-scale uplift and downwarping by as much as 15 m (50 ft) and intertidal scouring by seismic sea waves played a significant role in altering near-shore habitats (Feder and Bryson-Schwafel 1988; McRoy 1988). Many salmon spawning streams were impacted, as were regions supporting economically important razor clam resources. The salmon have adjusted, but the clams have never recovered to commercial abundance.

Meteorology and Physical Oceanography

Cycling seasonal weather and its influence on ocean temperatures, currents, and mixing is a principal factor shaping marine habitats in the Gulf of Alaska. Described as one of the most active meteorological regions in the northern hemisphere, the Gulf's huge swings between winter and summer atmospheric pressure fields create wind environments (oceanic and coastal) that influence the sea surface and water column in a variety of important ways (Wilson and Overland 1986).

During oceanographic "winter" (October to early March), the Aleutian Low Pressure system is characterized by a progression of gulf-wide storms (averaging three to five per month) traveling across the region from west to east (Weingartner 2007). Counterclockwise winds blowing around the centers of these storms facilitate the transport of warm moist air northward from lower latitudes along the eastern coastal Gulf of Alaska. These winds warm the coastal zone and deliver large amounts of precipitation as they lift over the adjacent mountains. Depending on the season, this precipitation quickly enters the coastal flow as runoff or is stored as snow for later release. Some studies suggest that this dominating weather pattern varies between weak and strong on decadal time scales. During groups of years when the Aleutian

Low Pressure system is well developed, winter in the eastern Gulf of Alaska is generally wetter and warmer than average. However, when the Aleutian Low Pressure system is weak and shifted to the west, the eastern coastal Gulf of Alaska tends to be cooler and drier in the winter. Described as the Pacific Decadal Oscillation, this cycling is believed to cause dominance shifts in fish, bird, and marine mammal populations lasting for many years.

Occasionally during the winter, very cold Arctic air originating in Siberia or interior Alaska floods south into the coastal zone, creating freezing conditions and icing. These events pose serious hazards to navigation and can also cause high mortalities in exposed intertidal seabed communities.

In the "summer" (June through September), the Aleutian Low Pressure system is replaced by weak high pressure in the Gulf of Alaska, termed the North Pacific High. This different condition is characterized by a modest reversal in average wind speed and direction, and a much more benign coastal weather pattern. During both "seasons," the Gulf is cloud covered more than 50 percent of the time.

An ecologically important air-sea interaction is the transfer of momentum from surface wind fields to upper layer waves and currents. As a general rule, only about 3 to 4 percent of the wind's momentum is transferred to and sustains surface currents, but this is enough to create along- and cross-shelf circulation patterns that significantly affect the living system by dispersing reproductive products (eggs, larvae, and other drifting life stages) across broad regions. The momentum transfer also creates a homogeneous surface "wind-mixed layer" that is shallow in the summer (less than 50 m), but deepens (more than 150 m) in the late fall, winter, and early spring. Coupled with cooling convective processes, the resulting deep winter overturn provides an important physical mechanism to replenish fertilizing elements in the surface waters from sources far below (Cooney 2005).

Another physical feature of the coastal Gulf of Alaska is the presence of a strong, nearshore coastal jet or current, the Alaska Coastal Current. Originating in northern British Columbia and southern Southeast Alaska, the current flows westward at speeds up to 1.0 ms^{-1} in a narrow band 20 nautical miles or so in width along the inner shelf (Royer 1981). Not to be confused with the much larger and sluggish Alaska shelf-break current, the coastal flow distributes sediments and plankton populations from east to west across 1,000 km of coastal habitat from Yakutat to Unimak Pass. Flow rates are highest in the fall, urged by runoff and seasonally increasing coastal winds. A portion of this flow enters Prince William Sound through Hinchinbrook Entrance and exits from Montague Strait and other passages in the southwestern corner of the Sound. During the summer, when the wind forcing is weak and freshwater inflow is pronounced, surface currents in the Sound can occasionally reverse—surface flow out at Hinchinbrook Entrance and in at Montague Strait.

The gulf-wide seasonally dynamic ocean climate described here is experienced in Prince William Sound to a greater or lesser degree each year, but modified somewhat by local topographic features producing gap winds, jets, and katabatic flows (downhill drainage winds) in some of the larger fjords and river valleys. Seasonally, mixed-layer marine temperatures are coldest (3°C–4°C) in early to mid-March, and salinities are highest (more than 32 ppt). At the same time, the depth of the wind-mixed layer is at its maximum (Vaughan et al. 2001). Springtime warming (increasing day length and sun angle) and the beginning of seasonal runoff provide the necessary conditions for stabilizing the upper layer, an important precursor to each year's cycle of marine plant production. The upper layer of the Sound is freshest (less than 28 ppt) and warmest (12°C–14°C) in mid- to late summer.

Continuing physical and biological studies of the Gulf of Alaska are demonstrating the presence of recently discovered large shelf eddies—designated Haida and Sitka (Crawford et al. 2007). Formed over the shelf and then propagating westward into the deep ocean, these features transport iron-rich waters and shelf plankton into the much less productive open ocean. Recent estimates suggest that more than 70 percent of surface phytoplankton stocks in the oceanic Gulf are actually found in these multiyear eddies. The ecological implications of these findings are currently being explored. It seems likely that these large and enduring eddies also play a role in transporting organic subsidies from the ocean onto the shelf and into coastal regions. This cross-shelf process links the open ocean production cycle with that of shallower systems, and provides a way for the bordering ocean to "feed" the shelf. Studies of the stable isotope composition of specific pelagic fauna sampled in the Sound clearly demonstrate this surprising conclusion (Kline 1999).

Chemical Cycles

Studies of the marine chemistry of the Gulf of Alaska, including Prince William Sound, have focused on distinguishing different water masses and their interactions—open ocean, shelf, and coastal waters, including those of fjords and estuaries (Reeburgh and Kipphut 1986). Two issues concern us here: (1) the seasonal cycle of deepwater oxygen renewal and (2) processes that contribute to the distribution and utilization of fertilizing elements like dissolved inorganic nitrogen, phosphorus, silicon, and iron. Both oxygen and these inorganic plant nutrients are termed "nonconservative" properties because, in addition to their concentrations being altered at the sea surface (by runoff and other sources), they are also subject to biological alteration throughout the water column and at the seabed.

In the absence of a mechanism to annually provide the essential inorganic nutrients for plants, and to renew oxygen in waters below the surface, the marine production cycle would eventually stagnate. Fortunately, the system has evolved a sophisticated means to address both of these problems. First, death in the living system continually produces a rain of particulate organic matter (detritus) that serves as a forage base for bacterial decomposition. This remineralizing process releases inorganic forms of the fertilizing elements back into the water column and at the seabed (Cooney, 2005). At depths below the productive surface layer (the photic zone), these elements increase their concentrations over time. In the Sound, the inorganic nutrients are returned to the surface waters each winter by wind and convective overturn.

Because the bacterial remineralizing process is oxygen dependent, without a periodic renewal, waters below the surface would quickly (in a few years) become devoid of oxygen (anoxic) and lifeless. In Prince William Sound and neighboring shelf waters, subsurface oxygen levels are replenished by the annual intrusion of seasonally available dense flows from the bordering ocean that replace the deepest shelf waters and refresh the dissolved oxygen (Weingartner 2007).

BIOLOGICAL RESOURCES

The subarctic marine ecosystem hosts a wide variety of species that occur together over a broad swath of the northeastern North Pacific Ocean and its coastal waters. A listing of common plant and animal populations is representative of collections and observations stretching from northern British Columbia to the eastern Bering Sea.

THE PLANKTON

The plankton community in coastal, shelf, and open ocean waters is composed of small, drifting or weakly swimming plant (phyto) and animal (zoo) species. Phytoplankton—the community of mostly tiny single-celled plants—represents an important primary producer base for the entire pelagic (water column) food web. Phytoplankters grow in the presence of sunlight in the upper 30 to 50 m using the fertilizing inorganic nutrients (nitrogen, phosphorus, silicon, and iron, among others) and dissolved carbon dioxide. During the subarctic winter (October to mid-March), low surface light and a deep wind-mixed layer inhibit this photosynthesis to barely measurable levels. However, in April and early May, increasing light, an abundance of inorganic nutrients supplied by the winter overturn, and a stabilizing mixed layer create a near-surface environment that supports explosive plant growth, usually most evident in April (Goering et al. 1973).

Lagging the peak of plant production by approximately 20 to 30 days, the zooplankton community (copepods, amphipods, krill, and many other taxa) eventually crop down the plant stocks, and by early to mid-June have reduced the primary producers to low levels (Cooney et al. 2001; Eslinger et al. 2001). The result of this feeding and growth is the development of a huge zooplankton layer in the upper 50 m that is composed of many different species but dominated in mass by the large-bodied copepod, *Neocalanus*. These tiny crustaceans swarm near the surface in densities approaching 1,000 individuals m^3 that attract many planktivores, including adult herring and pollock. The older juvenile stages of *Neocalanus* fill their bodies with lipid reserves to allow overwintering and spawning in the deep water (more than 400 m).

The resulting increased grazing, coupled with a strongly stratified water column that acts to inhibit further nutrient renewal, brings about the end of the spring production cycle. In some (but apparently not all) years, conditions in the fall may also promote a second plankton bloom.

Most planktonic taxa participate in daily (light/dark) vertical migrations; stocks that are deep during the day rise to the surface at night and then return at dawn. This migration increases the biomass of consumers in the upper layers for roughly half of each day. Krill, some copepods, amphipods, and pelagic shrimps are strong vertical migrators in Prince William Sound, presumably to feed and perhaps for reproductive purposes. In addition, a few taxa (interzonal species like *Neocalanus*) undergo a seasonal vertical displacement, growing in the productive upper layers in the spring and early summer, and then sinking to depth to reproduce and—like the salmon—die.

The reproductive strategy for the plankton is to respond quickly to optimal growth conditions, increasing their numbers by simple division (the phytoplankton) or by producing one or many large clutches of eggs (the zooplankton) that are dispersed by mixing and currents. Some zooplankters produce eggs continuously over an entire growing season as long as food is available. The resulting progeny generally peak by early July and then begin declining to what will be annual lows in February and early March (Cooney et al. 2001).

THE BENTHOS

The benthic or seabed community is comprised of plants and animals that live on or in the bottom and receive their primary nutrition from a combination of water column "fall out" and the production of shallow-water macroalgae and sea grasses. The large algae (kelps and others) and sea grasses are limited in distribution to the intertidal and shallow subtidal environments—the littoral zone—where light is sufficient to support photosynthesis for much of the year. This "bathtub ring" of plants is extremely productive locally, but makes up only a small portion of the overall plant production in the region. Benthic consumers include a variety of suspension and deposit feeders (snails, bivalves, sea stars, worms, cucumbers, shrimps, crabs, and barnacles, to mention a few; Feder and Jewett 1988). The deepwater benthos (more than 200 m) in Prince William Sound lives in and on a substrate of very fine-grained sediments (glacial flours) that are generally low in organic matter. At these depths, less than 20 percent of the organic matter synthesized in the surface photic zone reaches the seabed (Cooney and Coyle 1988).

Littoral stands of sea grasses and macroalgae provide critical nursery habitat for young fishes like juvenile Pacific herring and salmon. Kelp forests and other macroalgae stands are subject to destructive wave action during the winter months. Their standing stocks degrade and rebuild annually.

The reproductive strategy for the first order benthos is to release large numbers of mostly planktonic eggs and larvae that drift with tidal currents for days and sometimes weeks, seeking suitable settling habitat. Reproduction in seabed communities is extended in time from late winter to early fall, providing additional opportunities for larval forms to find and use critical habitat.

In addition to many demersal (bottom-dwelling) fishes, benthic invertebrate predators and filter feeders are represented by a number of commercial, sport, and subsistence shellfish stocks like king, tanner, and demersal crabs; northern pink and spot shrimps; little neck, butter, and razor clams; and mussels (Kruse 2007).

MARINE AND ANADROMOUS FISHES

The subarctic Pacific hosts a variety of fishes, including pelagic and benthic forms, as well as some species that require freshwater natal habitats (salmon, trout, chars, and some smelts). Dominant commercial species include Pacific salmon, Pacific herring, walleye pollock, Pacific cod, sablefish, several species of rockfishes, and many flatfishes including halibut and arrowtooth flounder. Sharks and rays are represented, as are sculpins. Of the nearly 300 different fishes reported for the Gulf of Alaska, a dominant few support sizable commercial, sport, and subsistence fisheries (Mundy and Hollowed 2005; Merrell 1988; and Rogers and Rogers 1986).

The majority of the marine fishes in the Gulf of Alaska reproduce by releasing large numbers (up to millions) of eggs and subsequent larvae into shelf and coastal waters each year. This bet-hedging strategy assures that while most will die from predation and starvation, some will almost certainly find suitable nursery conditions for growth and survival to adulthood. In addition, most of these fishes have prolonged life histories allowing them to seed the waters under a variety of different annually varying reproductive conditions over several years. Some, like a few of the rockfishes, may live in excess of 50 years.

The salmon—pink, chum, silver, king, and red—are anadromous, spawning in freshwater but growing on shelf and open-ocean forage resources (plankton, small fishes, squids, and krill). These highly desirable commercial, sport, and subsistence stocks spend one or two years in freshwater coastal rivers and lakes as juveniles, and one to five years in shelf and ocean feeding areas. In Prince William Sound, salmon stocks are dominated numerically by pinks and chums, followed by smaller numbers of kings, reds, and silvers. A large ocean ranching program, focusing primarily on pink salmon, produces annual returns of between 15 and 30 million adults. Nearby world-class populations of Copper River kings and reds support a highly sought after early-season fishery—one of the most lucrative in Alaska.

Among the most important forage resources for other fishes, birds, and mammals in the Sound are Pacific herring, capelin, eulachon, and sand lance—all species with high fat content. Other schooling juveniles like pollock are abundant but much less fatty and thus not nearly as desirable. Krill are suspected to be important, but there have been few local studies to confirm this.

MARINE BIRDS AND MAMMALS

Higher order consumers in the Prince William Sound food web include a large community of resident and seasonal birds and mammals. It is estimated that the Gulf of Alaska annually hosts nearly eight million seabirds representing 28 species (Springer 2005; Benowitz-Fredericks et al. 2007). Fork-tailed storm-petrels, black-legged kittiwakes, common murres, and tufted puffins are among the most common; all occur in the Sound and the nearshore coastal region at various times of the year. These birds are principally distinguished by their nesting preferences—cliff nesters or burrowers—and their feeding types—ranging from surface dippers to deep divers. The cliff nesters do so to avoid ground predators like rats and foxes. Most seabirds are fish eaters, focusing on the small schooling species mentioned above. Others, like the glaucous-winged gulls, are also predators and scavengers.

Studies of year-to-year variations in seabird production strongly suggest that food supplies limit population growth and that food kind and availability are the critical variables. However, predation and inclement weather also play important roles from time to time. Good and poor production years (chicks fledged per nesting adult) do not occur in synchrony over the broad region, but rather seem associated with localized conditions at the different colonies (Springer et al. 2007). However, in an overall sense, there is a suggestion in the observational data that seabirds in the Gulf of Alaska have been in a slow decline for the past two or three decades.

Relatively common marine mammal species in the Gulf of Alaska include 18 taxa in four major groups: baleen whales; toothed whales; seals and sea lions; and sea otters (Calkins 1986; Iverson et al. 2007; and Lowry and Bodkin 2005). Of the baleen species, the humpback is common in all seasons, feeding on a variety of forage including krill, herring, and occasionally juvenile salmon. It appears that overwintering humpback whales are increasing their numbers in the Sound, and may be targeting adult herring.

Of the toothed whales, killers and belugas are the most easily distinguished by their shapes, colors, and behaviors. Killer whales come in two varieties: transient and resident. Residents occur in pods at given locations and are fish eaters, whereas the transient variety are far ranging and feed primarily on seals, sea lions, and sea otters. In 2000, there were nine resident pods and one transient group in the Gulf of Alaska. Harbor and Dall's porpoise are common small-toothed coastal species. Little is known about them, their feeding habits, or reproductive strategies.

Harbor seals and sea lions are common residents of Prince William Sound. The former live more solitary lives hauling out, resting, giving birth, suckling young, and molting in the edge zone of the region. Little is known of their habits and feeding requirements while they are at sea. Harbor seals are commonly found on ice floes near the face of calving tidewater glaciers, where they presumably find food and may be protected from predators like killer whales.

Sea lions are gregarious, hauling out on rocky shorelines and other protective sites. These rookeries can host hundreds of individuals. Unlike the secretive harbor seals, sea lions are noisy and generally curious, readily approaching fishing vessels and tourist boats. Steeply declining numbers of sea lions in the northern Gulf of Alaska over the past two decades have been attributed, in part, to competition with industrial-scale fishing operations targeting primarily pollock. Marine pollution and disease are other possible causes.

Sea lions are generalists in their feeding habits, eating a range of fishes, invertebrates, birds, and other marine mammals. Studies in the Sound demonstrate night feeding on dielly (also known as diurnally) migrating Pacific herring which travel vertically in a 24-hour cycle during the winter as a major strategy. Herring schools, staging for their spring reproductive cycle, rise to the surface at night where they become easy prey for sea lions (Thomas and Thorne 2003).

Sea otters have enjoyed a checkered past in Prince William Sound. Nearly driven to extinction by early European fur traders and now protected by the Marine Mammal Protection Act, they flourish. Hunted only by the Alaska Native community, sea otters also fall prey to sea lions and killer whales. These small mammals are creatures of the edge zone, being limited to depths of 100 m or less (Springer et al. 2007). Also generalists in their feeding habits, otters take a variety of invertebrates, including clams, crabs, and small demersal fishes. In this regard, their trophic linkages are most strongly tied to a nearshore benthic food web, not one based primarily on plankton as are the other marine mammals—whales and seals. Sea otters are particularly fond of sea urchins and crabs. There is some suggestion that historically rebounding otter stocks in the Sound have been responsible, in part, for the recent disappearance of Dungeness crab stocks in the region.

Most recently, sea otters in Prince William Sound were particularly impacted by the *Exxon Valdez* oil spill of March 1989. Forced to rest in the surface and shallow waters, large numbers quickly became covered with crude. In attempts to clean themselves, the otters ingested oil, and many died from internal organ failure.

AN EMERGING ECOLOGICAL UNDERSTANDING

When taken together, the descriptions of the different components comprising the Prince William Sound marine ecosystem provide insights into the more holistic aspects of critical dependencies and responses over a range of different time and space scales. The lowest trophic levels—primary producers and first order consumers—are tightly coupled to the strident meteorology and ocean physics characteristic of the northeastern Gulf of Alaska, characterized by shifting seasonal light levels and the ephemeral availability of critical inorganic nutrients in the upper layers.

As a result, the base of the production cycle toggles back and forth between light-limited (oceanographic winter) and nutrient-exhausted states (oceanographic summer). During the short transitions between these broad limiting conditions, a time is reached each year when weak stability in the upper water column establishes the critical conditions for a short, explosive period of spring (and perhaps fall) photosynthesis. The spring production period is short because the physics are rapidly moving toward a more stable, highly stratified, but nutrient-poor summer condition. The burst of growth is immediate because most of the primary producers (phytoplankton) have population replacement times as short as hours and days. This rapid doubling quickly builds seasonal plant plankton highs that are at first slowed and then eventually reversed by losses to grazers and nutrient exhaustion.

At the leading edge of the spring plankton bloom, the zooplankton community is present in low abundance, composed of survivors of the winter production hiatus. However, within a month or less, the small-bodied grazing community builds its numbers, and in the process captures a large portion of the energy synthesized by phytoplankton. Much of this "capture" is carried forward into the less productive summer season to feed rapidly growing juvenile fishes and other pelagic consumers.

In all years, varying portions of the phytoplankton bloom escape the water column to help feed a seabed community. The amount sinking to the bottom is apparently related to the duration of the bloom and the depth of the water. When the bloom is short but intense (quickly limited by declining nutrients), coupling to water column grazers is diminished and more organic matter reaches the seabed than when conditions extend phytoplankton production, allowing the pelagic consumers to harvest a much greater share. The deeper the water, the more organic matter is utilized above the bottom. The energy captured by photosynthesis is transferred to consumers throughout the food web.

This story is less well understood for the summer-to-winter transition. In mid- to late September, the upper layers of the Sound begin shifting from the strongly stratified summer condition (minimal inorganic nutrients) back to a deepening mixed water column more characteristic of winter (plant nutrients once again building in the surface waters). Depending on the time course and the magnitude of the returning Aleutian Low Pressure system, and the fall storms and runoff that come with it, there may be a period of overlap during some years when declining light and inorganic nutrient levels are sufficient to support a fall plankton bloom. Unfortunately, there have not been sufficient observations in the fall to verify whether this bloom is a regular feature or only occasionally present. Ongoing studies of Prince William Sound Pacific herring and their fall nutritional needs should help resolve this issue.

The timing of the spring plankton bloom (April) signals the beginning of the annual integrated growth cycle for the entire Sound, to include (but not limited to) the following phenomena: (1) the rapid production of upper-layer zooplankton stocks following the plant bloom, (2) krill swarming and reproducing at the surface, (3) pollock spawning in the deep water, and (4) herring depositing hundreds of millions of eggs at specific littoral locations around the region. These massive reproduction events—particularly the krill and herring spawning—draw a combined feeding frenzy by other consumers that supports large numbers of fish, bird, and mammal predators, and when appropriate, a local herring fishery. At this same time, seabirds begin returning to their rookeries, and juvenile pink and chum salmon begin leaving the region's freshwater natal areas and hatcheries. All these populations demand large quantities of forage as they rebound from the relatively unproductive winter season. In the Prince William Sound ecosystem, much of this forage is provided by a handful of abundant small fishes: capelin, eulachon, juvenile herring, sand lance, and juvenile salmon being the most important. The massive near-surface zooplankton layer (krill and copepods) is also an important forage resource.

Shortly after the peak of the spring plankton bloom, millions of seasonally migrating shorebirds and thousands of waterfowl (geese and swans among them) follow the east Pacific Flyway stopover, or establish nesting sites on the productive Copper River flats and immense freshwater delta. Shorebirds apparently track herring spawning events in the eastern Gulf of Alaska, deriving a huge energy boost by feeding on the small lipid-filled eggs when they can be found (Bishop and Green 2001).

As the season progresses into late spring and early summer, a second and equally impressive wave of biological activity occurs: adult salmon begin returning to hatcheries, natal streams, and rivers, and a large fishery begins. From approximately 300 mt (metric tons) of juveniles entering the Sound each year, a returning stock of between 25,000 to 50,000 mt of adults rush into the region to spawn. This return is closely monitored by Alaska Department of Fish and Game to assure a sufficient escapement of wild adults for each species. Red and king salmon are the first to arrive, followed shortly by pinks and chums. By late September, all but some stocks of silver salmon have completed their runs.

Once again with the salmon, the edge zone of the region is the site of intense reproduction and prey-predator activity. Bears, river otters, gulls, and eagles (to mention a few) glut on the dead and dying carcasses of spawned-out adults and their eggs. Salmon sharks and killer whales patrol the nearshore waters, and the region is dotted with a variety of commercial, sport, and subsistence vessels, large and small. At the regional hatcheries, millions of salmon eggs are taken from returning adults to restock incubators. Adults not required for the "egg take" are sold to help defray operating costs. The bulk of the hatchery return enters a large Prince William Sound common property fishery where it can make up 80 to 85 percent of the annual salmon landings.

In the streams and rivers, the dead and decomposing wild salmon relinquish their marine-derived nutrients to both freshwater and terrestrial riparian ecosystems. Here the ocean "feeds" the edge zone composed of nursery lakes, rivers, and streams some hundreds of miles to the north in the Copper River watershed.

By late September, the Prince William Sound ecosystem begins sliding back into the low production regime that is characteristic of winter. Seabirds leave their rookeries, juvenile salmon join migrating stocks in the coastal current, and a few pelagic and benthic fishes and invertebrates begin moving to deeper water, presumably to avoid the cooling, turbulent upper layers. Juvenile herring migrate to the shallows at the heads of fjords and embayments, where they spend the winter months, sometimes under a thin sheet of ice. Adult herring begin massing in selective fjords and bays. A long period of rest settles over the Sound, and the seasonal cycle is complete.

AN EMERGING UNDERSTANDING OF OUR PLACE IN THE ECOSYSTEM

The review and general summary of the biogeophysical and ecological settings of Prince William Sound and the coastal Gulf of Alaska present a picture of how the seasonally dominant weather patterns and longer meteorological cycles affect the region's living resources. This information is expected to become increasingly important as the global phenomenon of climate change plays out. The relationships described here between the living and nonliving parts of the system portray a complicated set of dependent populations and processes that result in the annually pulsed production of organic matter, which is then moved through an immense food web, with consequences for both marine and freshwater environments. As our views of the complexities of the ecology of the Sound become more sophisticated and accepted over time, the resulting information will enter related fields of knowledge where decisions about resource use and sustainability are debated and resolved.

In this regard, the author looks forward to a time when the findings of marine and terrestrial science and the questions of stakeholders, the general public, and resource managers come together in discussions crafted to produce an increasingly more thoughtful use of the natural world. The insights that follow are provided in the spirit of encouraging those kinds of interactions.

Coming to know the natural world through a variety of different recreational activities introduces each of us to the grandeur of nature and informs our ideas about how nature "works." These insights provide the personal avenues that citizens navigate to encounter the important political arenas where multiple-use decisions are debated and crafted. Prince William Sound and the surrounding landscapes and seascapes represent a world-class natural laboratory within which residents and "visitors" can encounter, appreciate, and enjoy the splendor of nature.

Coastal marine and freshwater habitats, though somewhat less well understood than their terrestrial counterparts, similarly require shared management efforts. The example of wild salmon life histories demonstrates dependencies that play important roles in the protection of both reproductive habitat and a resource with huge sport, subsistence, and commercial value. Understanding how nature "works" in this case is key to providing resource protections that bridge critical freshwater and marine environments.

Sharing information and its interpretation is key to trust building, idea generation, and conflict resolution. It is only in recent decades that marine ecosystems have emerged from "black-box" status and become contributors to the resolution of serious resource issues. In community engagements between stakeholders, the public, marine and terrestrial scientists, and educators and resource managers, it is incumbent upon the science community to take the lead—when appropriate—in assisting others to appreciate what "understanding" ecosystem form and function may mean, both generally and in the case of local resource use. For many in the science community, this is not a task that is easily shouldered. However, to be "part of the solution," plain speaking and clarity go a long way toward useful applications of findings and discussions of important theory. In a recent instance where a diverse community of users focused on questions of marine ecosystem restoration—Puget Sound, Washington—the approach was significantly assisted by clearly stated descriptions of how the Sound was configured by nature to support the populations that were the focus of restoration efforts. This success underscores the importance of including the results of marine ecosystem science in forums debating the use, enhancement, and regulation of natural systems.

STAKEHOLDER ESSAY

LOCAL KNOWLEDGE, LONG-DISTANCE JOURNEYS

NANCY BIRD

I WENT TO THE SLIDE SHOW in January 1977 attracted not only to the topic "The Undersea World of Prince William Sound" but also because this chamber of commerce luncheon meeting would help me meet Cordova "adult" residents. I'd come to town to work at the new youth center focused on providing healthy after-school activities for the 10- to 18-year-old crowd. The daylight hours were short, and the rain (and wind!) had rarely abated in the three weeks I'd been in Cordova.

Rick Rosenthal, a biologist with Alaska Department of Fish and Game, surprised me that day with an amazing world of colorful and diverse marine life. I'd snorkeled in the Red Sea and, while Prince William Sound is totally different, I'd never expected cold waters to be so rich and beautiful. Now, almost 35 years later, we've learned much about this region but still know too little about the rockfish and lingcod that Rick first showed me.

Understanding the basics, like where and when the currents flow, is an area where gains in knowledge have been made. In 1981, the tragic crash of a Coast Guard helicopter—while trying to assist a Copper River fisherman—graphically demonstrated the flow of currents from the Gulf of Alaska into Prince William Sound. The chopper went down near Hook Point on Hinchinbrook Island, and it was two days before it was seen again, bobbing in the mid-Sound waters near Naked Island.

Today, we'd have a lot better idea of where to look for surface objects, though we're still a long way from understanding currents at various depths. There is recognition of the complex current structures and gyres at different places and depths. The flow of these currents determines a lot about herring, salmon, and other fishes' survival. Residents here depend on these resources, and many are highly observant about their environment.

It was one drifter buoy deployed in May 1997 that corroborated the "anecdotal" though scientifically collected knowledge of 82 randomly chosen mariners and fishermen. Interviewed about the currents in the Copper River Delta, they maintained that an eastward flow occurred almost every year. That flow pattern is contrary to the prevailing east-to-west flow banked on by regulators as they assessed the risk to the Copper River Delta region of an oil tanker spill.

The 1997 drifter buoy first traveled southwest in the Sound, following a common flow pattern, but then turned north and exited the Sound east through Hinchinbrook Entrance, spending weeks bobbing eastward to Kayak Island before finally reversing itself. After drifting southwest, it reentered the Sound through Montague Strait and again flowed north, circumnavigating Montague Island before finally beaching itself near Seward in September.

It was the combined evidence of the local residents' knowledge and science that changed the tone of discussions about the risks of an oil spill staining the Copper River Delta and its world-renowned salmon. The result is that regulators do require oil spill drills for the Delta and recognize that we need to be prepared for responses to conditions of both prevailing and nonprevailing currents.

Since I made Cordova home in 1976, the world has seen dramatic technological shifts. I sometimes hesitate to call them "advances," but it is amazing that we can now implant an acoustic transmitter into the belly of a lingcod in Port Gravina and learn from an elephant seal where that lingcod traveled.

The story of the elephant seal and lingcod actually brings me back to Rick Rosenthal, who eventually left Alaska to focus more on his photographic skills. He produced a documentary on elephant seals and may well have photographed the seal from Santa Cruz, California, that was tagged in early 2010 with a combination transmitting and receiving acoustic tag. This relatively new "business card" tag records other transmitters the animal's tag hears as it travels throughout the oceans and also sends out its own signal to be recorded on receiving devices it may encounter.

It was not a surprise that this male elephant seal swam thousands of miles north to near Yakutat, Alaska. What was surprising was to learn through the elephant seal's tag—retrieved on a California beach in April 2011—where the lingcod from Prince William Sound had crossed paths with the elephant seal, at that spot southeast of Yakutat on April 17, 2010. The lingcod had traveled over 300 miles from Port Gravina in the year since it was tagged in April 2009. That was exciting new data on lingcod, as we've thought this species stayed relatively close to its "home" turf.

How exactly did we learn all this? The elephant seal had a satellite tag as well as the acoustic tag, and its movements were tracked via satellite. We know it returned close to its breeding colony by Santa Cruz before disappearing from the satellite data records. Its satellite tag washed up on a beach, and a few weeks later, in April 2011, its acoustic recorders were found in the intertidal zone on a low tide. Data from its recorders—including water temperature and depth as well as the encounter with the lingcod—were downloaded and shared with scientists and others through Internet websites.

It's certainly my hope that observations by local residents, scientists, and visitors will continue being collected and analyzed to keep improving our understanding, management, and use of Prince William Sound's resources. It's my home, and I want it to remain as beautiful and abundant as I first found it.

CHAPTER 3

A BRIEF HISTORY OF HUMANS IN PRINCE WILLIAM SOUND

PAUL TWARDOCK

INTRODUCTION

PRINCE WILLIAM SOUND'S natural beauty, abundant fish and wildlife, relatively pristine environment, and unpopulated landscape attract recreational users in growing numbers. However, the concept of Prince William Sound as an uninhabited wilderness is new for an area populated for thousands of years. Alaska, "The Last Frontier," and the Sound have seen dramatic changes in the size and type of population since Europeans made contact with native indigenous groups. The hypothesis of this chapter is that the Sound's population, starting with the Alaska Natives, ebbs and flows with social and economic changes, and that during the late 1960s, the population was at a low point when modern-day wilderness-seeking recreationists first visited the Sound. The attraction of uninhabited wilderness was a powerful one in America and Europe. As Catton (1997) points out, the idea of wilderness existing, even as an unvisited place, is appealing for many Americans.

EARLY HISTORY

The Sound's Alutiiq people inhabited the Sound for at least 3,000 years before contact with Captain James Cook in 1778 (Lethcoe and Lethcoe 1994). Early European explorers estimated the Prince William Sound Alaska Native population at around 700 to 1,000 in eight clans, though that estimate was probably low (Lethcoe and Lethcoe 1994; Wooley and Haggerty 1993). With the establishment of the Russian trading post at Port Constantine in 1793 (Chugach Alaska Corporation 2011), Europeans started the first commercial venture in the Sound that forever changed human use patterns. Alaska Natives were quickly affected by the Russians' presence. For a variety of possible reasons—disease, trading, and others—they slowly abandoned their summer camps and villages, moving to a few larger villages. The native population dropped to less than 300 by 1880 (Lethcoe and Lethcoe 1994). One of the first "tours" of the Sound occurred when Corporal John Ledyard of Cook's expedition was stuffed headfirst into a kayak for a trip to shore (Twardock 2004). The Europeans' interest in the Sound was not diminished by Ledyard's experience, and today recreation is one of the Sound's principle human uses. During the 1800s, the population of sea otters, the Russians' primary interest, declined. In 1867, the United States bought Alaska from Russia (Naske and Slotnick 1987), a prelude to the next great population shift in the Sound.

EARLY AMERICAN DECADES

A series of American military and scientific expeditions explored the Sound in the late 1880s. The Abercrombie and Allen expeditions pioneered routes into the Interior, while others explored the Sound itself. The information they gathered generated interest in the Sound's resources, including salmon and minerals. The first cannery was opened around 1888 near Cordova, and the first fox farm was established on Seal Island in 1894. With the decline of sea otters and the 1911 hunting ban, the fox farming industry would eventually establish farms on nearly every island in the Sound (Lethcoe and Lethcoe 1994; Wooley and Haggerty 1993).

Gold was discovered at Sunrise in Cook Inlet in 1895, and in 1896, when the famous Klondike gold rush started, the SS *Dora* unloaded miners at the head of Passage Canal for the trek over Portage Pass and the Hope mining district (Taylor 2000). In 1897, copper was discovered in the Sound, and the news of Klondike gold reached the public. Miners flooded north. Valdez was promoted as an All-American route to the Klondike, and in 1898, 3,000 to 4,000 people came to Valdez headed for the goldfields. Valdez was created, and many stayed behind, prospecting and making the Sound their home. In the early 1900s, mining was prevalent throughout the Sound. Lethcoe and Lethcoe (1994) report that 48 gold mines had been established between Valdez and Columbia Bay by 1912, nearly all of Knight Island had been staked and explored, and the Latouche town site had around 500 residents by 1920. A mail boat made a regular trip through the western Sound, serving 21 stops and over 350 people. The Port Wells district boasted a post office and the second largest gold mine in the Sound: the Granite Mine, with 50 men.

The mining era came and went relatively quickly. After World War I, the collapse of copper prices, played-out mines, and the enactment of the Jones Act requiring ships to be registered in the United States all contributed to the demise of mining. By the 1930s, most of the mines had shut down: Ellamar in '29; Latouche in '30. The big mines—Kennicott, Latouche, Ellamar, and Latouche—were completely or partially destroyed by bulldozers or nature by the early 1970s.

The Chugach National Forest was established in 1907, with the first ranger based in Cordova in 1908 (Lethcoe and Lethcoe 1994). The national forest encompassed the entire Sound, the Copper River Delta, and much of the Kenai Peninsula. Land ownership became uncertain as many longtime residents discovered they had no title to their fox farm, cannery, or home. This started a period of complex land negotiations that continue today.

Early tourism consisted primarily of the Alaska Steamship Company, the Alaska Railway, and the Copper River and Northwestern Railway, offering tours between the Sound and Fairbanks. The first commercial air traffic between Seattle and Anchorage started in 1946, and the Richardson Highway to Valdez was paved and open year-round in 1956. Chuck West restarted the "Golden Circle" tour with the Alaska Steamship Company in 1947, adapting the tour to include air travel out of Anchorage in the 1950s. In 1957, the *Jim Alice* started tours out of Valdez to Columbia Glacier and offered fishing trips. In 1958, Brad Phillips and Wally Hickel started day tours in the *Gypsy* out of Whittier, which shifted to Valdez when the army stopped rail service to Whittier in 1960 (Lethcoe and Lethcoe 1994). Also in the postwar era, vagabond Europeans toured the world with their folding kayaks, some making it to Alaska (McNamara 2010).

AFTER STATEHOOD

In 1959, Alaska was granted statehood and 103,350,000 acres, to be selected from federal lands (Naske and Slotnick 1987). By 2009, the state had title to 89,500,000 acres (ADNR 2009). Additionally, the University of Alaska was granted land as part of its land grant university status; 2,438 acres in six units were eventually transferred to the university (ADNR 1988). Ultimately, the state would select land in the Sound for 14 state marine parks, primarily near Valdez and Whittier.

On March 27, 1964, the Good Friday earthquake devastated the Sound. What was left of the canneries, mines, and fox farms was all but wiped out, either by the quake itself or the ensuing tsunamis. The village of Chenega was destroyed, with a loss of 26 lives. Valdez's waterfront was destroyed. Whittier dropped over five feet, and its small boat harbor and rail terminal were damaged. (Smelcher 2006; Taylor 2000). The cannery at McClure Bay was destroyed, as were smaller fishing operations throughout the Sound. The survivors of the tsunami at Chenega were scattered: some moved to Tatitlek, others to Valdez, Cordova, or Anchorage. For the first time in thousands of years, few year-round residents lived in the Sound outside of Cordova, Tatitlek, and Valdez.

With the tragedy, however, came the seeds of renewal. Some of the hundreds of millions of federal dollars that came to rebuild Alaska went to the Sound (Erskine 1999). Valdez was relocated to stable ground. Whittier's waterfront and rail line were rebuilt with tourism and shipping in mind. Brad Phillips started tours between Whittier and Valdez in 1964 with the reopening of railroad service postearthquake. Also in 1968, and far to the north, oil was discovered in Prudhoe Bay (Naske and Slotnick 1987), leading to dramatic changes in the Sound.

Partly to enable the Trans-Alaska Pipeline to be built from the North Slope to Valdez, the Alaska Native Claims Settlement Act (ANCSA) was passed in 1971. ANCSA created the Chugach Native Corporation (CNC) and allowed CNC and village corporations to select lands in the Chugach National Forest. CNC has also selected around 300 historical/cultural sites, many at the prime camping/recreational sites in the Sound. A few have been conveyed to CNC (Lethcoe and Lethcoe 1994). During the 1960s and 1970s, the salmon returns were inconsistent, with fishery closures common. In the early 1970s, the state took steps to more actively manage the salmon fisheries. Flush with oil revenues, the state built or facilitated building hatcheries on Evans Island (1975), Cannery Creek in Unakwik Inlet (1978), Main Bay (1981), and Esther Island (1985) (PWSAC 2017). The hatcheries stabilized salmon runs and led to the development of a strong fishing community. Statehood and oil revenues also led to the development of the Alaska Marine Highway and increased tourism (Catton 1997). The M/V *Tustumena* started ferry service between Seward, Valdez, and Cordova in 1964. The M/V *Bartlett* served Whittier, Cordova, and Valdez starting in 1970 (Johannsen and Johannsen 1975).

Kayaking, particularly in Folbot and Klepper kayaks, became popular and many paddlers found their way to the Sound. In *Fabulous Folbot Holidays*, printed circa 1973, Loyette Goodell described paddling a Folbot with her husband John from Whittier to Port Nellie Juan (Kissner 1973). The National Outdoor Leadership School (NOLS) started month-long kayaking courses in 1971 (Johannsen and Johannsen 1975; J. Niggemyer, personal communication). In the early 1970s, the first charter boat also started to operate out of Whittier and Valdez, shuttling hunters and kayakers to remote camps. Also in 1971, Stan Stephens started boat tours from Valdez, and in 1974, Jim and Nancy Lethcoe started Alaska Sailing Safaris (Lethcoe and Lethcoe 1994). By the mid-1970s, tourism had become well established. Johannsen and Johannsen reported multiple tour and charter boats operating out of Whittier and Valdez by 1975. There was even a backcountry lodge, the Prince William Sound Inn, at Thumb Bay on Knight Island. There were at least six USDA Forest Service public-use cabins for rent at five dollars per day (Johannsen and Johannsen 1975). The Bear Brothers Outfitters was started in 1974 by Ted West and Eric Singer, both former National Outdoor Leadership School (NOLS) instructors, and was based at the south end of

Culross Island. Kelley Weaverling took West's place and helped run it until 1981. The Bear Brothers lasted until 1982. In the meantime, NOLS continued to increase the number of kayaking courses during the 1980s, as did other outfitters, such as Prince William Sound Kayak Center. The first cruise ships started using Whittier in 1983 (Lethcoe and Lethcoe 1994).

The Alaska Native Claims Settlement Act allowed the federal government to permanently withdraw lands for certain purposes such as parks and wilderness designation. After almost a decade of negotiations, Congress passed the Alaska National Interest Lands Conservation Act (ANILCA) of 1980. As part of ANILCA, the 800,000 ha Nellie Juan Wilderness Study Area (WSA) was established. Covering most of western Prince William Sound, the Nellie Juan WSA is managed by the Chugach National Forest, which manages it primarily as de facto wilderness, though successive management plans recommend less wilderness than exists in the wilderness study area.

In 1984, the village of Chenega was relocated to Chenega Bay on Evans Island with 21 new homes built (ADCCED 2011). By the late 1980s, the Sound was, if not populated, at least rebuilding after decades of relative human absence. Then, on March 24, 1989, the *Exxon Valdez* grounded and spilled more than 11 million gallons of crude oil. The Sound was once again suddenly altered.

AFTER THE OIL SPILL

The immediate effects on the environment are well documented: thousands of animals and birds perished, and thousands of miles of shoreline were oiled. The cleanup, however, also brought impacts. Over 10,000 people worked on the cleanup, resulting in a long-term increase in the population of Valdez (Wooley and Haggerty 1993). Many cleanup workers experienced the Sound for the first time, some to return on their own. Exxon settled with the state and federal governments, paying $100 million in criminal charges and $900 million in a civil settlement. The money funded extensive studies and infrastructure improvements such as state park cabins. State money also went to improve access to Whittier, helping create a one-lane road.

The Whittier access project was projected to bring one million users to the Sound. Though that has never occurred, businesses responded by expanding, not only their capacity but their marketing. The increased publicity, easier access, and better facilities have contributed to a steady increase in recreational use. The civil settlement also funded habitat acquisition, in some places providing access to formerly private lands. Despite the spill, or because of it, one study showed an annual increase of recreational use of 7.5 percent between 1987 and 1998 (Twardock and Monz 2000).

Since the spill and opening of the Whittier Tunnel in 2000, recreational use has continued to climb. However, the future of human use in the region is less certain.

TABLE 3.1. Economic and demographic characteristics for Prince William Sound Communities (Chugach census subarea) relative to the adjacent Anchorage Municipality and Kenai Peninsula Borough

FEATURE	MUNICIPALITY OF ANCHORAGE	KENAI PENINSULA BOROUGH	CHUGACH CENSUS SUBAREA
Population and housing			
Population in 2014	300,549	57,212	**6,707**
Population in 2011	295,920	56,623	**6,733**
Average annual growth, 2011–2014	0.5%	0.3%	**-0.1%**
Occupied housing units in 2010	107,332	22,161	**2,676**
Employment and income			
Residents employed in 2013	130,673	23,909	**3,152**
Private sector	85%	80%	**74%**
Local government	8%	14%	**20%**
State government	7%	6%	**6%**
Median household income	$77,454	$61,793	**$91,338**
Fishing and subsistence			
Number of limited entry permit holders who fished	338	1,097	**334**
Estimated ex-vessel value of fish harvested	$46,630,382	$136,807,046	**$62,137,013**

Source: Alaska Department of Commerce, Community, and Economic Development (ADCCED 2011)

The global economic downturn and energy prices resulted in a three-year decline in total tourism to Alaska (ADCCED 2010). This report did, however, identify potential near-term increases in visits from foreign travelers and highlighted the importance of small coastal towns to destination travelers to the state. How this translates to tourism visits to the Sound is difficult to say, but it's safe to assume that the Sound's natural beauty and wildlife will continue to attract people: to play, live, and subsist in this unique wildland setting (see table 3.1).

STAKEHOLDER ESSAY

PERSONAL JOURNEYS, TRANSIENT TRACES

NANCY LETHCOE

JIM AND I FIRST VISITED PRINCE WILLIAM SOUND in 1969, taking a month to sail from Whittier to Valdez and back in our 21-foot boat with our daughter, a friend, and a German shepherd. The only chart still indicated that the mouth of Passage Canal was a closed military zone. The only place to launch a boat in Whittier was Smitty's Cove. And, there was, of course, no harbor.

At month's end, our lives were changed forever. Instead of pursuing our academic professions, we decided that Prince William Sound would be our careers. The following winter we visited the offices of the Chugach National Forest, attended meetings about clear-cutting in the Sound, and met a few kindred spirits who saw the recreational and tourism potential of the Sound's wilderness characteristics—the unlimited opportunities for tranquility, solitude, and silence; the intriguing signs of the presence, activities, and passing of former inhabitants; and the unimaginably spectacular scenery and wildlife watching opportunities.

Between 1969 and 2007, when Jim became ill, we spent about five months in the Sound every year, and for four years, we also overwintered in various parts of the Sound. We estimate that during those 38 years, we spent over 4,500 nights swinging on our anchor. We tried to make our presence in the Sound as trackless as the flight of birds.

The more time we spent exploring the beaches and uplands, the more we wanted to learn about the area's history. Here, on the beach, is a half-uncovered rusty chain. We follow it into the woods and find alder growing through the planks of a shipwreck. The steering wheel is still upright, entwined in alder branches. Whose boat? What happened?

On another beach, we find the wreck of an old dock rolled up above the high tide line. In the woods are the moss-covered outlines of old cabins. A two-hour walk inland, we find a small adit, or mining tunnel. What is the story here? The adit is too small to justify such a large and expensive dock.

Kayaking along the shore, we explore the sea caves. But this isn't a sea cave, it's an old mining adit. The rails and an ore cart are still in it. For how long, if ever, was it profitable?

We have just made a steep switchbacking ascent towards a ridgetop using lots of slide alders as handholds. As we stop to catch our breaths, we notice mine tailings, cables, and a vanishing cabin. We wonder how the miner managed to get so much stuff up the mountain and his ore down.

Walking along a retreating shoreline, I find a flat rock. Nearly a century ago, someone carved a memorial on it to infant Rose. The remains of a fox farmer's cabin are in a hollow near the shore. A room containing winter wood protects the southeastern wall.

Elsewhere, we find fox feeding pens, wooden dams, strawberry gardens, collapsed cabins, old fish trap anchors, cables around trees, a barge moved inland by the 1964 tsunami, bricks, docks, pilings, marine rails, sunken boats that catch our anchor, a child's weathered bones leaning upright against an overhanging cliff. All are as transient as the glaciologists' photographic stations.

It is good to feel that we are also part of the Sound's long human history. As recreationists and tourists, may we be trackless. May we only look and not destroy. May everyone leave historical artifacts in place for all to discover and enjoy. May those who come after us have the good fortune to see and contemplate what we have seen. For our transient human history, too, is part of the wilderness experience.

CHAPTER 4

THE SOCIAL AND ECONOMIC IMPACTS OF THE *EXXON VALDEZ* OIL SPILL AND A PATH TOWARD RECOVERY

AARON J. POE AND RANDY GIMBLETT

INTRODUCTION

IN 1989, THE HEART of Prince William Sound was severely impacted by the *Exxon Valdez* oil spill (EVOS), which disrupted species and habitats vital to the biodiversity of the region and thus disrupted social and economic services within the Sound's human communities. In the years since the spill, the Sound has experienced numerous changes, some of which were certainly set in motion by that catastrophe. Others, like changes in the way people access the area, changes in ocean uses, and fisheries and tourism practices, as well as environmental changes in coastal systems (e.g., climate change), clearly affect this region as well. It is not the intent of this volume to document the entirety of damage caused by the spill, but almost nothing can be said about Prince William Sound without considering the context and the continued social and biological impacts of this disaster.

An estimated 11 million gallons of North Slope crude oil spilled into the marine waters of Prince William Sound following the grounding of the tanker *Exxon Valdez* on Bligh Reef, some twenty miles from the terminus of the Trans-Alaska Pipeline in Valdez, Alaska. In March 1989, oil immediately began to spread south and west through the Sound, then out along the western coast of the Gulf of Alaska (figure 4.1). In total, some 1,400 miles of coastline were impacted by the spill, with oil spreading 600 miles from Bligh Reef as far west as Kodiak and the Alaska Peninsula (EVOS Trustee Council 1994).

FIGURE 4.1. Map produced by Alaska Department of Natural Resources, Land Records Information Service, showing the track of the *Exxon Valdez* oil spill.

It is estimated that approximately 40 to 45 percent of the oil washed ashore in the intertidal zone, with heavy oiling taking place along some 220 miles of shoreline, much of it within the Sound. Oil contamination had devastating impacts on natural resources throughout the region. Some effects were unintentionally further exacerbated by cleanup activities that took place for months at a time in 1989, 1990, and 1991. The impact to resources had compounding devastating effects on the human communities dependent on the region (EVOS Trustee Council 2010), some of which were felt most acutely in the Sound. Much of the oil has diminished, but pockets of unweathered oil still persists within the intertidal zone of the Sound, and its presence continues to potentially harm organisms (EVOS Trustee Council 2010) and serve as a reminder to many residents of the Sound the great blow their communities were dealt a generation prior.

Though much has been documented with respect to the effects on biological communities and resources impacted by the spill, impacts to the culture and economies as well as the sociological well-being of Sound communities are less well understood. Community studies from spill-affected areas found increases in substance abuse, divorce rates, reports of domestic violence, and mental illness following the spill that have persisted many years since the event (Picou and Martin 2007). Some long-term Cordova residents still say that the oil spill served to divide the community into groups of those who ultimately may have profited from lucrative spill cleanup contracts and those whose livelihood was destroyed by the damage done to commercial fisheries, with this division still resonating a generation later.

In the aftermath of the spill, the State of Alaska and the United States were successfully awarded $900 million in claims against Exxon Corporation. The *Exxon Valdez* Oil Spill Trustee Council (the Trustee Council) was formed in 1991 and entrusted with overseeing the use of these funds to restore the damage done by the spill. In order to guide restoration efforts, the Trustee Council adopted a restoration plan which helped define what was needed in terms of research, monitoring, habitat protection, and other general restoration activities (EVOS Trustee Council 2010). The Trustee Council adopted an official list of resources and services injured by the spill to help guide efforts under its restoration plan. The Injured Resources and Services List, or "the List," as it came to be known, served three main purposes: (1) identify natural resource and human service injuries caused by the oil spill and its subsequent cleanup efforts; (2) guide the restoration plan when it was first adopted, in terms of expenditure of public restoration funds, to assist the trustees and the public in ensuring that money was expended on resources that needed attention; and (3) become the source of recovery status for injured resources so that the trustees and public could monitor recovery of ecological function and human services.

Under the guidance of the recovery plan and the List, significant investment has been made toward resource recovery. More than $180 million has supported numerous research efforts to assess damage and recovery of the Sound's ecological systems. Another $375 million has funded habitat protection—often in the form of large-scale land acquisitions. An additional $65 million is committed to other previously approved, ongoing projects and habitat purchases. Overall administrative costs for the Trustee Council's efforts have accounted for $45 million, with another projected $18 million to $22 million needed to see the council through 2032. As of 2010, the Trustee Council holds approximately $65 million to $70 million available for further research, monitoring, and restoration, as well as a remaining $24 million for habitat acquisition and protection (NOAA 2010a).

The Trustee Council classifies recovery status for these natural resources based on the results of the most recent research into the categories described in table 4.1. As of 2012, an approximately equal number of resources are classified in the still recovering and recovered categories. Two species, pacific herring and pigeon guillemots, remain classified as not recovering. Table 4.2 shows recovery status in 2012 as described on the Trustee Council's website.

TABLE 4.1. Definitions established by the Trustee Council to describe recovery status of injured natural resources

Not recovering: These resources continue to show little or no clear improvement from injuries stemming from the oil spill. Recovery objectives have not been met.
Recovering: Recovering resources are demonstrating substantive progress toward recovery objectives but are still adversely affected by residual impacts of the spill or are currently being exposed to lingering oil. The amount of progress and time needed to attain full recovery varies depending on the species.
Very likely recovered: While there has been limited scientific research on the recovery status of these resources in recent years, prior studies suggest that there had been substantial progress toward recovery in the decade following the spill. In addition, so much time has passed since any indications of some spill injury, including exposure to oil, it is unlikely that there are any residual effects of the spill.
Recovered: Recovery objectives have been met, and the current condition of the resource is not related to residual effects of the oil spill.

TABLE 4.2. Recovery status of resources harmed by the *Exxon Valdez* oil spill as determined by the Trustee Council as of May 2012

RECOVERED	RECOVERING	VERY LIKELY RECOVERED	NOT RECOVERING
Archaeological resources	Barrow's goldeneyes	Cutthroat trout	Pacific herring
Bald eagles	Black oystercatchers	Rockfish	Pigeon guillemots
Common loons	Clams	Subtidal communities	
Common murres	Designated wilderness areas		
Cormorants	Harlequin ducks		
Dolly Varden	Intertidal communities		
Harbor seals	Killer whales		
Pink salmon	Mussels		
River otters	Sea otters		
Sockeye salmon	Sediments		

INJURY TO HUMAN SERVICES

Less research has been done to address the four dependent human services: passive use, recreation/tourism, subsistence, and commercial fishing. According to the restoration plan, the following impacts to these services took place as a result of the oil spill: (1) reduced the physical or biological functions performed by natural resources that support human services or (2) reduced aesthetic and intrinsic values or other indirect uses provided by natural resources.

In the immediate aftermath, recreation and commercial fishing use avoided areas affected by the spill. This led managers at the time (e.g., Hennig and Menefee 1995) to speculate that use patterns had likely shifted and people might be spending time in parts of the Sound that had been previously less used. Subsistence harvest efforts were dramatically curtailed throughout the spill-affected region, with many households reducing their overall harvest and having to shift from traditionally used areas. In the years immediately following 1989, the public perception was that the Sound was still reeling from damage done by the spill. More recently, there has been an increase in use of the Sound, possibly because of people's desire to see the area they have heard so much about and to experience it for themselves. This, combined with the fact that thousands of people "lived" in the Sound while working on cleanup efforts elevated awareness of the region, even locally within Alaska, and resulted in new interest in exploring its shorelines (chapter 3).

All four human services are still described by the Trustee Council as recovering from the effects of the spill. This is based on the assumption that each injured service is inextricably linked to the state of other injured resources in the region, and thus full recovery of the spill area cannot occur until both resources and services are restored (EVOS Trustee Council 2010). This assumption is somewhat dubious given that the recreation and tourism industry in the Sound, as described by Colt et al. (2002) and Fay (2008), has certainly increased beyond 1989 levels, with many more people participating in recreation activities in the region. It is easy to imagine that the tourism industry or recreation potential for such a spectacular land- and seascape as the Sound is not dependent upon a dozen species still described as impacted by the spill. However, it is less clear how the commercial fishing industry has recovered, given its former large dependence on target species like Pacific herring.

Certainly, subsistence harvest practices were impacted for many years following the spill for fear that contamination from lingering oil had altered the availability of safe subsistence resources in the Sound. For example, in Chenega Bay prior to the spill, marine mammals made up about 40 percent of the subsistence harvest and only 3 percent of subsistence harvest 10 years later, in 1998. Ten years after the spill, 63 percent of the households in spill-affected areas believed that subsistence resources had not recovered from the effects of the spill. Some subsistence community members reported having to increase effort (traveling farther, spending more time and money) to achieve comparable harvests to those before the spill and reported an increasing reliance on fish (Fall and Utermohle 1999). Even in the household harvest survey effort described in this volume (chapter 9), conducted 20 years after the spill, a few respondents expressed concern about harvesting in areas that had been damaged by the spill.

To this day, people who haven't visited the region or are otherwise unfamiliar with south coastal Alaska closely associate Prince William Sound with this disastrous event. This speaks to the recovery status of passive use, evaluated as the broad association made by society that the Sound still suffers from injury from the spill. The Trustee Council ties this association with the presence of lingering oil: as long as the public is aware of lingering oil and its potential effects on resources in the region, there are impacts to passive use. For example, in chapter 7, when recreation users were asked whether they had seen evidence of lingering oil on their trips into the Sound, 2 of 171 individuals responded in the affirmative. Whether or not they had definitively encountered lingering oil (relatively easy to do by turning over beach rocks on a few of the Sound's mostly heavily oiled sections of shoreline) is difficult to know, but at least in stated perceptions, it still exists.

What follows is a brief status synthesis for each of the human services according to the most recent recovery status evaluation completed by the Trustee Council (EVOS Trustee Council 2010). Complete details for recovery status as well as reports documenting the recovery of these services and the resources impacted by the spill can be found on the Trustee Council's website: http://www.evostc.state.ak.us/.

COMMERCIAL FISHING

Recovery Objective: Commercial fishing will have recovered when the commercially important fish species it depends upon have recovered and opportunities to catch these species are not lost or reduced by the effects of the oil spill. The strategy for restoring commercial fishing includes funding projects that accelerate fish population recovery, protect and purchase important habitat, and monitor recovery progress.

Recovery Status: By 2002, the Trustee Council considered pink salmon and sockeye salmon to be recovered from the oil spill, but the Pacific herring remains not recovering; therefore, restoration activities will continue to focus on this resource.

Commercial fishing was injured as a result of the spill's direct impacts to commercial fish species and through subsequent emergency fishing closures to protect stocks that had been significantly damaged by the spill. Commercial harvest of salmon, herring, crab, shrimp, rockfish, and sablefish was closed in 1989 throughout the Sound, and as a result, income from commercial fishing dramatically declined immediately following the spill. Disruptions to income from commercial fishing continue today, as evidenced by changes in average earnings, ex-vessel prices, and limited entry permit values. For example, commercial harvest of herring in the Sound has not reopened following the spill. Though other economic and environmental changes are certainly at play, many still see the spill as being the tipping point for the loss of these valuable stocks. This is particularly the case for Pacific herring.

The Trustee Council recognizes that economic changes in the industry, like increased world supply of salmon from farmed fish and corresponding reduced prices, have ripple effects on fisheries in the spill-affected region. Similarly, management restrictions, like individual fishing quotas (or IFQs) for halibut and sablefish, and allocation changes in the amount of salmon available to commercial fishermen versus that reserved for subsistence or recreational fishers, also affect the industry.

SUBSISTENCE

Recovery Objective: Subsistence will have recovered when injured resources used for subsistence are healthy and productive and exist at prespill levels. In addition, there is recognition that people must be confident that the resources are safe to eat and that the cultural values provided by gathering, preparing, and sharing food need to be reintegrated into community life.

Recovery Status: Fears about food safety have diminished since the spill, but it is still a concern for some users. Additionally, harvest levels from villages in the spill area are comparable to other Alaskan communities. However, many subsistence resources injured by the spill, including clams, mussels, and harbor seals, have still not recovered from the effects of the spill. For these reasons, subsistence continues to recover from the effects of the oil spill, but has not yet recovered.

Fifteen predominantly Alaska Native communities—with a total population of about 2,200 people in the oil spill area—relied heavily on harvests of subsistence resources, such as fish, shellfish, seals, and waterfowl. In addition, the spill impacted another 13,000 subsistence permit holders throughout the spill area. Subsistence harvesters from the Alaska Native communities of Chenega Bay and Tatitlek, as well as the rural communities of Cordova and Whittier, were hard-hit by the spill. Oil affected harvests through a variety of mechanisms beginning with an immediate and continued reduction in the availability of key fish and wildlife resources. This was made worse by concern about possible health effects of eating oiled fish and wildlife from areas even after they had been declared "safe" for harvest. Finally, and perhaps less broadly recognized initially by the Trustee Council, was the disruption of the traditional lifestyle due to participation in cleanup and related activities and the presence of the spill cleanup industry in these formerly more isolated communities.

Subsistence use is a central way of life for many of the communities affected by the spill, thus the "value" of what was lost cannot be measured by reduction in harvest success. The subsistence lifestyle is a cultural value of traditional and customary use of natural resources to support the household and the community at large. It is perhaps best described as an annual cycle (see chapter 9 essay), the components of which reinforce the societal fabric of these often isolated communities. The spill disrupted opportunities for young people to learn these integral cultural practices and as a result, fear exists that this tradition may be lost as this generation grows into the adults who would have been the primary harvesters in their communities. A 2004 survey found that 83 percent of respondents felt the "traditional way of life" had been injured, and 74 percent still felt that recovery had not occurred.

Harvest levels have generally rebounded in many communities in Cook Inlet and Lower Kodiak, but Sound communities (save Cordova) continued to decrease as of 2003. Harvest composition was also altered. In the first few years following the spill, people harvested more fish and shellfish than marine mammals. Target species of marine mammals like seals, sea lions, and sea otters had declined immediately following the spill, and the common perception was that these mammals, along with shellfish like clams and mussels, were contaminated and unsafe to eat. From 1989 to 1994, subsistence foods were tested for oil contamination, with no or very low concentrations of petroleum hydrocarbons found in most, but perceptions continued to drive reduced overall harvest for several years after the spill. By 2003, most subsistence users expressed confidence in foods like seals, fish, and chitons, while the safety of shellfish such as clams was still met with skepticism.

Many other factors likely contribute to the changes observed in subsistence harvest and the cultural lifestyle it supports. Demographic changes in these small rural communities, ocean warming, increased competition for subsistence resources by other harvesters in some areas, changes in predator-prey regimes (like the increase in sea otters observed in the Sound in recent years), and increased concern about other contaminants all may play a role in decreased harvest. The Trustee Council recognized these complicating factors in their more recent assessment, and separating which among them may be primary drivers makes for complicated analysis of both the environmental and human systems.

RECREATION AND TOURISM

Recovery Objective: Recreation and tourism will have recovered, in large part, when the fish and wildlife resources on which they depend have recovered, and recreation use of oiled beaches is no longer impaired.

Recovery Status: Even though visitation has increased since the oil spill, the Trustee Council's recovery objective requires that the injured resources important to recreation be recovered and recreational use of oiled beaches is unimpaired. Given that several natural resources have not recovered from the effects of the spill the Council finds recreation to be recovering, but not yet recovered.

In the immediate years following the spill, recreation use was displaced from a number of spill-contaminated areas (Hennig and Menefee 1995). Since this time, recreation and tourism has increased in the Sound and is a key component of the economy in the region (Colt et al. 2002). Similarly, the number of visitors to Alaska has increased since the spill. By 2020, it is expected that the recreation and tourism industry in Southcentral Alaska will grow approximately 2.8 percent each year. In 1999 and 2002, telephone interviews of people who used the spill area for recreation before and after the spill indicated that although oil remained on beaches, it did not deter their use. They continued to report diminished wildlife sightings in the Sound, particularly in heavily oiled areas like Knight Island. They also reported seeing fewer seabirds, killer whales, sea lions, seals, and sea otters, though some also reported observing increases in the number of seabirds in the last several years. As of 2001, over $10 million from settlement funds has also been spent by the Trustee Council on repair and restoration of recreational facilities that had been damaged by the spill and subsequent cleanup efforts.

Recreation and tourism rely on what are traditionally described as consumptive (e.g., fishing, hunting, and gathering) and non-consumptive (e.g., wildlife watching, hiking, camping, etc.) uses of natural resources. Although these activities have increased in the Sound, some beaches used for recreation still contain lingering oil—though it's unclear what level of overlap is actually taking place. Of greater concern to the Trustee Council are the resources important to recreation and tourism that are still not considered recovered from the spill. Iconic species sought by wildlife watchers, like killer whales, sea otters, Kittlitz's and marbled murrelets, harlequin ducks, Barrow's goldeneyes, and black oystercatchers are still in recovery; pigeon guillemots are not recovering. There are also still species sought by harvesters, including clams and mussels, which are recovering, and fish, including rockfish and cutthroat trout, with a status of very likely recovered.

PASSIVE USE

Recovery Objective: Passive uses will have recovered when people perceive that aesthetic and intrinsic values associated with the spill area are no longer diminished by the oil spill.

Recovery Status: Until the public no longer perceives that lingering oil is adversely affecting the aesthetics and intrinsic value of the spill area, it cannot be considered recovered. Because recovery of a number of injured resources is incomplete, the Trustee Council considers services related to passive use to be recovering from the effects of the spill.

The Trustee Council determined that passive use injuries occurred from the spill because impacts to scenic shorelines, wilderness areas, and popular wildlife species were easily observed by the public, both in person and through media coverage throughout the United States and the world. The Trustee Council believes that the key to the recovery of passive use is providing the public with current information on the status of injured natural resources and human communities and work to show the progress being made toward their recovery. Toward that end, restoration efforts in research, monitoring, and other restoration programs as well as habitat protection and acquisition have been made available to the public, and accomplishments are promoted by the Trustee Council. An example promoted by the Trustee Council is the success of their land acquisition program, which as of 2006 protected more than 630,000 acres of habitat, including more than 1,400 miles of coastline with over 300 streams valuable for salmon spawning and rearing.

All program efforts are funded through a transparent process driven by an annual work plan, which documents the projects selected to implement restoration of injured resources and services. Through the process of inviting proposals, publishing an annual program of work, and ensuring that all project reports are archived through their website and a network of libraries, the Trustee Council seeks to make information about recovery available to the public. In 1996, 1999, 2002, 2006, and 2010, they offered a synthesis of the new information on recovery status of the injured resources and services. The Trustee Council also supports researchers with publishing project results in peer-reviewed scientific literature to engage a larger professional audience.

Additionally, the Trustee Council supports an annual marine science symposium held in Anchorage, Alaska, which is open to the public. It is attended by hundreds of marine scientists, managers, and members of the public, and it provides a venue in which to report the progress of restoration in the spill area. Public input comes from a 15-member advisory committee comprised of regional stakeholders with the intent that they will help others stay informed of the progress of restoration as well as provide public opinion to the Trustee Council as decisions are made. Public meetings are also held periodically throughout the spill area where public comment on Trustee Council progress is requested.

LOOKING BEYOND THE LIST—TOWARD AN ECOSYSTEM APPROACH

Research continues to advance the understanding of recovery status for injured resources and services, and the Trustee Council is still pursuing strategic land acquisition and protection efforts. Immediately following the spill, the majority of effort focused on understanding the effects of exposure to oil on those species and habitats that came in direct contact with it, as well as those that continued to have indirect exposure. In recent years, understanding of the complexities of these systems has evolved, in part due to extensive Trustee Council–supported research. Restoration efforts have also aimed to identify potential impacts to the recovery of these injured resources and services as a result of other environmental stressors, including increased human use.

The importance of managing human use within the spill region was recognized in the original 1994 restoration plan. The management of human use was identified as one of three types of general restoration projects that could be funded. The Trustee Council believed that projects which redirected human users and reduced human disturbances to resources were important to the overall recovery effort. In 2006, the Trustee Council more clearly stated that "stressors other than oil," such as increasing human use, may be affecting resources still recovering from the spill. The Trustee Council encouraged the identification and use of tools/strategies to manage these other stressors such that they might help resource recovery. The Trustee Council sees this as an essential part of a holistic, ecosystem approach to restoration for both the resources and their dependent services.

Following this logic, the Trustee Council emphasized an ecosystem approach, which included addressing human use. They felt such projects may ultimately protect resources and their habitat as well as protect human services from impacting themselves (i.e., "being loved to death"). In their 2010 assessment, they recognized the increasing difficulty in distinguishing between spill impacts and other environmental and societal changes occurring within the region. They have prioritized the use of remaining research funds toward efforts that provide information critical to the support of a "healthy functioning ecosystem" with the hope that by evaluating these other system components they can make better decisions on restoration efforts.

Key to a systems approach evaluating the condition of recovering resources and their inextricably linked human services is an assessment of the very nature of human use patterns and user experience in the region. For example, increasing use in the Sound presents something of a paradox. High numbers translate into economic recovery of the recreation/tourism service but if not managed effectively could contribute to a decline in the quality of visitor experience, result in conflicts with subsistence users, and possibly further the degradation of recovering resources that visitors come to the region to see—thus hindering the long-term recovery of human services in the Sound. A dominant assumption in natural resource management is that more use equates with impacted natural systems, yet finding a way to allow and even promote use in the spill area may be necessary for recovery of the affected human services.

This conundrum captured the attention of human use researchers working for the Chugach National Forest in the late 1990s (e.g., chapter 5). They began the first work on integrating human use assessment into resource recovery evaluations (chapter 11) in the Sound. This early work came in response to the proposed tunnel that, for the first time, would allow vehicle, and thus trailered boat, access to the western Sound port of Whittier. This work alongside apparently increasing use in the western Sound, including new and more requests for commercial recreation permits in the early 2000s, captured the attention of the Chugach leadership. Furthermore, some stakeholders in the region began to look to the dominant land manager in the Sound (approximately 80 percent of the uplands, including the 2.1-million-acre Nellie Juan Wilderness Study Area) to address perceived changes and potential problems that may arise from increasing use.

THE CHUGACH NATIONAL FOREST'S HUMAN USE MANAGEMENT

The Chugach National Forest, as the major Prince William Sound land-managing federal representative on the Trustee Council, is dedicated to playing an important role in the recovery process. Forest leadership saw a key aspect of this recovery to be the distribution, behavior, and experience of recreation/tourism users throughout the Sound. They saw this not only as a potential impact on recovering natural resources but also as a possible threat to the recovery of human services, including the recreation/tourism human service itself, and subsistence. Like recreation and tourism, subsistence harvest is a vital part of the Alaskan lifestyle (see essay by Kate McLaughlin in this chapter). It is widely recognized in Alaskan land and resource management that subsistence harvest provides irreplaceable cultural, spiritual, personal, and sustenance values. The Alaska National Interest Lands Conservation Act (ANILCA) requires the Forest Service to consider the effect of management activities on subsistence. The Chugach National Forest was concerned about the increase in commercial recreation activity interfering with subsistence practices.

Increased access from the opening of the Whittier Tunnel and pressure from recreational activities on the already injured and recovering resources was also seen as a potential issue. There was also concern among some who use the Sound as to whether the very wilderness experiences that Alaska residents and visitors seek are being threatened. As recreational use levels increase in the Sound (e.g., kayaking, wildlife viewing, pleasure boating, hunting, fishing, and camping), it seemed inevitable that encounter levels within and among these groups would also increase. Some within Chugach National Forest leadership, as well as some stakeholders, felt that user conflicts were inevitable and thus the quality of recreation as a human service might be degraded.

The forest leadership saw that rigorous characterization of recreation and tourism had received less attention from the Trustee Council in the past and that this would be a critical first step for managers in the Sound to be able to address concerns about increasing human use. They also recognized shortcomings in their own management planning efforts (see chapter 19) in their 2002 Revised Land and Resource Management Plan or "Forest Plan." For example, the Forest Plan established fundamental values by which the Chugach uplands in the Sound were to be managed, but in many cases failed to take into consideration activity *already* occurring in adjacent areas, including the salt water that supports the vast majority of human use activity in the region. Furthermore, many management prescriptions defined by the Forest Plan were generalizations that pertained to large spatial areas and did not specifically attempt to address the variation in uses between specific areas.

In order to refine its management approach, the Chugach National Forest produced two carrying capacity documents for the region: one covering the eastern and the other covering the western portion of the Sound. These were only used to allocate commercial use and had similar functional limitations, as they were based on the Recreation Opportunity Spectrum or ROS (Clark and Stanky 1979). They relied heavily on calculating the appropriate number of recreation groups at one time based on either the available acreage of uplands in one case (the eastern Sound) or numbers of available shoreline campsites (the western Sound). They were also developed in the absence of specific information about existing patterns of private use within the region (we now know this private use is approximately 90 percent of recreation use in the Sound; see chapter 7) and did not represent differences in use types very well when it came to permit allocation.

Unfortunately, the assumptions and intent behind each were different, with the Eastern Prince William Sound Capacity Analysis being developed primarily to address commercial hunting guides and used to keep permit numbers low. This tack was in response to concerns from big game guides in the 1990s about "crowding" with other guides. This concern about crowding, which originated between competing hunters, was carried over, perhaps unintentionally, to other commercial activities, even if the proposed use was actually kayaking or shore hiking. The Western Prince William Sound Capacity Analysis began in 2001 in response to experiences of backcountry kayak crews working for the Chugach National Forest in the field, as well as anecdotal information gathered at the public contact yurt in Whittier.

Over several years both carrying capacity analyses were updated as new data and information came in about use numbers in the Sound or concerns about social or resource impacts. This resulted in each evolving independently, and by the mid-2000s, the Western Prince William Sound Capacity Analysis, which covered the proposed wilderness study area, was more permissive of increased commercial use than the Eastern Prince William Sound Capacity Analysis.

ADAPTING TOWARD A SUSTAINABLE HUMAN USE MANAGEMENT APPROACH

Forest leadership understood that the existing carrying capacity tools framed within the 2002 Forest Plan did not adequately address human use concerns with respect to forest management but that this also had profound implications for recovery of resources and services injured by the oil spill. At approximately the same time, the Trustee Council was increasingly recognizing the need to change their approach and develop a better understanding of *other* stressors in the spill area, including human use. These two recognized needs converged to become the focus of a 2006 workshop for Chugach National Forest staff to synthesize data and professional judgments on human use management issues facing the forest in the Sound.

This workshop identified leading issues for the Chugach National Forest in the Sound and ultimately led to three broad goals: (1) evaluate the Chugach National Forest's current recreation management approach for the Sound; (2) recommend management and monitoring strategies that improve Chugach National Forest recreation management and provide for the recovery of *Exxon Valdez* oil spill–injured natural resources and human services; and (3) identify partnership, collaboration, and stakeholder engagement opportunities that promote sustainable use of the Sound.

Based on these goals, a suite of projects was conceived to better inform management objectives for both the Chugach National Forest and the recovering resources and services. These efforts include many of those described in this volume and implemented through a combination of funds administered by the *Exxon Valdez* Oil Spill Trustee Council and the Chugach National Forest. Collectively these projects became known as the Prince William Sound Framework, the results of which aimed to guide the management of existing human use and facilitate future recreation demand while simultaneously protecting and enhancing the ecological values of the Sound. The framework is based on four foundational planning principles: integration of social and ecological systems, multiple scales of analysis (ranging from regional- to landscape- to site-specific levels), issue-based planning, and collaboration.

This endeavor to better understand the dynamics of human use and potential for user conflict as well as species and habitat impacts—coupled with increased efforts in stakeholder engagement—was seen as a way to adapt Chugach National Forest management approach to changing conditions in the region. The resulting recommended management approach (chapter 19) is in line with national discussions regarding visitor management approaches, recognizing that protecting resources and visitor experiences requires a more sophisticated approach than simply specifying how much visitor use is too much (i.e., visitor carrying capacity analysis). It was the hope of the Chugach National Forest that enhanced understanding of use dynamics and increased stakeholder engagement would ultimately translate into improvements in human use management and in recovery of resources and services injured by the oil spill.

STAKEHOLDER ESSAY

A SUBSISTENCE WAY OF LIFE

KATE MCLAUGHLIN

SUBSISTENCE IS AN IMPORTANT WORD in Alaska, the meaning of which is not always fully understood or appreciated by those who don't live by it. For my grandparents, who lived through the Great Depression, subsistence is a dirty word. It conjures up images of poverty and barely scraping by. Subsistence food gathering is something to rise above, to put behind you and hopefully never have to think about again.

For rural and Native Alaskans who live by it, subsistence is something else entirely. It is the right to provide quality food for your family, to continue a cultural experience that has gone on for millennia, and to fulfill a spiritual need to be part and parcel of nature, not something outside of it.

The *Exxon Valdez* oil spill (EVOS) radically affected the Alutiiq people in western Prince William Sound in a sudden and profound way that even more than 20 years later is still impacting the social and cultural fabric of the communities here. Immediately after the spill, Exxon representatives came to Chenega Bay and pledged to the elders that they would make them "whole again." On the backs of this promise, Exxon brought in hundreds of pounds of store-bought food. At the time, many people were unaccustomed to frozen chicken or soda pop. With over 11 million gallons of oil washing up on the beaches, all commercial and subsistence fishing stopped. No longer could mussels or clams be gathered on the beaches. Seal meat (a staple of the Alutiiq diet) would be potentially contaminated. Diets radically changed within the space of days, along with a way of life.

Fears still linger regarding the potential health and safety of marine foods, with continued issues of lingering oil from EVOS still very much present in the Sound. Children who have grown up since the spill were raised on store-bought foods and away from the fishing culture their parents engaged in. These children don't know how to gather, clean, and prepare these foods anymore. And too many play video games instead of putting on rubber boots and going out onto the land and sea. With this loss of knowledge come great ills. The radical change in diet and loss of nutritional quality has caused heart disease, obesity, and diabetes rates to soar.

There is no denying that subsistence food gathering is hard work, and sometimes dangerous. Yet with those challenges come great benefits, such as an understanding of and a deep personal linkage with the natural environment. The loss of a sense of stewardship between the environment and the people who live within it can have profound effects upon communities. Children are growing up without having learned the valuable life lessons that hunting, fishing, and gathering teach. The innate spiritual connection between the people and the land and waters which surround them has been broken.

Much effort is being made to regain this precious knowledge. Traditional camps have been established in communities across the Sound where children and elders come together each year to learn the old ways and reconnect with their environment. Those of us who still participate in subsistence gathering have an obligation to our children to ensure that they have the opportunity to learn about the true nature of subsistence and why such knowledge is valuable to them and their future success in our modern world.

CHAPTER 5

IDENTIFYING LANDSCAPE VALUES IN PRINCE WILLIAM SOUND WITH PUBLIC PARTICIPATION GEOGRAPHIC INFORMATION SYSTEMS (PPGIS)

GREG BROWN

IN 1998 AND 2000, two public participation geographic information systems or PPGIS studies were completed, asking Alaska residents to identify what they valued about the Sound. The purpose of the 1998 study was to inform the Chugach National Forest Land and Resource Management Plan, a 10- to 15-year plan that guides management of the 5.5-million-acre national forest that cradles the Sound. The 2000 study aimed to assist nongovernmental organizations or NGOs to develop a conservation strategy for protection of the Sound by identifying conservation hot spots and to examine policy issues such as shoreline development, tourism activity, and cruise ship regulation in the Sound. The two PPGIS studies, in combination, provide a reasonably comprehensive view of *what* Alaskans value about the Sound and *where* they perceive these values. Full details of this work are published in Brown et al. (2004), but the following offers a useful insight into landscape values and perspectives held by stakeholders following the 1989 oil spill as they looked ahead to the opening of the Whittier Tunnel.

The term *public participation geographic information systems* or PPGIS was conceived in 1996 at the meeting of the National Center for Geographic Information and Analysis. The concept describes the process of using GIS technologies to produce local knowledge with the goal of empowering marginalized populations (see Sieber 2006; Brown 2005; and Sawicki and Peterman 2002 for a review of PPGIS applications and methods). An implicit, normative assumption of the PPGIS approach is that public land management *should* be guided by public values for the region. The studies were guided by the following assumptions: (1) the public has significant place-based knowledge and values that are essential to an inclusive and effective planning process; (2) the values of the general public that comprise "the silent majority" are seldom explicit in a public planning process; and (3) methods can be developed to measure the place-based information from the general public in a systematic and unbiased way.

The 1998 Chugach National Forest study was a mail survey of Alaskan residents using Dillman's (1978) total design method. The sampling methodology consisted of randomly selecting individuals from households in 12 communities (Anchorage, Cooper Landing, Cordova, Girdwood, Hope/Sunrise, Kenai, Moose Pass, Seward, Soldotna, Sterling, Valdez, and Whittier). Sampling was limited to one individual per household. Participants were sent a map of the Chugach National Forest and four sticker dots associated with 13 categories of landscape values. They were instructed to place the sticker dots "directly on the map over those locations on the forest that you feel best represent those values."

The 2000 Prince William Sound study used similar methods to the 1998 Chugach National Forest study. Questionnaires were mailed to randomly selected households in the Prince William Sound communities of Cordova, Valdez, Whittier, the villages of Tatitlek and Chenega Bay, and the city of Anchorage, as well as a random statewide sample of Alaskan residents. In addition to the sticker dots indicating landscape values, the 2000 study contained a series of survey questions that asked opinions about potential threats to Prince William Sound values, such as oil transport, shoreline development, increased small tour operator activity, and large cruise ships. These questions were measured on a five-point Likert scale ranging from "No impact" to "Very large impact."

The sticker dots representing the 13 landscape values from the two studies were digitized into a GIS as point features. The landscape value codes and additional attributes relating to respondent characteristics (e.g., community of residence) were joined to the points. To describe the general distribution of the landscape values within the study area, the number of landscape value points were summed and ranked for each value. Hot spot maps of the point distributions for the thirteen landscape values were generated using kernel density estimation techniques in ArcGIS Spatial Analyst.

The overall survey response rate for the two studies was about 31 percent, with a total of 13,895 landscape value points identified in the study area. Respondent characteristics showed more male respondents (62 percent) who were middle-aged (mean 47 years) and long-time Alaska residents (mean of 26 years). The distribution of landscape values for the Prince William Sound communities generally reflects the spatial proximity of the respondent community. For example, the landscape values for Cordova (figure 5.1a), Valdez (figure 5.1b), and Whittier (figure 5.1c) tend to concentrate

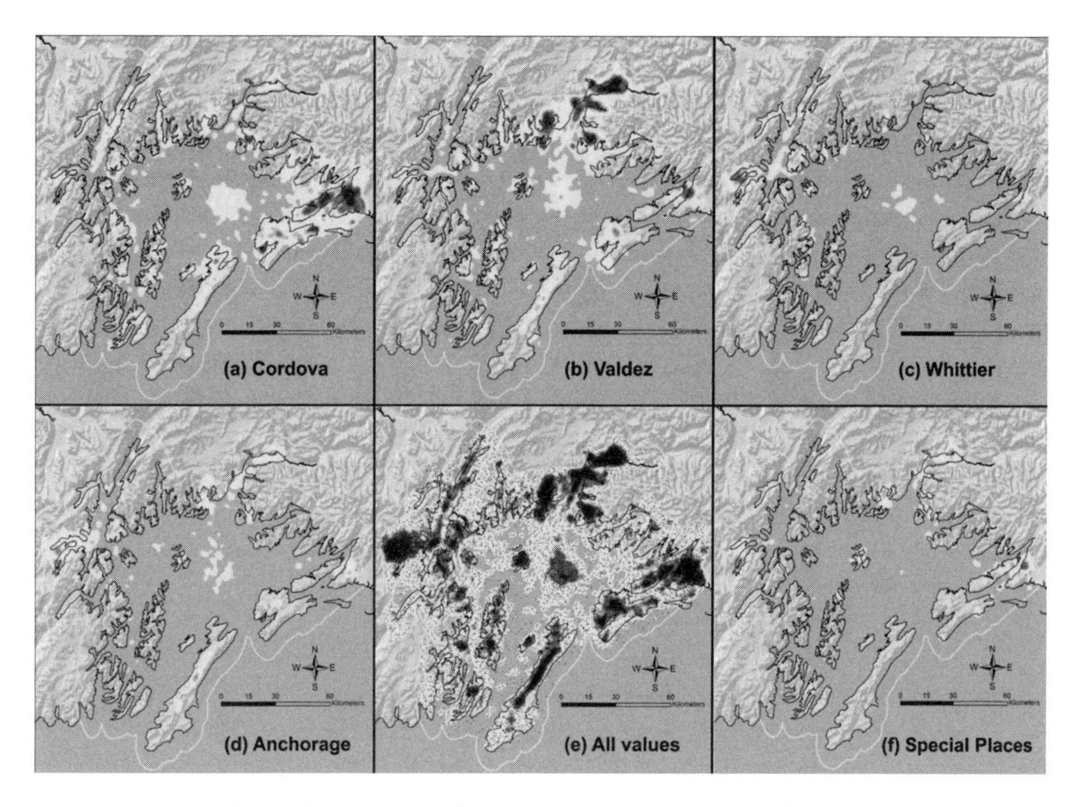

FIGURE 5.1. Spatial distribution of all landscape values by community: (a) Cordova, (b) Valdez, (c) Whittier, (d) Anchorage, (e) all values in all communities, and (f) special places.

along the coast within 20 to 30 km of the community. The spatial distribution of landscape values for Anchorage residents (figure 5.1d) are more dispersed throughout the Sound, although the highest concentration is near Whittier, the closest access point for Anchorage residents. The distribution of responses by all respondents from all communities surveyed (figure 5.1e) reveals that despite the relative remoteness of the Sound, there are few places in the Sound that don't hold significant value.

The results also show that Alaskans know the Sound intimately and value it for many reasons, but of greatest importance are its scenic beauty, recreational opportunities, biological richness, and economic opportunities. Subsistence values are especially important to residents of Cordova/Eyak and the villages of Tatitlek and Chenega Bay. These values are strengthened through participating in outdoor activities in the Sound, the most common being recreational fishing, use of the Alaska Marine Highway, subsistence gathering, and touring/sightseeing in private boats.

Despite the ecological disaster resulting from the 1989 *Exxon Valdez* oil spill, respondents perceived the Sound to be in relatively good ecological condition and

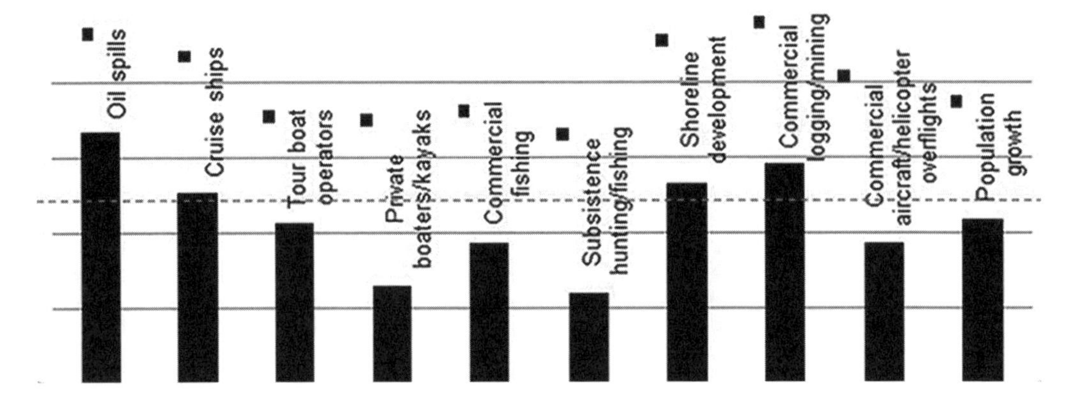

FIGURE 5.2. Alaskans' perceptions of future potential impacts to Prince William Sound. Mean scores on 1 to 5 scale, where 1 = no impact, 2 = small impact, 3 = medium impact, 4 = large impact, 5 = very large impact. Activities with a mean value above 3.5 are perceived to have potentially large, negative impacts on the Sound. Perceived large impacts appear above the dotted line.

wanted it to remain that way. They believed shoreline development should be limited and targeted to protecting the marine environment and enhancing primitive recreation opportunities. Resort development and floating lodges are viewed as incompatible with existing values for the Sound. Prince William Sound residents expressed awareness of the potential for significant growth in marine-based tourism in the Sound. They were open to small, incremental growth in cruise ship activity, but all communities opposed large increases in cruise ship numbers.

When asked to evaluate activities with the greatest potential impact to Prince William Sound, the largest potential impacts were related to industrial activity in the Sound: oil spills, mining, and logging activity (figure 5.2). Respondents were concerned about the potential catastrophic nature of these large-scale activities. Large cruise ships and shoreline development were evaluated as having between medium and large potential impact on the Sound. This 2005 study concludes with the recommendation for "Prince William Sound communities, state and federal interests, interest groups, and visitors to the Sound to engage in a comprehensive, proactive planning process that seeks to protect the current values in the Sound. Without a serious dialogue about the limits of acceptable change in the Sound, incremental changes may overwhelm the very attributes of the Sound that make it so special to its residents and visitors."

STAKEHOLDER ESSAY

OPPORTUNITY OR HOSTAGE?

HAROLD BLEHM

CHUGACH NATIONAL FOREST and Prince William Sound are as inseparable as picture and frame. It is a vast area and the second-largest national forest in the system. Three portal communities—Seward, Whittier, and Valdez—from west to east, are on the highway system and account for most of the recreational users entering the Sound. The entire Chugach National Forest has limited access with a mere 90 miles of Forest Service road.

Though on the highway system, Valdez is on the eastern margin and not within the national forest, yet the eastern third of the state of Alaska accesses the Sound through Valdez. The next closest opportunity is Whittier, some 360 highway miles west of Valdez. All three portal communities have small boat harbors that are at capacity. The average wait for a slip is six to seven years. All three harbors hope to expand their capacity, but it is a long and expensive process.

The city of Cordova, home of Cordova Ranger District, is not on the highway system, and few recreationists originate from there. Whether by chance or design, Cordova Ranger District is sited in an isolated community. The only access is by sea or air. By investing heavily in infrastructure in an isolated community, Cordova Ranger District has already drawn criticism and may continue to do so as use and demand increase. Those who can afford to ferry or fly will continue to travel to Cordova. Unfortunately, a casual, impromptu weekend for the general public in the eastern side of Chugach National Forest is not possible unless you own a boat or plane. A trip to Cordova via public transportation requires planning, reservations, and cash, and is weather dependent and subject to cancellation. In other words, it's difficult and expensive to get there.

Most of Cordova Ranger District is in storage and will remain so. The uplands are, for the most part, unused. Is the eastern Sound intentionally held hostage in the name of preservation? Is this part of a management strategy or a happy accident? Are plans underway to provide access to the eastern sound, or is it to be held hostage indefinitely? Demand exists for connecting Cordova Ranger District to the highway system. The U.S. Forest Service needs to make the forest more accessible to justify continued expenditure of public funds.

The vast majority of the users entering the Sound by boat do not intend to set foot on dry land until they return to port, using the forest as a visual resource only. The uplands are vast and quiet. From a boat, even heavily used beaches appear pristine. Once you set foot on them, it becomes evident that they are not. An additional detriment to the viewshed is the accumulation of marine debris. Though it is largely inert, it will persist indefinitely.

A great deal of data has been generated since the 1989 *Exxon Valdez* oil spill, indicating which beaches are recovering, may recover, or are not likely to recover. As with any management unit, some areas must be designated "sacrificial" to protect others. These exist to take some of the burden from more fragile areas. These areas can be hardened to accommodate trampling and yet must retain their character, to blend with the mission of the managing agency. Forest managers should site these most heavily used areas on oil-impacted beaches, as long as they are not toxic to users.

Development has always followed transportation or trade routes, with areas immediately adjacent—whether roads, rivers, lakes, and canals—most heavily affected. This natural pattern should be taken advantage of, and users should be directed to appropriate areas along well-used transportation corridors. Managers should identify and appropriately develop resources, areas of interest, camping areas, and private property, using maps, brochures and, for safety's sake, GPS coordinates. For as long as possible, the most far-flung areas of the Sound should be left alone. Wilderness enthusiasts will seek them out on their own.

A number of organizations, including the U.S. Forest Service, State of Alaska Parks and Recreation, Alaska Geographic, the National Wildlife Federation, and private consultants, have joined an effort to create the Prince William Sound Marine Trail, which basically follows the route of the Alaska Marine Highway: this is a monumental task, but a good start has been made.

Coincidentally, the areas of the Sound most heavily oiled by the 1989 disaster also tend to follow the Alaska Marine Highway route, the primary artery in the Sound, from Bligh Reef southwest to Kodiak.

The bottom line is that use will increase. It is still possible for the Forest Service to manage use in the Sound and to direct it toward appropriate areas. They should not wait too long. Managers of Prince William Sound have an opportunity to do it right. In this case, opportunity is a moving target. With the opening of the Whittier railroad tunnel to private motor vehicles, that target is gaining speed. It will become harder to hit.

Management strategies for sustainability could include reduction of use of sensitive areas by steering users to more appropriate sites, restriction (almost impossible in an area this large), stabilization and site hardening to accommodate an increase of users, or the ever-present skeleton at the feast, the "no action" or do-nothing alternative. All strategies have inherent risks and challenges are to be expected.

The Sound is used and use will increase. Hopefully there will always be a secluded cove to provide the wilderness that we need. Appropriate management is required. It's too important to be left to chance.

CHAPTER 6

INTEGRATING LEGACY DATA TO ESTABLISH A SPATIAL AND SEASONAL FRAME FOR PREDICTING HUMAN USE HOT SPOTS IN PRINCE WILLIAM SOUND

CHANDRA B. POE AND SAMANTHA GREENWOOD

SOUND BITES

THIS WORK DEMONSTRATES an approach to summarizing human use data at multiple spatial scales from system wide to regional and down to the site level across seasons.

We show that a collaborative effort to synthesize use and permit data from multiple managers is essential for sustainable human use management.

The results of this work can be used to frame Sound-wide analyses of ecological and human use questions for a variety of land and resource managers in the region.

INTRODUCTION

Human use is not evenly distributed in Prince William Sound. Certain locations are more desirable for a variety of reasons, including distance from communities, presence of glaciers and postglacial landscapes, availability of landing areas, protected anchorages, sport fish streams, cabins and other recreation infrastructure, and wild game concentrations (Murphy et al. 2004). Human use "hot spots" are important areas in the Sound where human use is more concentrated relative to surrounding areas. In many cases, these locations are physiographic bottlenecks, restricting access to desirable intertidal areas and their associated upland opportunities for recreation activities. They also vary seasonally based on wildlife distribution (important for both harvest and viewing activities) and weather conditions.

In desirable areas, competing uses (e.g., wildlife viewing and hunting) are likely to converge, potentially resulting in detrimental effects upon one another and creating a larger cumulative impact on the associated resources. It is critical for the sustainable management of human use in an area hosting a number of *Exxon Valdez* oil spill–injured resources (both recovered and recovering) that the location, timing, and nature of these human use hot spots be well understood by resource managers. Several disparate data sources characterizing human use in the Sound are currently available, but these have not been compiled into a single comprehensive database; nor have they been evaluated relative to their impacts upon one another, or upon *Exxon Valdez* oil spill–injured resources.

This chapter describes the effort to first define a seasonal and spatial framework that was then used to combine contemporary GIS and tabular data from a variety of sources for Prince William Sound to create a database characterizing known concentrations of human use and their seasonal patterns of use prior to 2008. Through this work, we were able to conduct a comprehensive spatial analysis to identify concentrations of human activity in the Sound (hot spots) and to define a framework of analysis for future human use efforts in the Sound.

METHODS

IDENTIFICATION OF POTENTIAL DATA SOURCES

The major categories of human uses of Prince William Sound were identified as sportfishing; commercial fishing; recreational hunting; subsistence harvesting; kayaking, including day trips and overnight camping; recreational boating; commercial outfitter-guide operations; commercial sightseeing tours; land permits, such as communication sites; public use cabins; private cabins; and beach use, including hiking, beachcombing, berry picking, and camping.

Examples of data sources reviewed included reported use numbers, such as reported use by special use permit holders or Alaska Department of Fish and Game (ADF&G) guided sportfishing trips; infrastructure, such as cabins, hardened campsites, and hatcheries; and attraction points, such as tidewater glaciers, wildlife concentration areas, and trails. Data reviewed included published and unpublished reports, guidebooks, existing maps, and spatial data and tabular data maintained by the Forest Service and ADF&G.

Throughout, we collaborated with other landowners, such as the State of Alaska and Alaska Native Corporations, to obtain human use data. Data received from other agencies were reviewed for both spatial and tabular content by the biologists who provided the data. Public meetings were also used to better define our data and find other sources of human use data in Prince William Sound. Public input and knowledge were sought during three open-house meetings in Valdez, Cordova, and Anchorage. Although attendance was minimal, these meetings led to some helpful changes and updates, as well as new discoveries of data sources.

SEASONS

Human use of the Sound is strongly seasonal due to variable weather and extreme winter conditions. All data were analyzed as three distinct data sets separated by seasons, as described below. Recreational use of uplands (including hunting, recreational boating, and sportfishing) occurs mainly from May through September. Previous studies have used a recreational season of May 1 through September 30 in eastern Prince William Sound (USDA Forest Service 2008) and May 15 through September 15 in western Prince William Sound (USDA Forest Service 2005). Our seasons were refined based on a review of existing recreational use studies of the Sound, as well as hunting seasons established by ADF&G-harvested species in the Sound, including brown bear, black bear, deer, and mountain goats.

SPATIAL SCALE

Due to the large geographic extent of the study area (4.7 million acres), spatial scale of data organization is an important component of this analysis. The Chugach National Forest currently uses existing administrative units, known as Analysis Areas, for management of special permits in the Sound. These Analysis Areas range in size from over 380,000 acres (Montague Island Area) to just under 11,000 acres (Perry/Lone Islands Area). Open-water areas of the Sound had not previously been defined into management or specific analysis units by the Chugach National Forest. Due to the broad objectives of this study to define human use concentration areas throughout the Sound, the large expanses of salt water not included in existing Analysis Areas were further defined. To accomplish this, we incorporated existing salt water polygons established by Alaska Department of Fish and Game Sport Fish Division as reporting units for commercial sportfishing guides in Prince William Sound.

The Analysis Areas created were too large to give a resolution useful for defining concentration areas of human use or for analyses needed at the bay or subwatershed level. As a result, General Areas were created as a lower-level spatial unit useful for land and resource managers. These General Areas are large enough for ease of display yet small enough to capture the spatial distribution of various landscape attributes, including human use. The General Areas are defined mainly by named geographic features, such as individual bays as shown on standard USGS topographic quads. General Areas were created using physical landscape features and established place-names to allow ease of use by multiple agencies and the public. They were created without consideration of management designations or land-ownership boundaries.

Not every named bay or cove was made into a General Area, and several General Areas were defined based on local knowledge rather than officially named areas on USGS maps. Several reference sources were used in determining which areas should be designated as a General Area. Jim and Nancy Lethcoe's *A Cruising Guide to Prince William Sound* (1998) was used with the authors' permission as a valuable source of data on recreation features and potential anchorages and was used in several instances to define a General Area not labeled on USGS maps or marine charts. Paul Twardock's *Guide to Camping and Kayaking in Prince William Sound* (2004) was used as another reference to determine popular recreation areas, particularly around Valdez, where Forest Service data is severely limited. Additionally, a map of areas permitted for use by commercial outfitter and guide companies under permit with the Forest Service was used to ensure that popular commercially used sites were accounted for. Data points from a 2005 human use study (Gimblett et al. 2007) were also considered to ensure that areas of public use concentration were adequately defined.

BUFFERING OF GENERAL AREAS

The majority of General Areas are defined by saltwater features. However, because much recreation use occurs onshore in tidelands and uplands, it was necessary to associate use assigned to any given bay with the surrounding land areas. The Buffer tool in ArcGIS (ESRI 2009) was used to expand the borders of each saltwater General Area outward by 100 m on all sides, so that they overlapped the adjacent land area. This step was also critical to ensure that subsequent investigations of potential user conflict and overlap with sensitive resources could analyze human use across tideline boundaries.

DATA COMPILATION

All spatial data used in this project and contained in the geodatabase are in the coordinate system North American Datum of 1983 (NAD83) Universal Transverse Mercator, zone 6N (UTM 6N). GIS software used was the ArcGIS 9.3 suite (ESRI 2009). After a review of identified data sources, we determined a need for two separate types of spatial data: raster and vector. Existing data fell into two broad categories: (1) measurable data with a metric of relative use (for example, number of nights a cabin was reserved or number of permits fished in a given area); and (2) locational/descriptive data with no associated measure of relative use (for instance, ferry routes). Data with measures of use were included as raster data sets, while descriptive data were included as vector data sets. Raster data include hunting, fishing, beach use, and ranked anchorages. Descriptive data include ferry and tour routes, campsite locations, land use permit locations, potential attractions, inhabited areas, aircraft landing site, and boat anchorage point locations.

Many data sets were in tabular formats rather than spatial formats. For example, hunting and fishing data obtained from various ADF&G departments were in the form of spreadsheets with no associated spatial data set. Where spatial units existed to which these data could be linked, the spatial units were digitized and spatially joined to the tabular data so that it could be mapped.

A set of vector data sets was created that contained ferry routes, daily scenic tour routes, public use cabin locations, documented campsite locations, land use permit locations, potential attractions, inhabited areas, aircraft landing site point locations, and anchorage point locations.

As defined in ArcGIS, a raster "consists of a matrix of cells organized into rows and columns (or a grid) where each cell contains a value representing information." Rasters can be powerful data representations useful for many types of spatial analyses. Data sets that do not share the same spatial boundaries can be converted to rasters and combined using various statistics. Converting data to a raster can be visualized as laying a piece of graph paper over the top of a shape, then assigning the majority value under each square of the graph paper to that square, resulting in a "pixelated" representation of the shape. Distinct seasonal rasters were created for measured use data—for example, number of nights reserved at a cabin or number of fishing trips recorded in a given fishing use area. A summary of created rasters is given in table 6.1, followed by sections providing detailed methods for each data set.

DATA COMBINATION

Overall ranks of human use were calculated for each season independently; in other words, a rank of "highest" use in summer season reflects different absolute use numbers than a rank of "highest" in fall season, when there are lower overall numbers of people using the Sound and also a different distribution of people than in the summer season.

To determine areas of relatively high overall human use, raster data sets were overlaid to find areas where all the various uses were concentrated. In order to overlay disparate data sources (for example the various polygon boundaries determined by Alaska Department of Fish and Game and Forest Service administrative units), raster data were combined using Spatial Analyst tools in ArcMap. All rasters were classified using the Jenks Natural Breaks algorithm in ArcMap, yielding values of 1 (lowest use) to 4 (highest use). All measurable data were classified from 1 to 4, creating a series of data sets presented in order of relative importance or rank rather than actual values. This is a necessary step due to the highly variable metrics in the data considered.

Because hunting data is unique among other data sets in its focus on reporting use to upland areas, additional steps were taken following the creation of the raster data sets of hunting rank. Hunting data collected by ADF&G are reported to Uniform Coding Units (UCUs), which are delineated only to the shoreline and are based on

TABLE 6.1. Summary of raster datasets

DATASET	YEARS OF DATA INCLUDED	SEASONAL GRIDS CREATED	UNIT OF MEASURE	SPATIAL UNIT	RANGE OF VALUES (AVERAGE ANNUAL)
ADF&G commercial fishing	2003–2007	spring, summer, fall	permits fished	comm. fish statistical area polygon	0.2–666 permits fished per area
ADF&G sportfish (reported from commercial fishing guides)	2005–2007	spring, summer, fall	number of trips	sportfish statistical area polygon	0.3–658 trips per area
Human use study and user experience trip diary points (divided by fishing or beach use)	2005; 2008	spring, summer, fall	number of points mapped by survey respondents	summarized by General Area	beach use: 1–20 points per GA; fishing: 1–70 points per GA
Special use permit reported use (categorized either as hunting or beach use)	2005–2007	spring, summer, fall	service days (one client on NFS lands for any part of one 24-hr day)	summarized by General Area	0.3–49 service days per GA (hunting); 0.3–358 service days per GA (beach use)
Forest Service and Alaska State Marine Park cabin reservations	2004–2006 (except Decision Pt., opened in 2005)	spring, summer, fall	number of nights reserved	summarized to General Area in which cabin located	2–116 nights per cabin
ADF&G hunting data for black bear, brown bear, mtn. goats	2002–2006	spring, fall	hunter days	ADF&G Uniform Coding Units	0.2–95 hunter days per UCU
ADF&G hunting data for deer	2006	summer, fall	hunted days (estimated)	ADF&G deer reporting units	1–880 estimated hunter days per deer reporting unit
Water taxi drop-off records	2005–2007	spring, summer, fall	number of clients dropped off	summarized to General Area	1–648 clients per GA

upland drainages rather than bays or fjords. In order to overlay these data with the other raster data sets of fishing and beach use, hunting use in these upland areas was spatially associated with the bays and coves adjacent to each UCU.

Individual ranked rasters were overlaid using the cell statistics function in Spatial Analyst to give the total sum of all ranks for each 100 m^2 pixel. Summed rank values range from 0 to 21 (the maximum potential value for a single pixel would be 44 if it had received the maximum rank possible for every input layer). The total rank sum for each season was classified using Jenks Natural Breaks to yield an overall relative use rank of 1 to 4, with 1 being the lowest relative use and 4 being the highest relative use. Using the zonal statistics function in the Spatial Analyst extension of ArcMap (ESRI 2009), the majority value for each General Area was calculated and assigned to all pixels in the General Area. This results in some loss of spatial variation, particularly in larger General Area polygons, but is necessary to be able to compare and summarize results across uses and seasons.

VALIDATION OF RELATIVE RANKS

To test the assumption that our calculated overall use levels had a reasonable level of accuracy, we compared our results to an independent human use data set. Vessel observations collected in 2008 by the User Experience Project (chapter 7) were used to create a density surface for the salt water of the Sound using the Kernel Density tool in ArcMap Spatial Analyst. The density surface was then sampled with zonal statistics, using our General Area polygons as the analysis zones. This resulted in a sum of all pixel values contained within each General Area. These sums were normalized by dividing by the area of each General Area and then classified into four classes using the Jenks Natural Breaks algorithm in ArcMap.

OVERLAPPING USES

We also wanted to identify locations in which there was a convergence of a variety of human uses. In order to summarize those areas with the greatest potential for desired use by multiple and possibly incompatible user groups, we examined the relative contribution of each use type to the overall rank of each General Area. Those areas with a relative ranking of higher or highest were considered to have the greatest potential for overlap of various user groups.

General Areas of higher or highest overall relative use rank were examined to determine which areas had multiple uses with similar contributions to the overall human presence in the area. The uses of hunting and beach recreation (kayaking, camping, hiking, etc.) have been previously identified by Forest Service managers as the greatest potential for conflict; thus, these ranks were evaluated, and areas with similar beach use and hunting ranks are identified as having a higher potential for

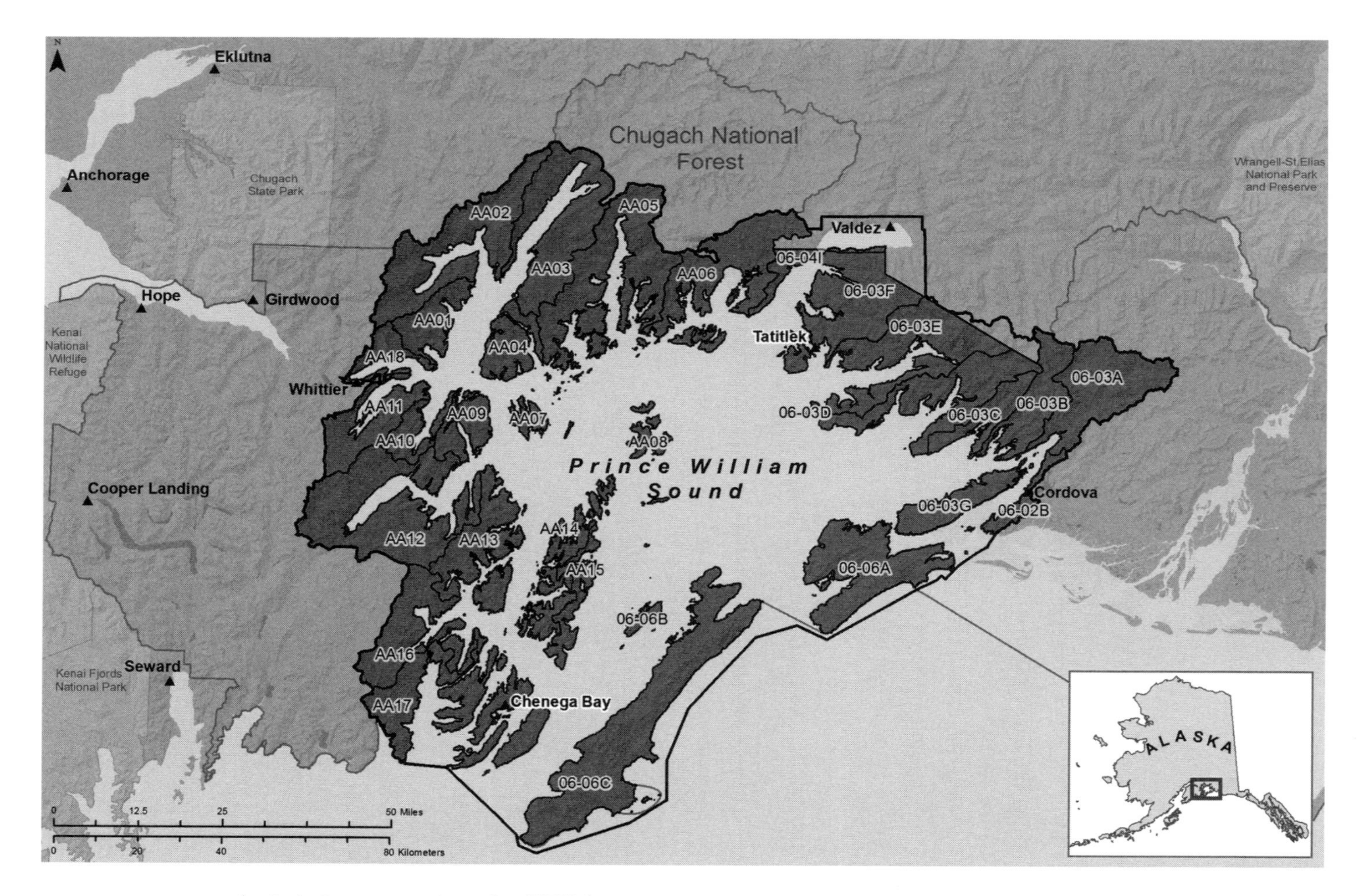

FIGURE 6.1. Analysis Areas currently used in PWS for special use permit administration. Note that the portion of AA 06-02B represented adjacent to Cordova is a small fraction of the larger area, which is managed as part of the West Delta rather than PWS.

overlap of multiple user groups onshore. Note: we do not assume that this potential overlap of use has led to actual conflict.

RESULTS

Based on evaluation of available data sources, four seasons were defined for this study:

- Spring = April 1 through June 14
- Summer = June 15 through August 31
- Fall = September 1 through December 31
- Winter = January 1 through March 31

An overview of 31 existing Analysis Areas in the Sound is shown in figure 6.2. These spatial units are used primarily by Forest Service permit administrators to manage the amount of commercial use in the Sound and are currently the scale to which capacity recommendations for commercial use have been established. They are too large to meet the objectives of this study, and no analyses were conducted at the Analysis Area scale. However, data are presented according to established Analysis Areas to provide a logical geographic grouping.

The General Areas created for this study vary widely in size, ranging from 3 to 200,000 acres. The trend is for smaller General Areas to have a higher known human use concentration while larger areas have less concentrated use. This reflects the current understanding of human use concentration. Areas of known higher use can be better defined for management purposes than areas where use patterns are less understood or where use is not concentrated.

OVERALL HUMAN USE LEVELS

Human use levels were summarized for each General Area relative to all the General Areas in the Sound, including beach use data, fishing data, and hunting data. Use was evaluated seasonally. Overall, use levels are higher in the western half of the Sound than the eastern half. This is likely due to relative ease of access from Whittier, which is road accessible and in close proximity to Alaska's major population center. Valdez is also a major access port, accessible from the Richardson Highway, which connects Prince William Sound to Fairbanks and other interior Alaska communities.

SEASONAL VARIATION IN USE DISTRIBUTION

Analysis of available data confirms the importance of seasonal variation in the distribution of human use in Prince William Sound. Overall use and human presence is highest in the summer months, followed by spring, then fall, with the least amount

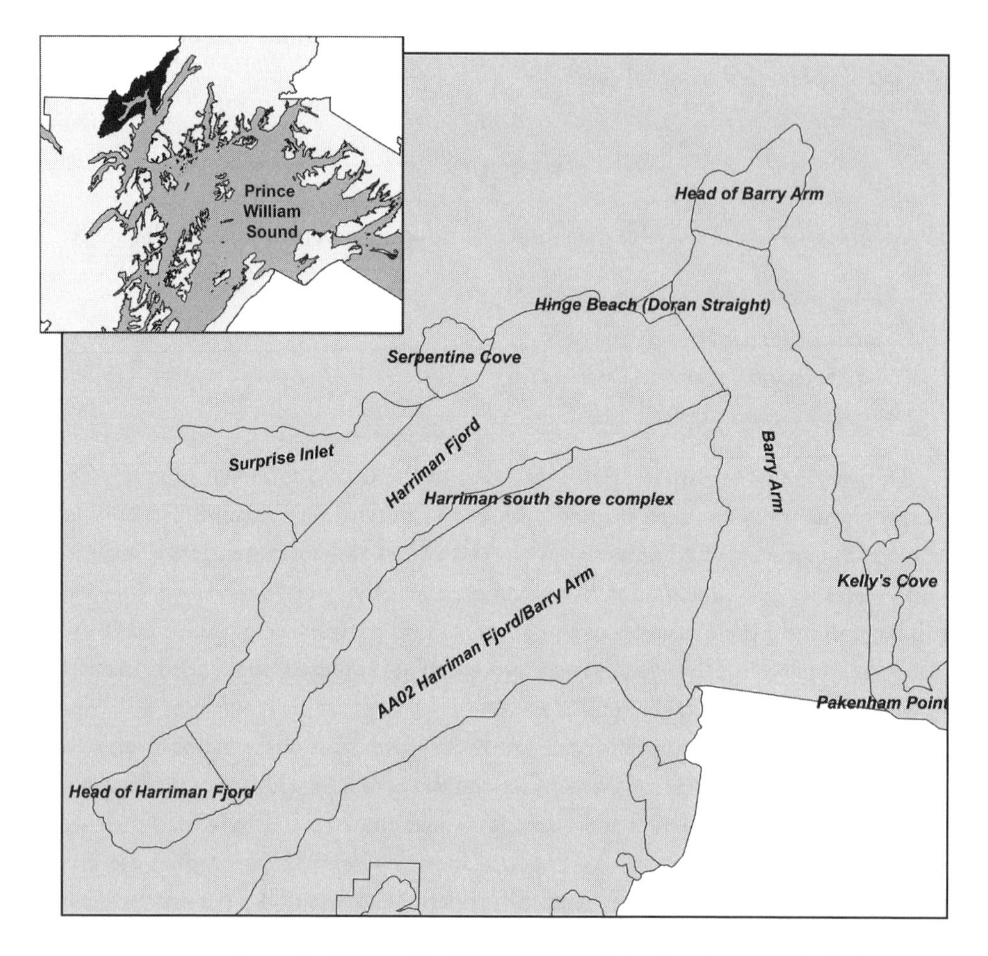

FIGURE 6.2. Example of the smaller General Areas defined within a single Analysis Area. Harriman Fjord/Barry Arm, Analysis Area 02, is the shaded area in the inset above; this Analysis Area was divided into a total of 11 General Areas of variable size.

of human use documented in the winter. In fact, data were insufficient to develop valid relative ranks in the winter season, when human use of the Sound is low and mainly limited to commercial fishing and other commercial open-water activities. Spatially, human use appears to be more widespread in summer months and more concentrated in areas closer to Whittier and Valdez in the spring and fall. Hunting dominates the documented human use in the fall season.

Three categories of use were selected to demonstrate the variation in measurable data behind these ranks (table 6.2): (1) Forest Service commercial use permit holders beach use (total service days); (2) public use cabins (nights reserved); and (3) hunter days as summarized by ADF&G data for black bear, brown bear, mountain goats,

TABLE 6.2. Seasonal variation in measured values for three different data types

HUMAN USE DATA SOURCE AND METRIC	MAXIMUM VALUE PER SEASON		
	SPRING	SUMMER	FALL
Forest Service commercial permit for beach use (days)	142	1,074	39
Public use cabins (nights reserved)	276	424	356
ADF&G harvest data (days hunted)	381	76	1,226

and deer. The Forest Service special use permit data reflects a use that is directly managed by the Chugach National Forest, while cabin reservations reflect the most discrete quantitative public use data available. Hunting data reflects both commercially guided and private hunting effort.

The importance of seasonal variation is clearly shown in table 6.2, which indicates that the spring season may be the time of greatest potential for conflict between user groups. Beach use peaks in the summer when hunting is at its lowest level, while hunting peaks in the fall after beach use has dramatically declined; however, both uses have a presence in the spring.

VALIDATION OF RELATIVE RANKS

The rank of vessel density (1 to 4) summarized to General Area was compared with the calculated human use hot spot rank (also 1 to 4) using a Pearson's correlation coefficient, with a result of $r = .71$. This indicates a high level of correlation between our relative rank summaries of use and an empirical data set of locations of vessels mapped in by the efforts described in chapter 7.

AREAS WITH GREATEST POTENTIAL FOR OVERLAP OF BEACH USER GROUPS

Those areas with a relative ranking of higher or highest were considered to have the greatest potential for overlap of various user groups. Sites within the western Sound with the highest relative degree of overlap where *potential* user conflicts between hunters and others using the shoreline and uplands might occur include (1) Blackstone Bay: throughout; (2) College Fjord: specifically, Coghill and Golden areas; (3) Esther Island/Passage: Esther Passage; (4) Kings Bay/Nellie Juan: specifically, Derickson Bay and Nellie Juan Spit. The complete list of General Areas with simultaneous hunting and other beach uses, and the seasonality of that overlap, can be found in Poe and Greenwood (2010).

DISCUSSION AND CONCLUSIONS

Human use in Prince William Sound, Alaska, is not evenly distributed, spatially or temporally. Certain locations within this remote area are more desirable for a variety of reasons, including distance from communities, presence of glaciers and postglacial landscapes, availability of landing areas, protected anchorages, sportfish streams, cabins, and wild game concentrations (see chapter 11). Human use hot spots are important areas in the Sound where human use is more concentrated relative to surrounding areas. In many cases, these locations are physiographic bottlenecks, restricting access to desirable intertidal areas and their associated upland opportunities for recreation activities. They also vary seasonally, based on wildlife distribution (important for both harvest and viewing activities) and weather conditions.

Several disparate data sources describing various human uses of Prince William Sound exist but have not previously been collected and integrated to identify areas of relatively higher concentrations of overall human use. In such areas, competing uses (e.g., wildlife viewing and hunting) may converge at certain times, potentially resulting in detrimental effects upon one another and creating a larger cumulative impact on associated resources. It is critical for the sustainable management of human use in an area hosting a number of *Exxon Valdez* oil spill–injured resources (both recovered and recovering) that the location, timing, and nature of these human use hot spots be well understood by resource managers.

While the summing of cumulative ranks is not necessarily the ideal method for combining ordinal data, this method is suitable to provide a generalized depiction of human use patterns across the Sound. These results appear to accurately describe human use hot spots in the Sound based on feedback from longtime users of the region and were found to statistically correlate with overall boat density predictions made in chapter 7.

Overall human use levels, including beach use data, fishing data, and hunting data, were summarized for each General Area relative to the entire study area. Overall, use levels are higher in the western half of Prince William Sound than the eastern half. This is likely due to relative ease of access from Whittier, which is road accessible and in close proximity to Anchorage, which is Alaska's major population center. Valdez is also a major access port, accessible from the Richardson Highway, which connects the Sound to Fairbanks and other interior Alaska communities. A substantial portion of recreationists entering Prince William Sound from Valdez use the western Sound. Highest use levels were found at areas closest to Whittier (particularly Blackstone Bay) and Valdez. Forest Service and State of Alaska public use cabins act as nodes of concentrated use. Designation and developments at state marine parks also appear to increase overall human use.

Overall numbers of people using the Sound are the highest in summer; however, we believe the highest potential for cumulative impacts of various uses and highest potential for conflict between different user types occur in the spring season, when hunting, other beach use such as camping and day hiking, and fishing are all present in the Sound. In summer, use is dominated by beach use and fishing; in fall the dominant user group is hunters. Managers may be able to use calculated ranks of various uses within General Areas to pinpoint areas of potential user conflict. It is important to note, however, that this is not a definite indicator of perceived crowding or conflict between users; some research suggests there may be greater risk for perceived crowding and user conflict in areas that are the least crowded, as individuals seek out these areas with a goal of solitude and thereby may be least tolerant of any encounter with others (Blahna 2007).

This evaluation of human use allowed us to pursue another analysis of keen interest to those concerned about oil from the 1989 spill still embedded in shorelines of the Sound. Lingering oil data points were collected in 2001 across Prince William Sound. These data were randomly sampled points at beach segments stratified according to their oiling history and targeting those beaches known to have received higher levels of oiling from the *Exxon Valdez* oil spill (Pella and Maselko 2007). We intersected these sample locations with our General Area (GA) polygons to determine their coincidence with areas of human use.

We found that six of our larger Analysis Areas to date still contain beaches known to have lingering oil from the spill based on the distribution described by Pella and Maselko (2007). Lingering oil remains at East Knight Island, Green Island, Perry/Lone Islands, Port Bainbridge, West Knight Island, and West Knight Island Passage. Two of these larger areas (Green Island and West Knight Island) have small-scale GAs with a rank of highest human use when looking at patterns of use from the Sound as a whole. These data provide a glimpse of the distribution of human use relative to lingering oil in Prince William Sound. Each of the smaller General Areas where lingering oil was identified is included in table 6.3, ordered by relative level of human use that might be expected for that area. These data provide a glimpse of the distribution of human use overlap with lingering oil in the Sound. As research continues to refine predictions about the distribution of lingering oil, the human use distribution that we have created would allow for easy future comparisons.

LESSONS LEARNED

This effort aimed to produce a seasonal and spatial framework summarizing human use distribution across the Sound as a system using the diversity of data available. This integration of data from multiple managers represents a significant departure from previous attempts by the U.S. Forest Service to summarize use only within the jurisdictions of the forest boundary and is a critical aspect of sustainable recreation management. We see it as a potential model that other managers could use to evaluate

TABLE 6.3. List of General Areas containing one to several lingering oil points as identified by Pella and Maselko (2007)

ANALYSIS AREA	GENERAL AREA	RELATIVE HUMAN USE RANK
Green Island 06-06B	Gibbon Anchorage	Highest
West Knight Island AA14	Herring Bay	Highest
East Knight Island AA15	Bay of Isles	Higher
East Knight Island AA15	Hogan Bay	Higher
Green Island 06-06B	Green Island	Higher
West Knight Island AA14	West Knight Island	Higher
West Knight Island AA14	Ingot Island	Higher
West Knight Island AA14	Lower Passage	Higher
West Knight Island AA14	Northwest Bay	Higher
Perry Island AA07	Day Care Cove	Lower
Port Bainbridge AA17	Latouche Island	Lower
Port Bainbridge AA17	Shelter Bay/Evans Island	Lower
Port Bainbridge AA17	Sleepy Bay	Lower
West Knight Island AA14	Block Island/Eleanor Island	Lower
West Knight Island AA14	Disk Island	Lower
West Knight Island AA14	Herring Point	Lower
West Knight Island AA14	Ingot Cove	Lower
West Knight Island AA14	Knight Island Pass CS #1	Lower
West Knight Island AA14	Smith Island	Lower
West Knight Island AA14	Squire Island	Lower
West Knight Island Passage AA13	Crafton Island	Lower
East Knight Island AA15	East Knight Island	Lowest
East Knight Island AA15	Point Helen	Lowest
Port Bainbridge AA17	Bettle Island	Lowest
Port Bainbridge AA17	Elrington Island	Lowest
Port Bainbridge AA17	Evans Island	Lowest
West Knight Island AA14	Aguliak Island	Lowest

their land units within the context of larger human and ecological systems. Managers talk about the need to summarize use across landscapes to address complex management procedures like spatially dynamic zoning of use allocation (see chapter 19), but also to address large threats like climate change and invasive species. This approach establishes the study area framework for these types of analyses in the Sound.

Conducting this analysis in a systems approach also required us to engage with a diversity of other land and resource managers in the Prince William Sound region, often initiating relationships and finding common interests for the first time. In all cases, managers willingly provided information into what they saw as an effort that could potentially benefit them as well, especially given that the outputs are readily available as seasonally attributed summaries of human use in GIS layer format. This approach also resonated with individual stakeholders invited to offer feedback on our early draft summaries and help proof our map layers. Stakeholders often fail to see practical differences in jurisdictional boundaries or perspectives and found our approach of summarizing use across the Sound refreshing if somewhat surprising from a federal agency without specific jurisdiction over the water.

For example, one of the data products of this project is the creation of topologically clean spatial polygons representing Alaska Department of Fish and Game reporting areas for hunting and fishing data. We recommend continued updating of identified hunting and fishing data sources through partners at ADF&G. These data should be updated at regular semiannual intervals to monitor changes in private and commercial fishing and hunting use levels in the Sound. Not only does this allow the Forest Service to evaluate their permitted activities in the context of other commercial uses in the Sound, it also affords us a potential window into possible locations and seasons of user conflict (see chapter 10).

Empirical validation of predicted human use hot spot areas may be conducted as necessary through continuation of boat-based or aerial surveys similar to those conducted in chapter 7 and chapter 11, respectively. Such surveys are expensive to complete in the remote wilderness of the Sound and should be targeted as needed to respond to identified areas of concern and/or potential user conflicts. This analysis, combined with the predictions made in chapter 7, offers managers a place to start when designing this type of strategic monitoring effort.

Finally, the methods we developed for merging of land- and water-based sets of data allowed us to summarize use in the system such that managers could inform both land- and water-based management objectives regardless of jurisdiction. Though joint management objectives are unlikely to be agreed upon soon, by establishing a spatial and seasonal frame that is functional for other area land and resource managers, the foundation for possible collaborative analysis is laid. Furthermore, our summaries of use distribution by season allow managers to take into account seasonal shifts in their planning efforts relative to species or habitat distribution and contingency planning for responses to natural or human-caused disasters.

CHAPTER 7

CHARACTERIZING THE SPATIAL AND TEMPORAL PATTERNS OF RECREATION USE AND EXPERIENCE OPPORTUNITIES IN THE SOUND

Perceptions of the Stakeholders

AARON J. POE, RANDY GIMBLETT, AND ROBERT M. ITAMI

SOUND BITES

THE PERCEPTION OF "CROWDING," which can negatively impact recreation user experience, does not seem to be a widespread concern among recreationists in the Sound.

Visitors to remote regions of the Sound have a variety of motivations, which are not necessarily solitude dependent. These include wildlife and glacier viewing, fishing, and spending time with family and friends.

Our approach of describing the intensity of recreation use in mapped data layers allows managers to better understand interactions between recreation and other ecosystem components, such as sensitive species or other human uses.

METHODS

Thirty-one Analysis Areas currently identified by the Chugach National Forest for administration of special-use permits in Prince William Sound were used to summarize our data. Each Analysis Area ranges in total size from 841 km^2 (Montague Island) to 30 km^2 (Green Island) and in total contain 4,828 km (3,000 miles) of shoreline. An additional product from that analysis is a layer of smaller spatial units referred to as General Areas, defined as midlevel spatial units large enough for ease of display and small enough to capture the spatial distribution of specific landscape attributes, including human use features. A total of 537 General Areas were defined in the study area, ranging in size from 0.1 to 800 km^2 (averaging 36 km^2). The delineation of these units is described in chapter 6. Human use of Prince William Sound is strongly seasonal due to variable weather and extreme winter conditions. Recreational use of uplands (including hunting), recreational boating, and sportfishing occur mainly from May through September. Previous studies have used a recreational season of May 1 through September 30 in eastern Prince William Sound (USDA Forest Service 2008) and May 15 through September 15 in western Prince William Sound (USDA Forest Service 2005). Four seasons were defined for this study by the work in chapter 6, and are used here, including the following:

- Spring = April 1 through June 14
- Summer = June 15 through August 31
- Fall = September 1 through December 31
- Winter = January 1 through March 31

In order to complete an empirical inventory of boat use across Prince William Sound, water-based transects were operated between May and September of 2007 and 2008. These transects were operated using a combination of small motorized boat- and kayak-based observation crews. May through early June (spring) and late August through mid-October (fall) surveys were completed by motorized boat due to the need to cover vast distances in short windows of time around inclement weather patterns. Summer surveys (mid-June through mid-August) were completed by kayak-based crews, and these observations included at least three daily, shore-based observation periods of 30 minutes in duration.

Each transect began when the observer entered the water and was ready to record observations and ended when the observer left the kayak or stopped the boat for longer than 15 minutes or during challenging water conditions when safety was a concern and observations could not be recorded at the time they occurred (e.g., during a long crossing of open water or rough seas). Boats traveled within a range of 10 to 20 knots, and kayaks from 2 to 4 knots, during water-based surveys. For kayaks, transects occurred within approximately 100 m of shorelines and for motorized boats within approximately 250 m of shoreline. All observers were equipped with a survey datasheet, observation datasheet, map, and global positioning system (GPS) for recording waypoints for each transect. Waypoints were taken for the shore-based surveys to capture the location of land visit. For both modes of transportation, the total amount of time spent on each transect was recorded to allow for comparison of use density between Analysis Areas.

At least one visitor observation datasheet was used each day to record the location of encounters with boats or shore parties. A visitor observation began at the first sighting of a water- or land-based visitor and ended when the visitor was no longer in sight. The following were recorded with each observation: time of detection; party location on a 1:73,560 topographic map as well as a GPS waypoint for the observer; the vessel type (K = kayak, TC = tour/cruise, OS = onshore user, CF = commercial fishing, OST = onshore tent, FW = fixed wing aircraft, IN = inflatable or skiff, HE = helicopter, CC = cabin cruiser, S = sailboat, MY = motor yacht, OT = other [barges, oil rigs]); estimated categorical distance at first encounter and closest approach (<100 m, 101–500 m, 501 m–1 km, >1 km); and the duration of the observation, defined as the total time the vessel was visible to the observer, as well as whether the vessel was moving. Encounter dynamics of kayak-based observers were summarized for water and shore encounter separately, and focused on distance of earliest detection and closest distance of approach. The locations and times of encounters with small fixed-wing aircraft and helicopters were also recorded as part of this inventory process.

The locations of all vessels observed were stored in a GIS database, and observations were summarized to Analysis Areas. Those vessel classes likely indicative of a recreation user (all but CF, commercial fishing; and OT, other) were identified as recreation groups and removed from the set for summary analysis, predictive density modeling, and subsequent comparison to a GIS database of human use hot spots compiled in chapter 6.

Recreation locations identified on questionnaires distributed during 2005 and summarized by Wolfe et al. (2006), as well as those locations collected by the questionnaire distribution effort described below, were analyzed for spatial correlation with transect observations. This was accomplished through kernel density interpolation of mapped use points from 2005 and 2008 questionnaires in Spatial Analyst (ESRI 2009) using default settings, including the search radius of 5,538 m. This process generated a continuous raster surface of 250 m cells representing the density of recreation vessels in the Sound. This surface was then classified into five classes using quantile classification scheme in Spatial Analyst. The points observed on transect were then evaluated relative to class membership using a chi-square and subsequent Pearson's coefficient of correlation analysis.

Subsequent kernel density interpolations were used to generate density surfaces (250 m cells using default search radii settings) from a combination of both trip diary and transect observation locations for the spring, summer, and fall seasons. Using the zonal statistics tool in Spatial Analyst, summary recreation-vessel density values were predicted for the larger Analysis Areas and smaller General Areas, and categorized using quantile classification into areas of high, medium, and low overall vessel use. This was completed for each season of use.

In an effort to explore relative groups of recreation users at one time within Analysis Areas, further evaluation of summer transect observations was completed. This was accomplished by selecting all recreation groups observed on transect (including both water- and shore-based observations) within each Analysis Area, and dividing that number by the total hours of observation within that Analysis Area to get a relative observations/hour measure. These observations were compared to the number of permitted days for commercial recreation described in chapter 6.

In order to characterize visitors to Prince William Sound, a survey tool was constructed and on-site contacts of survey targets were made by Forest Service and contract personnel. The survey consisted of a carefully organized set of questions that were structured to elicit response from visitors on a number of experience-based questions, situational responses to encounters with others, and characteristics of their trips. Included in the survey was a trip diary to document spatial and temporal trip patterns. The trip diary was a map of the Sound with instructions on how to document a visitor's own trip from the time of departure, where the visitor camped, where the visitor stopped, the duration of these activities, routes taken, perceptions of place, and displacement.

Prior to beginning data collection, support was gathered for the project by contacting some recreation groups as well as local transporters (e.g., water and taxi services), harbor masters, and guides to seek input into the survey process and inform them about our intentions of sampling in 2008. Survey distribution was focused at the three primary harbors used to access Prince William Sound: Whittier, Valdez, and Cordova. Time of day was incorporated into sampling procedures such that distribution efforts focused on peak-use periods. Sampling was undertaken approximately four days a week for a six- to eight-hour period from mid-May through mid-September of 2008. Weekdays were sampled as randomized two-day blocks to ensure weekday use was captured. Both weekend days were consistently sampled throughout the survey period. Each of the surveys was coded by location (Whittier, Valdez, or Cordova) with a number associated with this coding.

Researchers distributing the surveys kept track of the survey number, number in party, mode of travel, and time of exit. They also kept track of refusals in an attempt to acquire an estimate of the total numbers of visitors entering the Sound from the respective locations for the sampling period. The instrument package included a return-addressed, stamped envelope and an enrollment postcard for a prize drawing (a participation incentive). Survey recipients were instructed to log the course of their activities throughout the duration of their trips, with an option to return completed surveys by mail or at collection sites at harbor distribution locations.

Concurrent with survey distribution, survey crews collected and recorded several data items from intercepted individuals. In order to determine whether those who declined to participate in the survey were statistically different from those who participated, all individuals were asked if they were Alaska residents, the primary purpose of their trip (cruising, kayaking, and hunting/fishing), the vessel type, and the anticipated length of their trip, and whether the respondent agreed or refused to accept and complete a survey. This information was useful in calculating several statistics, including overall response rate and response rate by class (purpose of trip, vessel type, and Alaska residency).

Questionnaire responses were entered into a tabular database and responses were tallied for the individual questions posed using a combination of ranks and summary proportions of response. Individual locations of recorded use mapped on the associated trip diary were digitized into GIS and summarized at the spatial resolution of Analysis Areas.

In order to explore temporal and spatial distribution of respondents, we used Recreation Behavior Simulator (RBSim), a software program fully integrated with ArcMap that has been specifically developed by researchers from GeoDimensions Pty. Ltd. and the University of Arizona for studying patterns of recreation use (e.g., Gimblett et al. 2001; Gimblett 2002; Itami 2003). RBSim has been used in the Sound previously by Gimblett and Itami (2006) to develop a simulation model for the Sound using data from Wolfe et al. (2006) to replicate the data collected on visitor use.

A trace simulation was developed from the trip diaries and survey responses to evaluate the distribution of use across the Sound by mode of travel, analyze popular destinations, and evaluate how long visitors spend at these sites and whom they encounter, both on water and on land. A series of simulations of existing peak use and forecast future use levels in the Sound were undertaken. The simulation was replicated 100 times to acquire a more accurate representation of distribution of visitor use. For each simulation replication, the following behavior variables were randomized: (1) selection of trip itineraries—a trip was randomly drawn from a pool of trips with the same port and the same day of the week; (2) launch times were randomized based on ranges from questionnaire data; (3) travel speeds were randomized around a mean for each vessel type (Gimblett and Itami 2006); (4) departure times from destinations were randomized within the time frame of one hour.

Simulation outputs were then averaged over the 100 replications. Since many trips are multiple-day trips, it was necessary to run the simulation a sufficient time in order to fully populate the simulation before outputs were calculated. In this simulation, the first two weeks of June were used as the test period and were discarded before simulation outputs were calculated on the remaining simulation days. The outputs of the simulations provided a spatial view of the changes in visitor densities and volumes at peak use levels by day of the week and by port of departure. Outputs were reported to the Analysis Area level by vessel class. Potential recreation use futures over the next 10 years using a theoretical 1.5 percent per year growth regime were also explored using this trace simulation.

RESULTS FROM TRANSECT SURVEYS

In total, 108 sampling days were accomplished between June 28, 2007, and October 4, 2008. Multiple crews of observers logged over 1,600 hours of combined transect and shore observations during these two years of sampling. Observers based in motorized boats were able to accomplish multiple surveys of Analysis Areas within Prince William Sound for a total of 4,172 km of transects and 4,205 km during spring and summer, respectively. Kayak-based observers conducted one-time inventories of Analysis Areas, accomplishing a total of 2,923 km of transects during summer, and in addition were able to complete 332 fixed-point shore observations. Spring and fall inventory efforts covered approximately 60 percent of Sound shoreline, with summer efforts covering approximately half. Sample effort occurred within all 31 Analysis Areas, as well as a substantial portion of open water areas of the Sound.

A total of n = 2,868 observations of recreation, commercial fishing, and aircraft groups were identified and mapped to individual Analysis Areas during all seasons of sampling combined. The relative composition of encounter type from least to greatest was aircraft (7 percent), commercial fishing (40 percent), and recreation groups (53 percent). Substantial variation occurs in the relative composition of these three classes of users. Within these broad categories, the relative proportions of specific vessel/user types encountered also varied seasonally. Throughout all seasons, recreation groups were the most common type encountered, and within that class, the specific type of cabin cruiser was the most common.

Further investigation of the seasonal variation in use reported for only those recreation groups with the potential to use uplands shows that cabin cruisers are consistently the most often encountered vessel type, followed by motor yachts. Perhaps somewhat surprisingly, kayak groups represented a small overall portion of the use in the Sound, though this may be due in part to difficulty detecting smaller boats without a motorized noise signature.

Numbers of encounters with only recreation parties varied widely by location, ranging from 0 around Hinchinbrook Island to 13.5/hour in Culross Passage during the summer. Observers on motorized boats encountered more individuals per hour (3.1 +/-3.3 during spring and 2.4 +/-3.6 detections/hour during fall) due to the faster rate of speed on transects, with high rates of variance of detection rates across Analysis Areas. Summer surveys from kayaks returned lower rates of detection, 1.0 +/-1.0 detections/hour, yet substantially more total detections. When controlling for the approximately tenfold increase in survey speed of boat-based observers (e.g., the threefold increase in days of sample effort), we found summer months support approximately three times the amount of total average monthly recreation traffic.

Two correlation analyses were completed using these transect data. The first evaluated the spatial relationship between 977 summertime recreation vessel observations to a classified predictive, raster surface generated from 3,703 trip diary locations collected in 2005 and 2008. Based on a five-class membership identity analysis, we found a strong positive relationship ($r = 0.77$) between the distribution of trip diary locations and observations compiled from kayak- and shore-based surveys. Based on this strong positive relationship between diary and transect locations, predictive surfaces were generated using combined vessel observation points from transect with diary locations from within the same season.

Based on these results, three raster surfaces were generated for spring, summer, and fall from 1,583, 4,680, and 483 input points, respectively. Using these raster surfaces, summary predictions of recreation use (high, medium, and low) were generated for Analysis Areas and General Areas. In this example, red crosshatch polygons represented the highest 20 percent of use summarized to General Areas, and darker-color landscape polygons indicated Analysis Areas supporting larger amounts of recreation vessel traffic. These results for Analysis Areas during summer are presented in table 7.1, alongside relative intensities of commercial operations that are permitted by the Forest Service summarized in chapter 6.

The second correlation analysis evaluated four classes of total vessel use (including both commercial fishing and recreation vessels) during summer months reported by the human use hot spots analysis from chapter 6 within 350 saltwater General Areas for the summer season. When summer density predictions for all vessel classes combined were similarly classified into four classes and summarized to the General Area level, they returned a correlation value of $r = 0.72$ with hot spot predictions of overall use. The class membership of transect observations is relative to predicted density class, 1 being highest and 5 being lowest. These two correlation findings suggest that when evaluated in generalities at the 4 or 5 class level, our transect data corroborates similar predictions of use documented by Wolfe et al. (2006) and in chapter 6.

We found commercial recreation use permitted by the Chugach National Forest (summarized in chapter 6) was overall a small proportion of total use occurring within each Analysis Area, estimated at less than 10 percent of total use. However, a few Analysis Areas likely support double-digit percentages of commercial use, including Harriman Fjord/Barry Arm, Icy Bay, and Columbia Bay. A few areas, like Port Fidalgo, Nelson Bay, and Sheep Bay, apparently host no summertime commercial use permitted by the Forest Service.

Results from encounter dynamics experienced by our observers relative to detection distance and proximity of approach were summarized for 2007 and 2008 transect

TABLE 7.1. Summer vessel density and special use permits by Analysis Area

ANALYSIS AREA	SUMMER RECREATION VESSEL DENSITY	SPECIAL USE PERMIT
Blackstone Bay AA11	High	High
Cochrane Bay AA10	High	Low
College Fjord AA03	Low	Medium
Columbia Bay AA06	Medium	High
Culross Island/Passage AA09	High	High
Eaglek/Unakwik AA05	Medium	High
East Knight Island AA15	Medium	Low
East Valdez Arm 06-03F	High	None
Esther Island/Passage AA04	Medium	Medium
Fidalgo Bay 06-03E	Low	None
Gravina Bay 06-03D	Low	Low
Green Island 06-06B	Low	Low
Harriman Fjord/Barry Arm AA02	Medium	High
Hawkins Island 06-03G	Medium	Low
Hinchinbrook Island 06-06A	Low	Low
Icy/Whale Bay AA16	Low	High
Kings Bay/Port Nellie Juan AA12	Medium	Medium
Montague Island 06-06C	Low	Low
Naked Island Group AA08	Medium	Low
Nelson Bay 06-03A	Low	None
Open Water	Low	None
Passage Canal AA18	High	High
Perry Island AA07	High	Medium
Port Bainbridge AA17	Low	Medium
Sheep Bay 06-03C	Low	None
Simpson Bay 06-03B	Medium	Low
Valdez Area Nonforest	High	High
West Delta 06-02B	Medium	None
West Knight Island AA14	Medium	Medium
West Knight Island Passage AA13	High	Medium
West Port Wells AA01	High	Medium
West Valdez Arm 06-04	High	Medium

observations. The vast majority of other parties (97 percent) detected occurred when those parties were on the water. Approximately 90 percent of detections were made when the other party was underway. The majority of detections of any other party happened at distances >1,000 m across all three types of sampling: shore based (68 percent), kayak based (80 percent) and boat based (70 percent). A similar pattern was found when evaluating the closest distance of approach, with 61 percent and 65 percent staying >1,000 m away from shore- and kayak-based parties, respectively. A different pattern emerged for boat-based observers, in which 38 percent of groups stayed >1,000 m, with 18 percent approaching within 100 m. This is likely due to the fact that encounters of other boats were the most frequent type, and boat-based observers were likely using similar travel routes farther out from shore when compared to kayak- and shore-based observers.

RESULTS FROM SURVEY AND TRIP DIARY

Table 7.2 describes the sampling process for the survey. A total of n = 1,377 surveys were distributed and a total return of n = 341 gave the research team an overall 25 percent return rate. We found that over 80 percent of those sampled were in-state residents; 20 percent were out-of-state visitors. Of those sampled, 45 percent typically spend two to three days in the Sound at a time, and over 68 percent go into the Sound two or more times a year. Of those going into the Sound two or more times a year, 98 percent are in-state residents.

The majority of respondents were accessing the Sound by small motorized boat (65 percent) followed by kayak and motor yacht. These three categories of users comprised 90 percent of survey respondents. The top three primary motivations for destination choice included good fishing (40 percent), glacier viewing (21 percent), and wildlife viewing (19 percent), with only 10 percent reporting that seeking solitude was a primary motivation for destination choice. When asked to identify the most sought-after recreation opportunities in the Sound, people ranked enjoying

TABLE 7.2. Survey response rate for Prince William Sound recreationists by port location

SURVEY LOCATION	TOTAL ACCEPTED	TOTAL REFUSED	TOTAL RETURNED	RESPONSE RATE (%)
Whittier	n = 667	n = 56	n = 176	26.39
Cordova	n = 95	n = 55	n = 25	26.32
Valdez	n = 615	n = 114	n = 140	22.76
Combined			n = 341	Average: 25.16

natural beauty most highly, followed by spending time with family and friends and fishing.

The locations of 3,703 summer use points were reported from map diaries from the questionnaires. When asked about the prime destinations that tend to be popular attraction sites, Blackstone Bay, Port Valdez, Jack Bay, Culross Passage, and Knight Island were highly rated. These same attraction sites strongly correlated with the results of the survey observation data. Respondents were overwhelmingly satisfied with their recreation experience, with 95 percent stating they would return and that their experience either met (23 percent) or exceeded (72 percent) their expectations. Over 90 percent of respondents reported no negative encounters with other groups, and no respondents reported being displaced from desired use locations as a result of encounters with other users.

Only 10 percent of respondents reported negative encounters with other users and none reported displacement due to encounters. But when asked "if you were to move to another location, what site qualities are you looking for?" the respondents indicated that within reasonable distance (35 percent), similar qualities to where we were (30 percent), and the next available location that can accommodate our group size (17 percent) were important. Most significantly, only two respondents reported seeing any sign of lingering oil from the spill.

RESULTS OF SIMULATION

Of the total of n = 341 trip diaries returned, only n = 4 trip itineraries were discarded because of incomplete or incorrect trip data, leaving a total of n = 338 valid itineraries. The distribution of these trips by launch site (port) strongly correlates with the observation data previously discussed. Trip itineraries, launches by port, day of week, and vessel type were analyzed by the day of week (Monday through Sunday), and vessel type (kayak, inflatable, small motorized boat, motor yacht, or tour boat) for the months of June through August. Once the day-of-week arrivals were calculated for each port for each vessel type, these values were used for the launch schedules for the full summer simulation for the months of June through August. A 10-year period at a 1.5 percent annual increase translates to a total projected use of 116 percent. An examination of the simulation outputs reveals that this figure does not represent a significant change in overall use, given the small levels of use reported earlier in this chapter.

DISCUSSION

This approach of empirical observations coupled with a user survey provides a means to characterize the users in Prince William Sound and to determine their spatial distribution across the Sound. This is much-needed information, not only for evaluating the status of oil spill recovery for recreation, but also for informing a revised management strategy for the long-term protection of the Sound (e.g., the recommendations in chapter 19).

Given our lack of control for vessel type, our encounters per hour number should be used only in terms of comparing relative levels of activity: for example, describing regions as high, medium, and low overall vessel use or evaluating relative proportions of total use compared to permitted use. Though detections per hour estimates can be predicted for both spring and fall boat surveys, these seasons of effort totaled only approximately one-tenth of observation time. As a result, they also show a higher overall degree of variability, which might be real as fall and spring use appears to be patchier in the region (chapter 6), or may be a result of lower sample effort. As such, the spring and fall surveys likely have less utility for making strong comparisons relative to overall vessel use.

Density surfaces produced for all seasons may prove more useful for spring and fall analyses, as they benefit from a point distribution set including several hundred user-identified locations from 2005 and 2008 questionnaires. Furthermore, our work comparing the transect observations to reported use locations from groups departing from ports is useful for describing relative use intensities within the Sound. This survey technique is useful for systems with relatively few entry points (Gimblett et al. 2005a) and should be regarded as a valuable tool for managers in the region.

When comparing our transect observations to commercial recreation use under special use permit by the Forest Service, we found these activities likely represent a small portion of the overall recreation use in the Sound. In Harriman Fjord and Barry Arm, the percentage permitted as commercial was estimated to be the highest, approaching 20 percent; but in most areas, these groups are single-digit percentages of overall use.

Blackstone Bay was consistently identified as a busy destination by survey responses to questionnaires and simulation results. This is a distinction from the three areas with the highest traffic—Passage Canal, Culross Passage, and Port Valdez—as these are more regularly used as corridors for travel than specific destinations. Other equivalent tidewater glacier areas with medium (e.g., Harriman Fjord) and low (College Fjord and Icy Bay) human use dynamics offer opportunities to study recreation and ecosystem interactions along a gradient in these unique environments.

Our density predictions and transect surveys demonstrate a clear relationship with port distance and regional variation in use intensity. This has been observed in an earlier study of western Prince William Sound (see chapter 11) relative to distance from Whittier. It is very clear from comments made by the boating communities in Valdez and Cordova that the distance from Whittier provides "some protection" from boating traffic (see chapter 8). Those users feel that the distance from Anchorage is an important factor in the lower use levels and the ability to attain feelings of solitude when paddling or boating in the Sound. Land and resource managers should keep this reality of the system in mind as they attempt to plan for solitude opportunities within the region.

Primary recreation activities reported by respondents were fishing (54 percent), kayaking (13 percent), and sightseeing (11 percent). When asked about desired recreational experiences during these activities, respondents prioritized based on (1) enjoying natural beauty; (2) spending time with family and friends; (3) fishing (primarily saltwater); and (4) being in a wild/undeveloped place. Specific choices about destinations were made based on (1) good fishing (in salt water); (2) glacier viewing; and (3) wildlife viewing. The ability to view wildlife was the only activity identified as "very important" to three categories of users (kayakers, small motorized boaters, yacht and sail boaters). This is perhaps unsurprising, as this finding mirrors larger-scale visitor use studies on motivations for visitors to the region (Hall and Cole 2007).

Although other studies frequently mention crowding or high number of encounters as being the reasons for displacement (Hall and Cole 2007), our study illustrates that most visitors to the Sound can still get to their planned destinations and obtain the type of solitude and other opportunities that they desire. Contrary to what might be considered a popular sentiment, too many visitors or crowding was barely mentioned by questionnaire respondents, and no respondents reported displacement as a result of interactions with others, though 10 percent reported negative encounters.

There are no significant differences between single-day and multiple-day trippers in their satisfaction, rate of negative encounters, displacement rates, or identification as either Alaska resident or out-of-state visitor. All users reported satisfaction levels above 90 to 95 percent, which seems exceedingly high for a visitor use satisfaction study. Our investigation failed to identify a clear link between user satisfaction and relative levels of use. However, sentiments shared by focus groups of hunters, and to a lesser degree kayakers, described in chapter 8 suggest that increasing amounts of specific user types can be troubling.

In terms of being able to achieve experiences associated with wildland experiences, respondents specifically reported they were able to experience solitude, with 78 percent selecting "4" or "5" on a five-point scale indicating their ability to achieve solitude. Similarly, solitude was a strong motivator for survey respondents, but only 10 percent identified it as a prominent reason for choosing their destination, suggesting that the ability to achieve solitude is not a limiting factor throughout the Sound as a whole. There was no correlation between longer trips and the desire for solitude. Similarly, when asked "if respondents were to feel crowded, what would they typically do," the overwhelming response (86 percent) was to relocate to another location, suggesting that solitude opportunities are not limited. Certainly, there are problems with such theoretical action questions, but results suggest that there are enough places still available to retreat from potential crowding experiences in the Sound.

As for the future, the simulation developed to provide a view over a 10-year period at a 1.5 percent annual increase translates to a total projected use of 116 percent. The results suggest that this does not represent a significant change in overall use. While these results should be viewed with caution, they do provide some insight into what increased visitation may have on the distribution of use into the future.

This effort has defined a useful baseline for future comparison of both recreation use distribution and quality of recreation experience. It also defines spring, summer, and fall patterns of use that can be used as a baseline to explore recreation use overlap with wildlife and other resources, as well as other human uses in the Sound. This effort has also established useful techniques for managers to consider in future monitoring of recreation dynamics in the Sound and similar environments.

LESSONS LEARNED

The work presented in this chapter underscores the importance of managing for a landscape where wilderness qualities (e.g., natural beauty, plentiful wildlife for viewing, fish and game for harvesting, and access to solitude) are available as part of the overall recreation experience. It also demonstrates that other experiences, such as simply viewing glaciers and spending time with family and friends, are key aspects to the recreation experience in Prince William Sound. The successful recovery of recreation in the Sound is likely dependent on recognizing and facilitating key recreation opportunities sought by users in the region while maintaining a spectrum of available wilderness experiences.

Given that there are some key experiences sought by recreationists in the Sound and the system is not in "crisis" mode in terms of conflict, crowding, or other social impacts, planning approaches should identify a few key issues and attempt to make systematic progress on them. Attempts at addressing key issues should focus primarily on assessing, managing, and engaging with private recreationists (those not supported by commercial guides), as they constitute the majority of use in the region. This is problematic, as these users typically have little connection or relationships with land and resource managers. Furthermore, in many cases, their direct use of the uplands may only account for a small portion of their trip, though certainly their activities have a potential social effect on those lands. Given the vast extent of the region, opportunities for managers to engage users are limited to ports of entry. This suggests that indirect management efforts in terms of education or management actions that passively affect, direct, or deflect use (i.e., campsite hardening or other low-impact facility options and outreach efforts identifying areas where more use is appropriate) are likely the best for this system.

Wilderness/wildland recreation research and management often focus on evaluating or limiting numbers of individuals using the landscape in order to predict and reduce the social impacts. Our work highlights some of the complexities that can be missed by approaches of singular focus. In the Sound, perceptions about encounters seem to have more to do with expectations and behaviors exhibited and witnessed, as well as the specific opportunities desired by different groups of recreationists. We suggest that future efforts attempting to assess quality of experience in Prince William Sound focus on understanding recreation behavioral norms and user expectations as approaches to further elucidating potential stressors on recreation experience.

STAKEHOLDER ESSAY

PRINCE WILLIAM SOUND REFLECTIONS

LISA JAEGER

I GREW UP AS THE DAUGHTER of a naval officer and have spent most of my life near the ocean: swimming, surfing, sailing, skiing, fishing, and gathering shellfish. Of all the seas and shores I have traveled, nothing compares to the magic of Prince William Sound. Every time I paddle a kayak in the Sound, I think of the native Alutiiq kayakers of long ago. I see the Sound as they saw it. We camp on the same beaches: ones with steep shores that will remain after high tide, with access to freshwater streams. We hike the same bear trails along salmon streams, eat blueberries and salmonberries, catch fish, and experience the rain. The sounds also remain the same: the splash of humpbacks breeching, sea otters on their backs munching on mussels, bald eagles calling to each other, the squawks of a kittiwake rookery, a calving tidal glacier, waves lapping on the beach.

Humans come to the Sound for a variety of reasons and using various modes of travel. Paddle a kayak, ride a tour boat, venture on a cruise ship, fly a plane, reel in a fish, take a ferry to another port, hunt a bear or a deer, beachcomb for treasure. I appreciate the diversity of the cultures that visit.

I have had many classic wildlife encounters in Prince William Sound and I hope to have many more. I can still remember how my heart jumped when a humpback came up near the stern of my kayak south of Knight Island, swimming with our kayak pod for a mile. The wildlife continues to give me joy. I never tire of viewing wildlife; that is what I live for!

Fortunately, the temperate rain forest ecosystem and the winter weather allow the beaches to recover until the next summer season. The difference in the beaches between May and August is telling. On trips in late May and early June, it looks as though no one has been on the beach. There are no human signs of flattened tent sites, and the vegetation in the woods is freshly bloomed. As the summer season progresses, more human impact is visible, and there are smashed plants, trash, rocks left in tent circles, and traces of campfires, including partially burnt wood.

Since the Whittier tunnel reopened for combined rail and vehicle traffic in 2000, more people travel in the Sound, and the level of impact is higher. The increased volume of motorboat traffic has decreased the number of marine mammals seen in waterways close to Whittier and Valdez. People need to recognize their impact. My wish is that we all may embrace the "leave little trace" ethic so that future Sound explorers can continue to experience the magic.

CHAPTER 8

LEVEL OF SUSTAINABLE ACTIVITY (LSA) IN PRINCE WILLIAM SOUND

Understanding and Managing Quality of Experience and Capacity in Wilderness Waterways

ROBERT M. ITAMI, RANDY GIMBLETT, AND AARON J. POE

SOUND BITES

LEVEL OF SUSTAINABLE ACTIVITY (LSA) provides an alternative and effective methodology to traditional "social carrying capacity" approaches in outdoor recreation.

An integrated planning tool such as LSA provides a robust way of obtaining insight into differing definitions of quality of experience, impact of a variety of vessel densities on experience, and management suggestions and directions from stakeholders.

The results of the LSA workshops lend support to more traditional social surveys and provide strategic and tactical guidance for managing visitor experience in wilderness settings.

INTRODUCTION

Recreational capacity decisions remain at the heart of many confrontational and highly scrutinized land use planning and management processes. Capacity issues are often ignored by land managers and recreationists until unacceptable conditions arise. When they do, resource managers have tended to settle on setting limits of use, which in many cases are arbitrarily derived or seek to retain the status quo of existing visitor use conditions. Social or visitor use/experience research that attempts to understand these unacceptable conditions has commonly only focused on social norms or normative research. Results from such studies argue for dominant visitor preferences to prevail in informing land use planning and management that facilitate a myopic set of values and attitudes. These approaches, commonly used for evaluating and limiting numbers of individuals using a landscape, have been shown by Cole (2001) and others to favor certain users and certain types of experiences, leading to decisions that may be inherently biased against visitors who seek an alternative experience from the "norm." This has ultimately led to the involvement of the courts in attempting to mandate and institute capacity determinations, a development that has oversimplified the complex issues surrounding visitor use patterns and the tools necessary to properly manage visitor landscapes. This has resulted in unsatisfactory solutions to visitor management and continues to be of great debate.

An alternative paradigm that accepts a more diverse population of users with different experience objectives, behavior, preferences, and tolerances to visitor density needs to be developed. A growing body of visitor-based research is now focusing on this area by employing normative theory and techniques (Pettebone et al. 2013) that move away from a single "normative metric" for defining social carrying capacity to one that explores the acceptability of a range of social and biophysical norms related to visitor use. In chapter 7, for example, we see that users come to the remote Prince William Sound region seeking a variety of experiences, several of which may not be dependent on numbers of other users. This chapter presents a framework called level of sustainable activity (LSA) that provides an alternative to traditional social carrying capacity approaches in visitor management of outdoor recreation. We use LSA to explore perceptions about crowding relative to use density, understand tolerance for other user groups, and identify key drivers of positive recreation experiences from the perspective of longtime recreationists in the Sound.

WATER-BASED RECREATION CARRYING CAPACITY

Carrying capacity is defined by Shelby and Heberlein (1986) as "the level of use beyond which impacts exceed levels specified by evaluative standards." In a recent monograph on carrying capacity, Whittaker et al. (2010) agree that capacity is the amount and type of use that is compatible with the management prescription for an area and is measured on a use level scale which includes (1) units of use, (2) timing, and (3) location components. Capacity can vary across an area for different uses, facilities, seasons, or other "management-relevant situations."

Bosley (2005) reviewed seven studies of boating carrying capacity for lakes and reservoirs in the United States. She found that most of these studies measured recreational boating capacity in terms of (1) boat density (acres per boat) derived by different methods for defining the useable or navigable water area by removing shallow water, water-near-shoreline facilities, and environmentally sensitive areas; (2) determining the number, type, and speed of watercraft (existing use); and (3) defining users' perception of crowding (social carrying capacity).

Environmental and safety issues are generally handled by assigning buffers around sensitive vegetation or erodible shorelines and facilities such as boating docks or swimming areas. A variety of recommended boating densities for single and mixed vessel types have been proposed in various studies as reported by Bosley (2005). Most of these studies do not document how these standards are developed or how they relate to recreation satisfaction. One study by Environmental Resources Management (ERM Inc. 2004) explicitly linked boating densities at Deep Creek Lake, Maryland, to social carrying capacity. Local residents and visitors were provided photographic simulations depicting mixed boating use at five levels of density.

Respondents were asked to make three judgments: (1) preferred boating use level, (2) the boating level that was so high they would not boat on the lake, and (3) the boating level at which some type of management action should be taken.

Using this methodology, residents were found to be more sensitive to crowding, as were canoeists and kayakers because of sensitivity to noise and wake conflicts. This suggests that boating densities and perceptions of crowding may differ between residents and visitors and also depend on interactions between users with differing vessel types. These types of dynamics inspired our attempt to understand complexities associated with user experience and potential user conflicts in the Sound from the perspective of longtime users and residents.

METHODS

Level of sustainable activity (LSA) is a framework developed to (1) define quality of experience objectives and criteria for competing user groups in outdoor recreation settings, (2) define visitor capacity levels within and between user groups, and (3) elicit management suggestions from users for attaining or maintaining quality of experience objectives.

The LSA concept builds upon the "level of service" concept developed by the Transportation Research Board (2000) to look at quality of experience relative to traffic density. Recreational capacity is different for each user group and varies in relation to distance from the nearest port, shoreline geometry, the provision of facilities, and the interaction with other users (Haas et al. 2004). Visitor management must therefore be based on a comprehensive framework that integrates all relevant factors in a format that is easy for users and decision-makers to understand and that can be adapted to a wide range of environments and travel modes.

In this example, LSA can be thought of as a scale of end-user experience. Selected bays or fjords will have a range of service levels defined for each vessel type, ranging from very low levels of use, with minimal environmental and social impacts, to high-density use, with high levels of user interaction, higher levels of potential environmental and social impacts, and more intensive facility and management requirements.

The LSA concept integrates (1) physical characteristics of the environmental setting, including navigable depth, width, and shoreline characteristics; (2) physical characteristics of different vessel types: their size, speed, and passenger capacity; (3) user preferences for levels of use for specific activities in specific management zones; (4) user attitudes toward competing traffic safety and environmental and social risk factors relating to increasing use densities; and (5) suggestions from users on management options for dealing with the above risks.

LSA WORKSHOP PARTICIPANTS

LSA workshops were conducted in three communities around Prince William Sound. Cordova is a small, remote community with a population of 2,242 (2008), accessible only by boat (including a ferry service) or plane. Valdez is a small, remote community with a population of 3,787 (2008), accessible by boat, road, and plane. Anchorage is Alaska's largest city with a population of 279,243 (2008), well serviced by all transportation modes and the source of most boat traffic into Prince William Sound. Within these three communities, three recreational user groups were selected with separate workshops for each group for a total of nine workshops: kayakers, recreational motor- and sailboaters, and hunters. Representatives of each group and each community were represented, with the exception of Valdez hunters, who were invited but did not attend. In total, 62 people participated in the workshops. Table 8.1 shows the number of people attending each workshop.

QUALITY OF EXPERIENCE OBJECTIVES

Quality of experience objectives are the expressed reasons why people pursue outdoor recreation, and the factors that contribute or detract from these experiences. Individuals in each group were asked to freely express their quality of experience criteria by answering the following questions: (1) What experience are you looking for? (2) What are the key factors that you look for to attain this experience?

TABLE 8.1. Workshop participation

COMMUNITY	KAYAKERS	HUNTERS	RECREATIONAL BOATERS
Anchorage	11	3	12
Cordova	7	6	6
Valdez	5	0	4

Note: Figures shown are the maximum attendance through a session; in a few cases, one person left before the workshop finished.

Quality of Experience—Kayakers

There is general agreement among kayakers about the type of experience they are looking for in Prince William Sound. As a group, kayakers are seeking contact with "wild places," a wilderness experience, solitude, quiet, enjoyment with nature, wildlife viewing, a feeling of independence, the sense that "you are the first one there," and a general feeling of self-sufficiency and independence. The feeling of camaraderie with other kayakers is very strong among the kayak community, especially with local kayakers as opposed to large commercial groups of kayakers. They also enjoy the great diversity of shoreline and the ability to explore new places. They are all aware of the unique combination of mountain, glacier, and ocean experiences and the ability to combine kayaking with hiking and wildlife viewing.

Quality of Experience—Recreational Motorboats and Sailboats

Like the kayakers, the focus of the recreational boating community (including motorized boats and sailboats) is the experience of nature. Activities associated with this group include wildlife viewing, glacier viewing, fishing, hiking, hunting, photography, beach picnics, beachcombing, clam digging, shrimping, exploring historic sites, and boating or sailing accompanied by whales or porpoises.

There is a much stronger social or family orientation expressed with this group as compared to kayakers, because with motorized boats and larger vessels it is possible to travel as a group with greater mobility and speed. Also, the larger vessels provide onboard facilities for sleeping and eating, making it possible to spend long periods out in comfort for everyone, including children and the elderly. The social orientation was expressed most strongly by the community groups represented but was not mentioned by Forest Service managers. Community respondents emphasized the importance of experiencing Prince William Sound as part of a social activity with family and friends. The Anchorage workshop attendees were particularly aware of the fact that Prince William Sound is their "backyard" and that easy access to wilderness from Anchorage was a special asset and privilege.

Quality of Experience—Hunters

The strong consensus among hunters is the focus on the availability of the resource. The availability of fish and game varies by location and season within Prince William Sound, and for residents of the three communities attending workshops, the quality of the hunting experience has to do with the abundance of the resource as it relates to hunting pressure on wildlife populations. Many of the hunters participating in the workshops consider themselves to be "meat" or subsistence hunters; that is, they eat the game that they shoot. As a result, many have had years of experience hunting in Prince William Sound and have noted changes in the wildlife populations.

Anchorage hunters specifically commented on the decline in the deer and bear populations.

Hunters at the workshops all agreed that solitude was an important part of the experience. One Anchorage hunter noted that there was a safety issue related to this as well, since you are less likely to be shot by another hunter if you have the place to yourself. Sport hunters are often brought in by guides, who may bring in hunting parties. There is a strong perception among longtime residents of the Sound that there is a difference between locals, who tend to be subsistence hunters, and visitors, who tend to be hunting for sport or recreation. Locals share the value of solitude and therefore will move on rather than intrude on another hunter in a bay. Sport hunters are perceived to have little or no respect for the importance of solitude and will often intrude on a bay without asking, thus diminishing the experience of the local hunter. This creates feelings that range from "uncomfortable" to animosity, largely due to lack of communication and knowledge about these "unwritten rules."

An important aspect of managing quality of experience in outdoor environments is determining the relationship between use levels and quality of experience. A standard measure of use levels is density: the number of people, or in this case, vessels or parties, per square kilometer. In the LSA approach, five levels are designated, from Level A (no use) to Level E (high use levels). These levels must be defined in the context of the level of traffic and environment under study. For example, it would be inappropriate to use the same density levels for an urbanized bay as for Prince William Sound, since the essential experience in the Sound is wilderness.

To determine realistic LSA levels for Prince William Sound, user survey data from 2005 was used. Daily use levels for kayaks, small motorized boats (<30 feet), and large motorized yachts and sail boats were examined for peak summer days for Blackstone, Unakwik Inlet, and Sheep and Simpson Bays. Table 8.2 shows the density of vessels across all three Analysis Areas (see chapter 6). Also note that the levels from B to E are in a geometric progression, with density doubling at each level. Once these density levels were determined, images were prepared for each Analysis Area using survey locations observed over the entire season in 2005, and the appropriate number of vessels was randomly selected for LSA Level E for each Analysis Area.

REPRESENTATIVE ANALYSIS AREAS

Figure 8.1 shows the location of the three Analysis Areas for the LSA workshops. Blackstone Bay is on the eastern side of Prince William Sound. Unakwik Inlet is north-central, and Sheep and Simpson Bays are north of Cordova. These three locations were selected for their geographic distribution, differences in size and shoreline configuration, level of use, and relative distances to Cordova, Valdez, and Anchorage. The rationale for selection is to determine if these factors affect local perceptions of quality of service.

TABLE 8.2. LSA density levels for three Analysis Areas

LSA LEVELS FOR PRINCE WILLIAM SOUND	
LSA level	*Vessels/sq km*
A	0.00
B	0.03
C	0.06
D	0.12
E	0.25

Blackstone Bay is on the western side of Prince William Sound and is the closest of the three Analysis Areas to Whittier and Anchorage. It therefore is closest to the largest urban population and consequently has heavier recreational use than the other two Analysis Areas in this study. Blackstone Bay is 64.1 sq km. Unakwik Inlet is 109.9 sq km and is the largest of the three Analysis Areas, with a distance of approximately 30 km from the northern end of the Inlet to the mouth. It also has the most diverse shoreline, with four smaller bays along the perimeter.

Sheep and Simpson Bays are three bays north of Cordova. Together they are 64 sq km and contain small islands and small coves.

LEVEL OF SUSTAINABLE ACTIVITY WORKSHOPS

A key concern in the management of outdoor recreation is an understanding of the motivations and quality of experience visitors are seeking. This includes personal objectives, expectations, the environmental setting, and interaction with other visitors. The LSA framework is designed to develop a user-based understanding of environmental and social capacity and ideas about how to optimally manage visitor experience. The basic premise of the LSA framework is that appropriate levels of recreational use density are specific to different user groups, their mode of travel, the geographic location of their destination, and the environmental context. All of these factors have an impact on quality of experience.

To gain an understanding of how these factors interact, users in Anchorage, Valdez, and Cordova representing three recreational groups—kayakers, recreational boaters, and hunters—were asked to evaluate five LSA density levels for three locations—Blackstone Bay, Unakwik Inlet, and Sheep and Simpson Bays—for three types of vessels: kayaks, small motorized boats (<30 feet), and large motorized yachts and sailboats (>30 feet). Participants in each user group were asked to make evaluations in three different contexts: Preferred or Ideal LSA, Expected Peak Season LSA at busy times, and Maximum Tolerable LSA. They were first asked to make the above evaluations for their own user group (kayakers, recreational boaters, or hunters) and

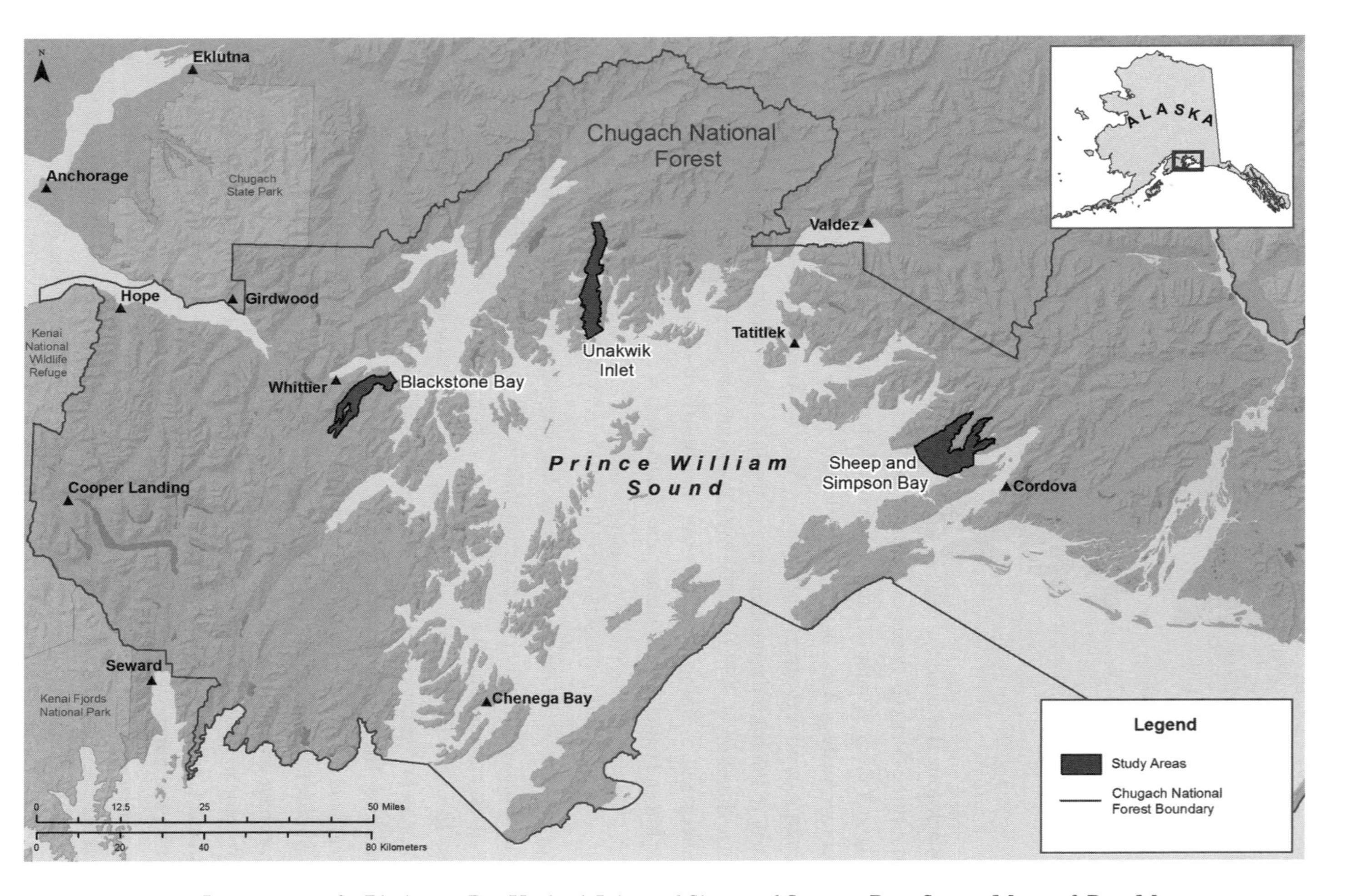

FIGURE 8.1. Location map for Blackstone Bay, Unakwik Inlet, and Sheep and Simpson Bays. Source: Microsoft Bing Maps 2009.

then asked to repeat the evaluations for competing user groups for a total of nine evaluations.

Workshop participants were instructed not to consider commercial traffic, such as fishing fleets or commercial tour boats and ferries. Individual evaluations were recorded and summarized by taking the median score for each group. After making these evaluations each user group was asked the following questions: (1) What are the impacts of other (types of) users on your experience? (2) What suggestions do you have for management to help you attain your desired experience? (3) Are there issues such as safety, noise, environmental impacts, or annoying behavior of other users that we haven't discussed? (4) Other issues/questions.

The results for each recreational group are discussed in the following sections.

RESULTS

LEVEL OF SUSTAINABLE ACTIVITY—KAYAKERS

Table 8.3 shows the median LSA values for kayakers in Anchorage, Valdez, and Cordova. The evaluations for Ideal LSA were consistent across all communities for all vessel types, with high value placed on low-density use (LSA level A—no other vessels or level B—0.03 vessels/sq km) across all three locations. These ratings are consistent with the high value that kayakers expressed for solitude and wilderness experience in the quality of experience discussion.

For Expected Peak Season LSA, the evaluations were more variable. The reason behind this variability is that some users had not visited some of the locations so opted out of the evaluation or made an evaluation based on "what they imagined" the traffic to be. The users who had visited the area may or may not have experienced the traffic during peak periods. Generally, however, the highest densities were assigned to Blackstone Bay and attributed to the close proximity to Whittier (with the easiest access to the greatest population), and Sheep and Simpson Bays because these bays are smaller, with relatively high use levels compared to the area of the bays. Unakwik Inlet had the lowest ratings, primarily due to the large size of the inlet and the relatively low use.

The median values for Maximum Tolerable LSA ratings were fairly consistent between communities for each bay. Also, the evaluations were fairly consistent by vessel type. Anchorage kayakers have the highest tolerance for other kayakers in Blackstone Bay (LSA level D). This is largely due to the large size of the bay and the fact that this level of use is already expected for Blackstone Bay. Valdez kayakers also had high tolerance for other kayakers in Sheep and Simpson Bays. Anchorage kayakers had the lowest tolerance for other vessel types at Unakwik Inlet (LSA level B for both small motorized boats and yachts and sailboats). Cordova kayakers also had

TABLE 8.3. Median LSA values for kayakers in the Anchorage, Valdez, and Cordova workshops

	Blackstone Bay	Unakwik Inlet	Sheep and Simpson Bays	Blackstone Bay	Unakwik Inlet	Sheep and Simpson Bays	Blackstone Bay	Unakwik Inlet	Sheep and Simpson Bays
	KAYAKERS RATING KAYAKS			KAYAKERS RATING SMALL MOTORIZED BOATS			KAYAKERS RATING MOTOR YACHTS AND SAILBOATS		
	Ideal LSA			*Ideal LSA*			*Ideal LSA*		
Anchorage kayakers	B	A	A	A	A	A	A	A	A
Valdez kayakers	A	A	B	A	A	A	B	B	A
Cordova kayakers	A	A	A	A	A	A	A	A	A
	Expected Peak Season LSA			*Expected Peak Season LSA*			*Expected Peak Season LSA*		
Anchorage kayakers	D	B	B	C	B	C	C	B	C
Valdez kayakers	D	B	C	D	B	C	C	B	C
Cordova kayakers	C	A	B	D	B	C	C	B	C
	Maximum Tolerable LSA			*Maximum Tolerable LSA*			*Maximum Tolerable LSA*		
Anchorage kayakers	D	B	C	C	B	C	C	B	C
Valdez kayakers	C	B–C	D	C	C	C	C	C	C
Cordova kayakers	C	C	C	C	C	D	C	B	C

Note: The density of vessels per square kilometer is described as A = 0, B = 0.03, C = 0.06, D = 0.12, and E = 0.25.

low tolerance for yachts and sailboats at Unakwik Inlet. Otherwise, there is a high level of consistency for other vessel types by kayakers in all three communities with an LSA rating of C for small motorized boats, and motorized yachts and sailboats, for all three locations.

When one compares the Expected Peak Level LSA for each vessel type and each location with the Maximum Tolerable LSA for the corresponding locations, it is evident that *Anchorage kayakers perceive that use levels across all locations for all vessel types are currently at capacity*. In other words, they would not tolerate heavier use during busy times. If confronted with these levels, they would either be displaced to quieter locations or they would try to schedule trips at alternative times. Some Anchorage kayakers indicated that there are already locations they avoid because of the unacceptably high use levels.

The picture for Valdez and Cordova kayakers in comparing Expected LSA to Maximum Tolerable LSA is more complex. Blackstone Bay is at or over capacity for all vessel types. Sheep and Simpson Bays are at capacity for motor yachts and sailboats and at capacity (Expected LSA = C, Maximum Tolerable LSA = C) for small motorized boats in the opinion of Valdez kayakers, or one LSA level under capacity in the opinion of Cordova kayakers (Expected LSA = C, Maximum Tolerable LSA = D). At Unakwik Inlet, Valdez and Cordova kayakers perceive that there is excess capacity for small motorized boats during peak periods (Expected LSA = B, Maximum Tolerable = C). Cordova kayakers feel that Unakwik Inlet is at capacity during peak periods for motorized yachts and sailboats (Expected LSA = B, Maximum Tolerable = C), while Valdez kayakers would tolerate LSA level C compared to Expected Peak Season LSA of B.

These conclusions should be regarded with caution given the variability of judgments for Expected Peak Season LSA, as described earlier. It is fairly clear that users perceive Blackstone to be busy by kayaker standards, and at or above capacity during peak periods. Unakwik Inlet, the largest body of water studied, is at the other extreme, with low perceived use levels at peak periods. Considering the large effort required to get to Unakwik Inlet (over 65 km from Whittier, over 75 km from Valdez, and over 110 km from Cordova), it appears that users not only expect low use but also would be annoyed if they arrived and found higher than expected use.

LEVEL OF SUSTAINABLE ACTIVITY—RECREATIONAL MOTORBOATS AND SAILBOATS

Table 8.4 shows the median LSA values for recreational boaters in Anchorage, Valdez, and Cordova. Most recreational boaters in the LSA workshops have vessels that are self-sufficient in that they provide facilities for sleeping, eating, and entertaining onboard. This allows for comfortable accommodation for all ages and physical abilities. This group tends not to camp onshore, so typically they do not compete with kayakers for campsites. They do, however, come ashore for other land-based activities like berry picking, hiking, and picnicking.

Generally, recreational boaters have a more social orientation than either kayakers or hunters; this is reflected in the generally higher LSA ratings for Ideal LSA than provided by the other user groups. The reasons expressed for the higher ratings include the concept that other boats in the area are good (in case of emergencies) and

TABLE 8.4. Median LSA values for recreational boaters in Anchorage, Valdez, and Cordova

	Blackstone Bay	Unakwik Inlet	Sheep and Simpson Bays	Blackstone Bay	Unakwik Inlet	Sheep and Simpson Bays	Blackstone Bay	Unakwik Inlet	Sheep and Simpson Bays
	RECREATIONAL BOATERS RATING KAYAKS			RECREATIONAL BOATERS RATING SMALL MOTORIZED BOATS			RECREATIONAL BOATERS RATING MOTOR YACHTS AND SAILBOATS		
	Ideal LSA			*Ideal LSA*			*Ideal LSA*		
Anchorage boaters	A	B	A	A	B	A	A	A	A
Valdez boaters	C	B	B	D	B	B	A	B	B
Cordova boaters	A	B	A	A	B	A	A	B	A
	Expected Peak Season LSA			*Expected Peak Season LSA*			*Expected Peak Season LSA*		
Anchorage boaters	D	C	B	C	C	C	C	B	C
Valdez boaters	C	B	B	D	B	C	A	B	C
Cordova boaters	E	D	C	D–E	D	D	E	C	C
	Maximum Tolerable LSA			*Maximum Tolerable LSA*			*Maximum Tolerable LSA*		
Anchorage boaters	D	D	D	D	C	D	D	C	C
Valdez boaters	D	D	C	E	C	E	B	C	C
Cordova boaters	E	D	D	C	D	C–D	D	C	D

Note: Forest Service managers pretested the LSA workshop format posing as recreational boaters. The density of vessels per square kilometer is described as A = 0; B = 0.03; C = 0.06; D = 0.12; and E = 0.25.

that kayaks cause little or no conflict. Valdez boaters had higher Ideal LSA ratings for kayaks and small motorized boats at Blackstone Bay. Their reasoning was that this is the busiest bay, they expect it to be busy, and it may be appropriate to have heavy traffic in this setting. The perception that kayaks have little impact on the feeling of wilderness is also reflected in the high LSA ratings for kayaks by all three recreational boating communities for allocations. Motor yachts and sailboats are seen to have more impact than motorboats of the same size; however, recreational boaters as a group are much more tolerant of other boats than kayakers and hunters across all vessel types and all locations.

Maximum Tolerable LSA ratings for kayakers are generally high for Blackstone Bay, as recreational boaters do not see them as having a big impact either on traffic or on sense of solitude. Also, there is an expectation that Blackstone Bay will be busy because of the proximity to Whittier and the opportunity to view glaciers. Respondents tend to be more tolerant of higher use levels in areas where higher use is expected and are known destinations for commercial tour boats and recreational boats. One of the key factors determining LSA levels for large boats is the availability of safe anchorages. Local knowledge and maps and guidebooks that show anchorages are valuable resources for recreational boaters in Prince William Sound.

Table 8.5 shows the median LSA values for hunters in Anchorage and Cordova. No values are reported for Valdez hunters because invited members of the community failed to attend the workshop. Cordova hunters did not rate Blackstone Bay for Expected Peak Season LSA or Maximum Tolerable LSA, either because they had not been to Blackstone Bay or because they don't hunt there. For Anchorage and Cordova hunters, there is a high level of agreement on the Ideal LSA with level A and B ratings across all locations for all vessel types. This is consistent with the expressed value hunters place on solitude and their preference for low competition when hunting. Anchorage hunters expressed concern that the Sound was getting too much pressure from hunters and that they were already getting displaced by heavy hunting use or depletion of game stocks.

This is reinforced by the LSA scores that show Maximum Tolerable LSA consistently lower than Expected Peak Season LSA across all locations. Hunters generally hunt from small motorized boats. Because hunting season is generally in the spring and fall, hunters are generally not in competition with the heavy summer kayaking and recreational boating season. Hunters see little problem with kayakers camping onshore as they are quiet and don't interrupt game hunting. Therefore, the LSA ratings for small motorized boats are of most relevance when analyzing hunter perceptions. Cordova hunters had no opinion about conditions in Blackstone Bay for kayaks or for motor yachts and sailboats and did not express the same level of hunting pressure as Anchorage hunters. Their scores show only Blackstone Bay under unacceptable traffic levels for small motorized boats. Interestingly, LSA scores for Maximum Tolerable LSA were higher for Cordova hunters for Blackstone and Unakwik Inlet (Anchorage Maximum Tolerable LSA = B, Cordova Maximum Tolerable LSA = C). This may reflect the fact that Anchorage hunters have more experience with high levels of hunting traffic and are better able to judge the impact of vessel density on hunting quality of experience.

TABLE 8.5. Median LSA values for hunters in the Anchorage and Cordova workshops

	Blackstone Bay	Unakwik Inlet	Sheep and Simpson Bays	Blackstone Bay	Unakwik Inlet	Sheep and Simpson Bays	Blackstone Bay	Unakwik Inlet	Sheep and Simpson Bays
	HUNTERS RATING KAYAKS			HUNTERS RATING SMALL MOTORIZED BOATS			HUNTERS RATING MOTOR YACHTS AND SAILBOATS		
	Ideal LSA			*Ideal LSA*			*Ideal LSA*		
Anchorage hunters	A	A	A	B	A	A	A	A	A
Valdez hunters	N/A	N/A	N/A	N/A	N/A	N/A	N/A	N/A	N/A
Cordova hunters	A	B	A	A	B	A	A	B	A
	Expected Peak Season LSA			*Expected Peak Season LSA*			*Expected Peak Season LSA*		
Anchorage hunters	C	B	N/A	E	D	C	B	C	C
Valdez hunters	N/A	N/A	N/A	N/A	N/A	N/A	N/A	N/A	N/A
Cordova hunters	No rating	A	B	E	B	B	No rating	B	C
	Maximum Tolerable LSA			*Maximum Tolerable LSA*			*Maximum Tolerable LSA*		
Anchorage hunters	B	B	C	B	B	C	B	A	B
Valdez hunters	N/A	N/A	N/A	N/A	N/A	N/A	N/A	N/A	N/A
Cordova hunters	No rating	C	No rating	C	C	C	No rating	C	D

Note: The density of vessels per square kilometer is described as A = 0; B = 0.03; C = 0.06; D = 0.12; and E = 0.25.

DISCUSSION

The ratings of the three communities and three recreational user groups reveal some distinct similarities and differences. Generally, hunters have the highest requirements for solitude and the lowest tolerance for competition. Generally, they prefer "one boat per bay" and this "unwritten law" is generally acknowledged and respected among hunters in local communities like Cordova. However, conflict with other hunters occurs when this concept is not recognized—especially by commercial hunting guides. The key issues relating to the hunter groups have to do with the management of the hunting resource. Anchorage hunters feel that wildlife stocks were overhunted and that the practice of bearbaiting is undesirable. Hunting pressures, primarily from sports hunters as opposed to subsistence hunters, are considered to be intolerable during the peak hunting season, which causes local hunters to either move from traditional hunting areas or hunt during the shoulder seasons. Hunters are generally in the Sound in the spring and fall and therefore do not overlap with the main summer recreational boating season. Therefore, they perceive little conflict with kayaks and other recreational boats; their main competition is with other hunters.

Kayakers follow hunters in the high value they place on solitude. This group values "low impact" camping, self-sufficiency, camaraderie with other kayakers, and contact with nature. They are generally tolerant of other kayakers because of the quiet mode of transportation and strongly shared values of low-impact recreation. However, they are less tolerant of small motorized boats because of the noise, speed, wake, impact on quiet and solitude, and competition for campsites. They are more tolerant of larger motor yachts and sailboats because these boats are self-contained and do not compete for campsites. Valdez kayakers were particularly opposed to the practice of bearbaiting because of the visual impact of litter from human foods used as bait. They were also concerned about the potential danger posed by bears becoming more interested in human foods and attracted to shoreline campsites.

Recreational boaters attending the LSA workshops were generally sailing motor yachts and sailboats. Smaller motorized boats like cabin cruisers, runabouts, and trailer boats were not represented among the respondents. Though this group, like the others, values solitude and wilderness experience as the Ideal LSA, they are, as a group, more socially oriented. They see little conflict or competition with kayaks and have higher tolerance for small motorized boats than do hunters or kayakers. They have equal or higher tolerance for kayakers than for motor yachts and sailing boats. For this group, the availability of anchorages is critical. Because they are self-sufficient and not reliant on onshore campsites, they tend not to have conflicts with kayakers. When they travel, kayakers tend to hug the shoreline, while recreational boaters and especially motorized yachts and sailboats use the open water. This group is quite aware of safety issues and mentioned the dangers for inexperienced sailors in boats of any size running out of fuel, being unaware of where to find safe shelter in inclement weather, using poorly equipped or maintained boats, or having inadequate seamanship, given the remoteness of many of the destinations in the Sound.

When comparing all three communities and user groups, it is apparent that all groups value feelings of solitude. The median rating for Ideal LSA across all bays, for all communities for all recreational groups, was A (no boats) or B (0.03 vessels/sq km). Those assigning a rating of A liked the idea of "being the first one there" or the feeling of "having the place to yourself." Those assigning a value of B felt that the presence of a few other vessels was positive for safety reasons. Valdez recreation boaters rated Blackstone Bay LSA a value of C because they felt that Blackstone Bay was unique as a tourist destination because of its proximity to Anchorage and the opportunity to view glaciers and wildlife in an already busy bay.

In most cases, users' median ratings for Expected Peak Season LSA were higher, or much higher, than Ideal LSA. This is expected given that Ideal LSA ratings were either A or B. However, Maximum Tolerable LSA ratings varied considerably between recreation types. Kayakers mostly assigned LSA a value of C for Maximum Tolerable LSA across all bays and all vessel types, with a few B and D ratings. Hunters also had relatively low Maximum Tolerable LSA ratings with mostly LSA B and C ratings for all locations. Recreational boaters had higher ratings across all bays and vessel types with median Maximum Tolerable LSA ratings of C and with even a few E ratings. These ratings are consistent with what users expressed in the workshops for quality of experience criteria, with recreational boaters being more social than kayakers or hunters.

THE IMPACT OF DISTANCE FROM PORT, SIZE, AND SHAPE OF THE DESTINATION ON LSA

Blackstone Bay, Unakwik Inlet, and Sheep and Simpson Bays were selected because of differences in the distance from Anchorage, differences in use levels, differences in total area and shape, and complexity of shoreline. Workshop participants did consider these variables as they were assigning LSA levels, especially for Maximum Tolerable LSA; however, there was not a clear difference in LSA ratings for the three locations.

It is very clear from comments made by the boating communities in Valdez and Cordova that the distance from Anchorage provides "some protection" from boating traffic. These users feel that the distance from Anchorage is an important factor in the lower use levels and the ability to attain feelings of solitude when paddling or boating in Prince William Sound. The reverse of this sentiment is true for Anchorage boaters who feel that the western side of the Sound is very busy and, with the opening of the Whittier Tunnel, that there is increasing pressure from trailerable motorboats.

The level of sustainable activity framework has been previously applied in urban waterways (Itami 2008). This is the first implementation of the framework in a wilderness setting. Whereas the framework is consistent with the previous studies, the main difference is in the scale and context of Prince William Sound. As a wilderness waterway, the Sound is vast in comparison to the urban examples. The LSA vessel density levels are therefore much less dense than the urban examples. By basing the LSA levels on previous field work, the densities portrayed (table 8.5) were accepted without question by workshop participants. They related easily to the density levels and were able to make judgments about the impact of these density levels on their experiences.

LESSONS LEARNED

The LSA framework has proven to be a robust way of obtaining structured insights into differing definitions of quality of experience, impact of vessel densities on experience, and management suggestions from Prince William Sound users. For example, this application of LSA has aided in our understanding of hunters, who were shown to be the most sensitive to the presence of other visitors, have the highest requirement for solitude, and place high value on the availability and quantity of wildlife and fish resources. Perhaps somewhat ironically, kayakers also place high value on solitude but considered themselves to be impacted most by hunting from small motorized boats that compete for shoreline campsites.

Users of the increasing number of small motorized boats, including rental boats and privately owned trailerable boats, are not represented in this study. The opening of Whittier Tunnel has had a major impact on the increasing number of these types of boats in the Sound. The user experience dynamics of that group (captured, in part, in chapter 7) appear to differ from some of those expressed by longer-term users or residents from communities in the Sound. Hence the levels of satisfaction experienced by respondents to the survey described in chapter 7 may not extend as well to longer-term users and some residents of the region. The LSA tool adds additional depth of perspective for managers to consider as they look at planning for sustainable human use.

Furthermore, insights from this work suggest that there is a need to develop a more comprehensive understanding of hunting practices in the Sound. Understanding hunting use in relationship to peak seasons for various game species as well as hunting pressure for various species relative to the location of wildlife stocks is important information for both recreation and wildlife managers. There may also be a need to work with commercial fishing and hunting guides to develop a code of practice that integrates their activities better with local traditional hunting and fishing practices. Similarly, there is need to educate all types of recreation users that are new to the region, including new hunters, about traditional hunting practices in order to try to avoid conflict with local hunters.

Finally, managers may wish to explore the potential impacts and opportunities of providing more tourist facilities close to Whittier on distribution of recreational boating to more remote destinations. Facilities, particularly for small motorized boats, including moorings, hardened campsites and picnic sites, and toilets, were identified as desirable in *some* locations where there is already high use.

STAKEHOLDER ESSAY

CHANGING ACCESS, COMPLEX VALUES

LYNN HIGHLAND

THE TRAIN TUNNEL TO WHITTIER opened in 1943, courtesy of the United States Army. It was the only land access to Whittier until the road was built through the same tunnel in 2000. I began riding the train in 1978 when getting to Whittier was a real undertaking, involving scurrying around after work on Friday afternoon and driving to Portage over one of the most dangerous highways in the country behind a motor home going 40 miles an hour to be there whenever the train finally decided to show up. Once it did, I could either schlep my stuff from the vehicle onto the baggage car or pay the much higher fare to put my truck on the train. If I opted for the less expensive passenger fare, all the stuff had to somehow get to the boat once I arrived in Whittier. Driving the vehicle onto the train simplified the whole affair but seemed too expensive for most of us. The train operated several times a day during the summer and several times a week during the winter. It was a nuisance, even ignoring the frequent vandalism in the Portage parking lot and the less-than-exact schedule.

I loved the train. Nuisance that it was, it created camaraderie among the folks who traveled to Whittier. We got to know each other, our kids and dogs played in the aisles, and loading and unloading everybody's gear was a group effort. Groups of kayakers, who smelled like wet sheep, prattled on and on about the beauty of the Sound. I loved the atmosphere the train created; I still have friends that I made on the train, and the kids who played in the aisles have their own kids now. When there were no seats left, we'd ride in the baggage car with the doors open (except in the tunnels); I had some really great naps sacked out on a pile of baggage. For a while, the railroad even operated a bar car on the train—not a particularly splendid idea given that folks were driving boats at one end of the trip or vehicles at the other, but it certainly added to the convivial atmosphere.

The people willing to put up with the inconvenience of the train, the expense of keeping a boat in Whittier, and the idiosyncrasies of harbor management were committed to Prince William Sound and vested in keeping it a special place that we visited over and over again, to the exclusion of any other place. We caught a lot of fish and shot a lot of deer and the occasional bear, but there weren't very many of us. What I enjoyed most was the beauty, the wildness, and the solitude. Prince William Sound was very forgiving of our presence—it either didn't notice or didn't care. There was a sense of having the wildness and the splendor of Prince William Sound all to myself, and I loved it. It seems I was a believer in Landscape Values before I had heard of them—before I knew about the R statistic, CSR, and the Z score, even.

Then the longest highway tunnel in North America opened on June 7, 2000, at a cost of about $100 million (in case you wonder why there's a federal deficit problem), and suddenly it was easy to get to Whittier and Prince William Sound. (It's also the only tunnel in the country—maybe the world—in which trains and cars share the roadway. Who would have thought trains would share?) One of the most spectacular and wildest areas in the world suddenly became only an hour away from Alaska's largest city. Anybody could get there with anything that floats and head out into the Sound. And they did, by the thousands.

In 2010, there were over 200,000 vehicle trips through the tunnel, mainly in the summer. A few years after the tunnel opened, there were over 5,000 paid small boat launches in Whittier during the summer. The launch ramp is a zoo on weekends, and behavior is just plain unpleasant at times; Whittier now maintains a police presence in the launch area on busy weekends. Rather than the leisurely train ride, there emerged a determination to get the boat in the water and head out as quickly as possible and to catch as many fish as possible. Shortly after the highway tunnel opened up access, improvements in technology allowed outboard-powered boats to travel longer distances at higher speeds and access all corners of Prince William Sound even on a weekend trip.

Things are now very different than when some of the earliest human use studies were done. The pressures on the resources are greater, and user focus seems to have changed from enjoying the wildness to primarily consumption—except in the case of the kayakers who, thanks to polar fleece, now smell lots better. Perhaps the principle that says the harder it is to join a group the more loyal the members are is at work here. If all you need to do is load your stuff and drive to Whittier, you don't develop the bond formed by the shared difficulty of the train experience.

This new consumptive focus shows. Bear sightings are a fraction of what they were before the road opened; the nonpelagic rockfish population seems down (as the biologists predicted it would be), and some salmon runs are weaker. Regulatory changes and enforcement seem slow to catch up with current resource management requirements. Prince William Sound has paid a price for improved access. It is not as wild as it was.

When the focus of the majority of users seems to be filling the freezer, how do values such as aesthetic, biological, learning, intrinsic, spiritual, cultural, historic, and therapeutic fit in? How do they reconcile with the attitude that says I can catch as many fish as I want because there are no troopers around to catch me? If public lands and waters should not be managed for specific public uses but rather managed for more general public values, we have an increasingly complex dilemma, don't we?

But Prince William Sound is a public resource; everyone has just as much right as anyone else to be there. How to keep it wild? Restrict access—on what basis? Should boat drivers need a license? Should a fisherman have to pass a test to get a fishing license—maybe even (gasp) be able to tell the difference between species of fish? Hold a lottery?

So there's the dilemma. The Sound is a beautiful place and attracts thousands of people. Because it does, it becomes diminished. How should the landscape values shared by some of the Sound's earlier recreationists like me be preserved while enabling freer access for newcomers to the region? Or should they be preserved?

I hope so.

Or maybe I'm just an old guy whining about the way things used to be.

CHAPTER 9

EVALUATING THE SUBSISTENCE SERVICE RECOVERY

Spatial and Temporal Characterization of Prince William Sound Subsistence Harvest Activities

AARON J. POE, RANDY GIMBLETT, AND MILO BURCHAM

SOUND BITES

REDUCED HOUSEHOLD HARVEST EFFORT in Prince William Sound is reportedly caused more by individual lifestyle changes and the perceived lack of resources than competition or user conflict.

Competition does exist, mostly in near proximity to communities, and is perceived as coming primarily from other local harvesters and commercial/guided sport harvest.

Harvesting user groups are poorly understood, yet this widespread stakeholder group is capable of offering insightful information such as traditional ecological knowledge, as well as much needed insights into their own behavioral and social norms.

INTRODUCTION

Subsistence harvest of fish, wildlife, and plant resources is a critical part of the Alaskan lifestyle. It is of great value in terms of the vital food resources that harvest brings into households in rural and often very remote communities where receiving foodstuffs can be difficult and expensive (Brown et al. 2002). Of equal or perhaps greater importance is its value as a cultural practice that reinforces the traditions of people still very dependent on wild landscapes. Subsistence harvest and its associated activities of preparation, storing, and sharing of resources is central to Alaska Native cultural identity and vitality. It's widely recognized in Alaskan land and wildlife management that subsistence harvest provides irreplaceable cultural, spiritual, personal, and sustenance value, and the Alaska National Interest Lands Conservation Act (ANILCA) requires that federal land managers consider the effects of management on subsistence activities (e.g., USDA Forest Service 2002). Generally, managers give subsistence harvest priority over other types of sport or commercial harvest of fish or wildlife.

In 1989, Prince William Sound, the heart of the Chugach National Forest, was severely impacted by the *Exxon Valdez* oil spill. In the aftermath of the spill, federal and state trustees were awarded criminal and civil restitution funds to help with the recovery (and the evaluation of the recovery) of injured resources and human services, including traditional practices of subsistence harvest, which is still listed as "recovering" (EVOS Trustee Council 2010).

The original cause for concern relative to subsistence was that contamination from oil had altered the availability of safe wild resources in the Sound. For example, in Chenega Bay prior to the spill, marine mammals made up about 40 percent of the harvest of wild resources and only 3 percent nearly a decade later in 1998. Ten years after the spill, 63 percent of the households in oil spill–affected areas believed that subsistence resources had not recovered from the effects of the spill. Community members reported having to increase effort (traveling farther, spending more time and money) to achieve comparable harvests to those before the spill, and community members reported an increasing reliance on fish (Fall et al. 1996).

Residents in the predominantly Alaska Native communities of Chenega Bay and Tatitlek, as well as residents of Cordova and Whittier, partake in a variety of subsistence harvest activities in the Sound. Fish and marine mammals comprise the majority of subsistence resources taken, but there is also significant use of other species, including Sitka black-tailed deer, black bear, mountain goats, waterfowl, seabirds, river otters, and mink (Stratton and Chisum 1986; Fall and Utermohle 1999).

Although many of these species use terrestrial habitats, the majority of time spent during harvest activities occurs in direct proximity to marine, intertidal, and shoreline areas. Furthermore, subsistence activities related to upland species are relevant for the purpose of evaluating the resource use as it impacts intertidal areas that serve as access points to inland subsistence harvest and gathering activities.

Historically, subsistence communities have expressed concern that activities such as timber harvest, road building, and recreation development could impact fish and wildlife populations or increase competition for subsistence resources. According to James Fall, statewide program manager for research with the division of subsistence with the Alaska Department of Fish and Game, in the Sound, subsistence users are concerned about increased competition for wildlife and fish resources from increasing numbers of private urban users and commercial operations (J. Fall, personal communication).

Recreational use is not evenly distributed in Prince William Sound. Certain areas are more desirable for a variety of reasons, including distance from access communities, presence of tidewater glaciers and postglacial landscapes, availability of landing beaches, protected anchorages, sportfish streams, wildlife viewing opportunities, cabins, and wild game concentrations. Independent use is not well understood for the Sound, though the majority of use happens during June, July, and August (Murphy et al. 2004, chap. 11). Recent work completed in chapter 7 produced predictive GIS raster surfaces for recreation activity in the Sound during spring, summer, and fall.

Subsistence users are also concerned about increased commercial recreation activity interfering with their subsistence practices. For example, many of the traditional Chenega Bay residents' harvest areas characterized by Stratton and Chisum (1986) have become popular for recreation activities, and it is reasonable to assume that this may result in increased contact and competition between user groups. Given that some recreationists using Prince William Sound may not understand the harvest traditions and rights of subsistence users, the potential for conflict is significant. Conflicts between user groups have significant implications for *Exxon Valdez* oil spill–impacted resources and services. Conflicts can diminish quality of life/experience for both subsistence and recreation groups (each already harmed by the spill) and push harvest and recreation activities into previously unused areas (Hennig and Menefee 1995), potentially negatively affecting the 25 impacted and recovering resources.

Various subsistence harvest studies have been conducted by researchers from Alaska Department of Fish and Game's Division of Subsistence since the early 1980s (e.g., Fall et al. 1996; Fall 2006). The results of these surveys are used to update a community profile database maintained by the division (Scott et al. 2001). Unfortunately, the variables recorded in such surveys are limited for spatially explicit analysis of harvest effort characterized in this database. The database tracks numbers of individuals (e.g., marine mammals) or pounds of resources (e.g., salmon) harvested but does not describe the daily effort of harvest or location of harvest at any level of resolution greater than Prince William Sound. Alaska Department of Fish and Game wildlife managers do track the harvest of some individual big game species, including black and brown bears, deer, and mountain goats taken under sport harvest.

One study completed in the early 1980s following the resettlement of the village of Chenega Bay (the original village site was destroyed in Alaska's 1964 Good Friday earthquake) evaluated a change in distribution of harvest and did report some mapped results of harvest areas. According to James Fall, statewide program manager for research with the division of subsistence with the Alaska Department of Fish and Game, data focused only on households from Chenega, and results are considered to be out of date at this time based on reported changes in harvest location as a result of the oil spill (J. Fall, personal communication). Consultation with Prince William Sound Alaska Native communities and regional subsistence users around the issue of evaluating the distribution of harvest efforts have confirmed this is information that is indeed lacking for two Alaska Native communities: Chenega Bay (Chenega Corporation 2008) and Tatitlek (Tatitlek IRA Council 2003).

Though some characterization has been completed for recreational use in the Sound (e.g., Murphy et al. 2004; Gimblett and Itami 2006), little is known regarding the spatial and temporal nature of the wide variety of subsistence harvest activities in the Sound. Similarly, little is known about the dynamics around the harvest effort and potential for competition within harvesting groups. An earlier study compiled spatially explicit harvest information for spring black bear sport harvest in the western Sound (Gimblett and Lace 2005), but a comprehensive, spatially explicit effort has not been completed. Understanding the subsistence harvest patterns in the Sound adds critical depth to the few existing Prince William Sound human use studies by describing the exact nature of overlap between subsistence and recreation use in the Sound. This information will help managers better understand the dynamics around these potential interactions and anticipate potential conflicts between these two injured human services.

METHODS

The methodology of this project involved a two-step process to evaluate harvest by residents of Prince William Sound and the spatial overlap with recreational users. The first step involved face-to-face household interviews, and the second step was to translate the results into spatially and seasonally explicit characterizations of harvest effort. It should be noted that the following general ethical principles (established as the appropriate way to engage subsistence communities in Southcentral Alaska) guided our efforts: "1) review and approval of the research plans by community governments prior to fieldwork; 2) informed consent by household members selected for interviewing (participation in the research was voluntary); 3) confidentiality of individual and household-level responses; 4) review of study findings by the participating communities; and 5) providing study findings and reports to each study community" (Fall et al. 1996).

HOUSEHOLD SURVEYS

Household style interviews around the theme of subsistence harvests have proven to be successful at evaluating summary harvest of resources in the years prior to and following the 1989 *Exxon Valdez* oil spill (e.g., Fall et al. 1996; Fall and Utermohle 1999; Fall 2006). However, as a prudent first step, five members from three of the target communities (Cordova, Tatitlek, and Whittier) were engaged in a pretest of the interview questions and data summary tools. The intent of this pretest was to elucidate possible confusion relative to question clarity, appropriateness of categorical responses, general comments regarding the proposed interview approach, and the appropriateness of the associated map document used to elicit responses.

Following survey instrument improvement, residents from the four Sound communities who are subsistence eligible (Chenega Bay, Cordova, Tatitlek, and Whittier, which are resident zone communities of the Chugach National Forest and are eligible to harvest resources under subsistence regulations in the national forest and outside the state nonsubsistence area) were consulted through individual household interviews conducted by current community institutions (e.g., tribal or community councils) or trusted residents from those communities. In the case of the two Alutiiq communities of Tatitlek and Chenega, individuals were approved by tribal leadership. Interviews in Cordova were conducted by employees for the Native Village of Eyak, which represents the Alaska Native residents of Cordova. The remote nature of these communities made this approach the most effective for reaching all households.

Information in this study was collected through oral interviews, and results were recorded by interviewers with assistance from a map document. Interviews occurred between March and November of 2009 in all four communities. Respondents described their household's harvesting practices for each resource in terms of the number of years pursuing each resource, the month(s) of harvest, and the duration of days invested in each resource. They were asked to summarize these efforts relative to spatial polygons representing subregions of Prince William Sound. For land-based harvests, they reported harvest to the resolution 88 uniform code unit (UCU) boundaries designated by ADF&G for the uplands of the Prince William Sound region. Water-based harvest was summarized to 72 marine polygons used to summarize sportfish harvest by the National Marine Fisheries Service (NMFS). Respondents were also asked about any recent changes in the location and effort expended in their harvest activities due to other users or changes in resource availability.

Responses were summarized and analyzed primarily as a summary data set for Prince William Sound as a whole. The spatial and temporal distribution of harvest was summarized by seasons (spring = April 1 through June 14; summer = June 15 through August 31; fall = September 1 through December 31; winter = January 1 through March 31; see chapter 6) and compared to predicted distributions of recreation use by private individuals (chapter 7) during those same seasons.

INTERVIEW METHODS

Phone records from the municipality of Cordova were used to randomly select 340 households (~40 percent of the population) for interviews based on recommended sampling procedures for Alaskan communities of fewer than 2,000 households (State of Alaska 2008). In the smaller communities of Chenega Bay (21 households), Tatitlek

(38), and Whittier (86) (2000 census), a combination of phone lists and door-to-door visits was used to conduct a complete census of 146 households. Interviewers introduced the project and made an inquiry by phone or in person about the availability of the head of each household to meet for a face-to-face interview for approximately 30 minutes to answer questions about their subsistence harvesting patterns. Three separate attempts were made to contact each household. Interviewers arranged meeting locations amenable to the household with an emphasis placed on the convenience of the respondent.

A total of five categorical response questions, including one four-part question and two narrative response questions, were asked during the course of the interview. Interviews concluded with interviewers briefly reviewing all reported results with the respondent to capture any inconsistencies in response or transcription errors.

ANALYSIS

Harvest effort by season and species was summarized relative to region as a whole. Summaries included three primary indicators. The first measure is *overall household use*, which is defined as any use of an individual resource reported by a surveyed household. The second measure, indicating spatial extent of harvest by household, is the *record*, defined as the water or land subregion polygon where an individual resource harvest was reported. The third measure, representing intensity of effort, is *harvest days*, which is defined as any amount of time spent attempting to harvest an individual resource within any portion of a given subregion. Variance around these three indicators of harvest effort was summarized by season, resource, and community using descriptive statistics.

An analysis of household use over time was completed by comparing total history of use with use reported in the last five years. Additional summary analysis of variance in total harvest days invested by individual respondents was completed for each harvested resource. The total harvest days were compiled by resource and season for each polygon and divided by the total area (km^2) of each polygon to return an index of harvest density. The 71 marine polygons were used as a zonal data set in Spatial Analyst, Zonal Statistics (ESRI 2009) to sample density raster surfaces predicting water-based recreation activity in Prince William Sound by season (chapter 7). An evaluation of overall correlation between recreation density and harvest density was completed using a Pearson's coefficient of correlation analysis in Excel (Microsoft 2007). Additionally, a rank summary analysis of sum recreation density relative to summary days of harvest effort will return those polygon units with the most intensely harvested polygons. The results of this analysis predict zones of highest intersection with recreation activity by season as well as the harvesting activities which most commonly co-occur with recreation activity.

A ranked summary analysis of most commonly selected reasons for change in use and discontinuation of cells was also completed. A spatial summary of subregions (polygons) where respondents report changing their use was completed with results summarized by cited reason for change in activity. This was also completed for those cells where respondents reported the intention to discontinue use. These resulting layers are those subregions where potential conflict zones may occur. These were then compared with the resulting volume of intersection analyses for recreation data in order to cross validate those results.

RESULTS

HOUSEHOLD SURVEYS

A total of 88 households contributed 1,473 records indicating locations of subsistence harvest. Approximately 75 percent of those records were locations that had been used at least once during the period of 2004 to 2008. The proportion of households contributing data ranged from 45 percent for Whittier to 38 percent and 16 percent for Chenega Bay and Tatitlek, respectively.

Cordova was approached through a sample selection scheme, and attempts were made to contact ~90 percent of the 340 households originally targeted by the effort. Of this total population of 307 households, interviewers attempted to make contact with 62 percent of household heads. Of those 190 households that were successfully contacted, 25 percent declined to give any information regarding their household's harvest efforts. Another 44 percent stated that they did not harvest resources within the Prince William Sound region. A further 14 percent reported they were harvesters but refused the survey. Of the remaining households contacted, 34 completed surveys. This represents 18 percent of the original sampled population contacted, or 60 percent of those households approached that harvest in the Sound.

All 88 households responding to the survey in the four study communities reported recent use within the period of 2004 to 2008. When evaluating frequency of resource use by households, halibut, salmon, berries, rockfish, and deer are the most regularly used both throughout time and in recent years. Overall reported harvest appears to have decreased for a number of resources when comparing recent to total history of household use. The average decrease in use for the 24 resources is 2.65 percent (+/-2.25 percent). Decreases in household use of halibut, rockfish, deer, and shrimp are outside of the standard deviation expected in overall change in use and thus may represent actual declines in use of those resources (table 9.1).

Households reporting harvest in recent years (2004 to 2008) were asked about intended future use of individual subregions (i.e., number of records). Of this group, 54 percent of respondents reported the intent for the same amount of use, 37 percent reported decreasing use, and 8 percent reported increasing use.

When asked the reasons for decrease and abandonment of an area, the majority of respondents identified *other*, which was a mix of reasons including health and

TABLE 9.1. Proportion of 88 households reporting use of a resource through time and within the past five years (~2004–2008) in Prince William Sound

RESOURCE	NUMBER OF HOUSEHOLDS USING	HOUSEHOLDS USING (%)	NUMBER OF HOUSEHOLDS USING IN LAST 5 YEARS	HOUSEHOLDS USING IN LAST 5 YEARS (%)	CHANGE IN 5 YEARS (%)
Berries	56	64	52	59	-5
Black bear	8	9	7	8	-1
Brown bear	4	5	7	8	-1
Crabs	12	14	7	8	-6
Deer	44	50	39	44	-6
Eggs	8	9	5	6	-3
Furbearers	1	1	1	1	0
Halibut	66	75	60	68	-7
Harbor seal	9	10	8	9	-1
Herring spawn	1	1	1	1	0
Mountain goats	10	11	8	9	-2
Other	6	7	6	7	0
Other finfish	21	24	18	20	-3
Plants	19	22	18	20	-1
Rockfish	50	57	45	51	-6
Salmon	65	74	62	70	-3
Sea lions	2	2	2	2	0
Sea otter	4	5	4	5	0
Seabirds	1	1	1	1	0
Sea ducks	9	10	7	8	-2
Shellfish	27	31	23	26	-5
Shorebirds	2	2	1	1	-1
Shrimp	37	42	32	36	-6
Waterfowl	21	24	17	19	-5

aging, changes in lifestyle, or a concern about lingering oil from the spill. The second and third most stated reasons for decrease and abandonment were *general change in resource availability*, a number of reasons broadly summarized as competition with others, and *cost of transportation*.

When considering only those harvest efforts occurring within the past five years, 88 households reported investing 16,723 total harvest days throughout the year across 126 land and water units (~80 percent of the subregion polygons) to harvest one to

TABLE 9.2. Annual harvest days reported by respondents from four Prince William Sound communities by resource type, ~2004–2008

RESOURCE	SUM OF HARVEST DAYS	HARVEST DAYS (%)
Salmon	3,767	27
Halibut	2,466	15
Rockfish	2,116	13
Berries	1,629	10
Harbor seal	1,258	8
Deer	1,181	7
Shrimp	1,057	6
Waterfowl	607	4
Shellfish	440	3
Sea ducks	357	2
Plants	336	2
Crabs	232	1
Black bear	185	1
Other finfish	177	1
Other	174	1
Sea lions	162	1
Eggs	143	1
Seabirds	126	1
Sea otter	102	1
Furbearers	64	<1
Mountain goats	52	<1
Brown bear	50	<1
Herring spawn	21	<1
Shorebirds	21	<1

several of 24 different resources. Fish species including salmon, halibut, and rockfish represent ~55 percent of the effort invested in harvest within the Sound. Another ~23 percent of total effort centers around resources acquired in the marine or intertidal environments, likely meaning that over three-fourths of total effort spent on harvest in the region occurs on or near the salt water. Picking berries and hunting deer are the most common activities occurring in upland areas, making up 17 percent of the Prince William Sound household harvesting effort (table 9.2).

Chenega Bay respondents (n = 8) reported the most total harvest days (8,006) of any of the four Sound communities and also harvested the greatest diversity of individual resources. This was followed by Cordova (n = 35) with 4,125 harvest days invested in 17 different resources. Whittier respondents (*n* = 39) reported a total of 3,736 days harvesting 12 resources. Tatitlek (*n* = 6) reported the smallest total effort of 856 harvest days spent across 13 resources. When controlling for total numbers of respondents between communities, Chenega Bay has the most individual effort, followed by Tatitlek, then Cordova and Whittier.

Of the 88 households reporting harvest of 24 different resources, the average number of years of experience harvesting individual resources was 15.5 (+/-12.1) years. Tatitlek's respondents appeared to have the longest tradition of resource harvest compared to other Sound communities.

When history of use was evaluated relative to specific resources being used by more than one household in our respondent group, we found that harvests of brown bear and harbor seal had somewhat longer traditions of use. As with deviation in overall harvest history by community, there was substantial deviation around individual resources.

Harvesting activity happened throughout all months of the year in the Sound with respondents from all communities. When evaluated by individual month, harvest appears to peak in July, August, and September for all Prince William Sound respondents. The community of Chenega Bay appears to have the most regular seasonal harvesting presence in the Sound, with no single month having fewer than 100 harvest days.

Evaluation of seasonal harvest patterns shows some resources are consistently harvested throughout all seasons, including halibut, rockfish, salmon, and shrimp. No respondents reporting recent harvest (2004 to 2008) stated that they planned to abandon areas as a result of competition. However, 15 households (17 percent) did report they intended to use 55 areas less than they currently had in future years at least in part due to competition. Thirty-seven of these were unique, with no more than one respondent selecting the same area. The households identifying the 55 areas included representatives from all four sampled communities. Units identified included 17 upland and 20 marine areas. A total of 14 units was identified by two or more respondents as locations where they have decreased overall harvest efforts in recent years. Eight of those subregions identified were marine, and six were upland polygons (figure 9.1).

Respondents from this group further specified 11 resources they intended to harvest less due to competition. Of the 11 resources, all four communities identified deer and halibut as resources they intended to use less as a result of competition. The greatest amount of competition was reported during summer and fall seasons for harvesting halibut and deer. Respondents reported competition likely to result in less intended use during all seasons of the year from Chenega Bay. The remaining three communities only reported competition that resulted in less effort spent in areas

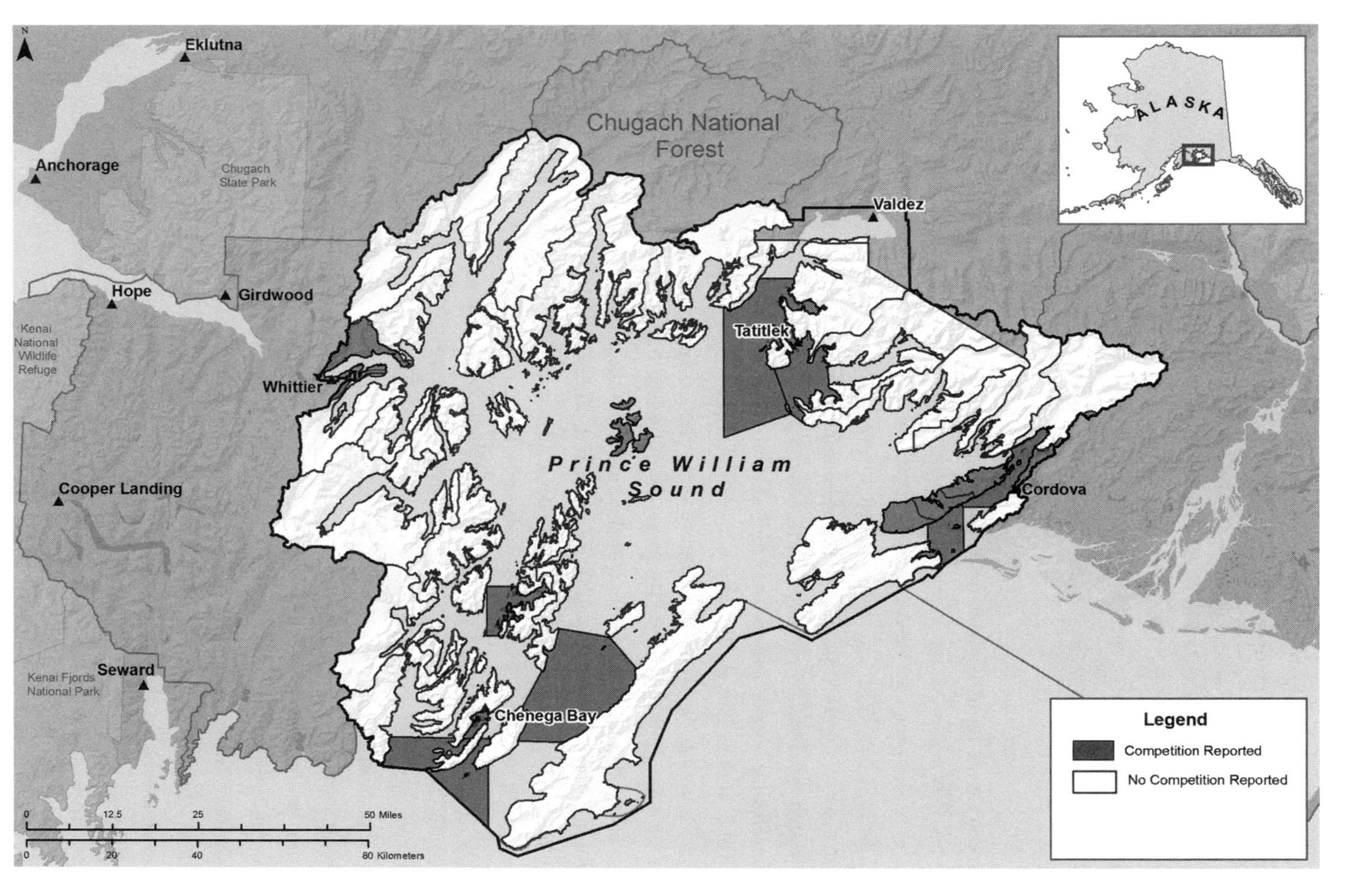

FIGURE 9.1 Fourteen subregions of Prince William Sound identified by two or more households as areas they intend to use less as a result of competition with others.

during spring, summer, and fall. Overall, households reported the greatest amount of competition overall during summer months, with Cordova and Chenega also identifying substantial competition during fall months.

When recent harvesters (2004 to 2008) were asked to identify all sources of competition and to rank those individuals most likely to be competing for resources, respondents reported the following ranging from most to least contributing to their sense of competition: (1) local harvesters; (2) commercial sport harvesters; (3) private harvesters from outside the Sound; (4) commercial harvesters; and (5) nonharvesters.

SPATIAL ANALYSIS

Harvests were reported from ~80 percent of subregion polygon units across the Sound. When harvest day effort for an entire year combined was evaluated, relative to size of individual polygons, the four most-used units were adjacent to the towns of Chenega Bay, Whittier, and Cordova. Overall harvest day density values ranged from 0.00 to 0.55 harvest days per km^2. Overall summer season is the most intensely harvested, followed by fall and spring, though variability in effort is certainly high between all subregion polygons during all three seasons.

Total annual harvest density is classified into four classes using the natural breaks function in ArcMap. It returns four classes that can be conceptualized as low, medium, higher, and highest in terms of harvest days/km^2. A number of areas adjacent to communities are used at the highest levels, including Sawmill Bay/mid-Elrington Passage unit as well as Evans Island, where the community of Chenega Bay is located, and two units adjacent to Cordova and Passage Canal (Whittier). Areas near Main and Eshamy Bays, west of Knight Island, also reveal higher use, along with Bainbridge Passage and Icy and Whale Bays (figure 9.2).

Spring harvest has the same top four highest-density harvest units as total annual harvests but with the addition of Knight Island Passage, as well as Derickson, McClure, and Kings Bays. During summer months, harvest is similar to the spring harvest, with three of the four top units being the same, but Main Bay (a key salmon fishing location) replaces a unit south of Cordova. Summer harvest is also generally more uniformly spread across numerous nearshore areas of the Sound, in contrast to a patchy distribution pattern displaying total annual and spring classifications of harvest effort. Fall harvest has similar overall patterns to summer, but increased effort is demonstrated in areas associated with waters and uplands around Cordova, causing Passage Canal to drop from the highest use class. A number of upland areas of importance appear in the medium density effort class across Hawkins, Latouche, Elrington, and Bligh Islands, as well as a pocket on the northern tip of Montague Island. Winter use is relatively low overall, only occurring in medium levels around Tatitlek and Cordova and low levels around Whittier. There is, however, significant

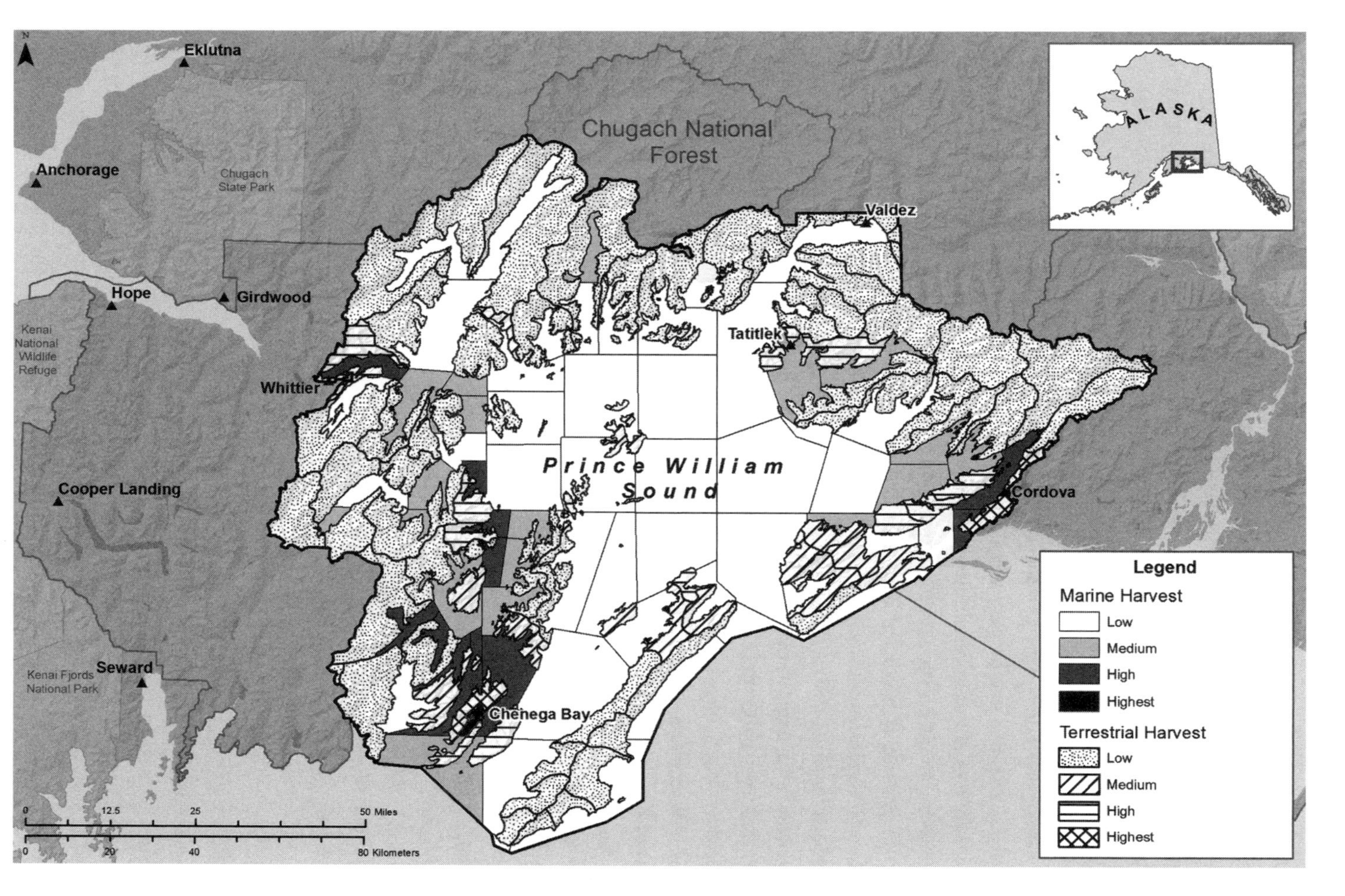

FIGURE 9.2. Total annual harvest day density for subregion polygon units broken into four classes from lighter to darker representing low, medium, higher, and highest, with highest being highlighted in black.

winter harvest activity in the vicinity of Chenega Bay, with the region's top four harvest units (Sawmill Bay/mid-Elrington Passage, Bainbridge Passage, Icy and Whale Bays) all occurring in close proximity to the village.

An overall correlation between harvest day density and recreation density (as compiled in chapter 7) shows a weak positive association between the two activities when summarized at the zonal level of subregion polygons (r = 0.20). When evaluated seasonally, the positive correlation between harvest and recreation activity becomes strongest during summer (r = 0.35) and substantially weaker during spring (r = 0.09) and fall (r = 0.05) respectively. Winter evaluations were not possible, as no data exists that can be used to predict distributions of winter recreation, which is generally thought of as very limited (chapter 6).

Specific water polygons where recreation intensity is relatively high (within the top 20 percent of predicted density range during summer) only overlap one area (Sawmill Bay) where harvest density is at its highest during summer. Other overlaps between high-density recreation areas and harvest areas of high importance occur in Passage Canal, Main Bay, and Eshamy Bay. A large number of medium importance and high recreation uses occur in the Sound, with most focused around Port Wells, Cochrane Bay, Culross Passage/Island, and Port Nellie Juan. Small pockets of this type of interaction with medium units also occur near Cordova and on the southwest part of Knight Island.

An evaluation of areas reporting high and highest levels of commercial sportfishing activity (chapter 6) showed overlap with one of medium importance to respondents on the northeast end of Hinchinbrook Island (Hawkins Island Passage). A polygon of high relative commercial sportfishing activity overlaps much of the area north of Hinchinbrook and extends out into Montague Island Passage. A similar polygon of highest commercial sportfishing encompasses much of southern ends of Prince of Wales, Elrington, and Latouche Passages south of Chenega Bay. This polygon abuts the highest use polygon for respondents during summer in Sawmill Bay/mid-Elrington Passage.

DISCUSSION

Harvest of a variety of resources from marine and terrestrial environments is obviously a key activity for residents of Prince William Sound. Subsistence harvest activities for Cordova, Chenega Bay, and Tatitlek have been well studied by the ADF&G Division of Subsistence in the aftermath of the oil spill. However, these studies failed to document harvests for the community of Whittier as well. The total numbers of days invested by households as well as the distances travelled by many to harvest resources clearly underscore the importance of this activity for all four communities.

It is important to note that the eight households from Chenega Bay participating in this study reported more total use in terms of harvest days invested than the other communities, although Cordova reported the most total records of harvest due to its larger sample size. When controlling for numbers of respondents per community, Chenega led substantially in terms of harvest days, followed by Tatitlek, Cordova, and Whittier. In a similar evaluation of extent of effort per household (i.e., number of records), Chenega also led, followed by Cordova, then Whittier and Tatitlek. A number of households in Tatitlek did not participate in our effort and likely resulted in this community not being fully represented in terms of harvest effort and extent of use. Even with reduced representation in the survey, Tatitlek led in terms of length of overall household history of harvesting in Prince William Sound. This difference with Chenega Bay is a likely consequence of the tragic destruction of the original village of Chenega following the 1964 earthquake and an essential 20-year absence from the southwest Sound until the community was reestablished in 1984.

Overall household use of wild resources in Prince William Sound appears likely to decline in terms of the numbers of resources harvested but also in terms of the spatial extent of harvest. Of the 24 resources harvested, households reported that they intended to harvest only 6 at similar levels in the future. Assuming respondents make good on intentions expressed in the survey, the remaining 18 resources can be expected to have less harvest from 1 to 7 percent of the households in our survey. Four resources (halibut, rockfish, shrimp, and deer) were identified by the most number of households (approximately six from our sampled group) as resources they intended to harvest less in the future. Certainly, harvest efforts are confounded by cyclical nature of species populations, but it is worth noting that these 4 resources have been strongly used in the past by harvesters in Prince William Sound, even following the oil spill, according to community harvest records maintained by Scott et al. (2001).

When use by all households is evaluated in terms of spatial extent of harvest (i.e., areas used by individual households), 37 percent of records showed decreased intended future use by respondents. Additionally, 12 percent of records reported that respondents intend to abandon use of areas entirely. Perceived competition with others does not appear to be the strongest motivator for declines in use or abandonment. The most often identified impetus was a number of personal reasons ("other") ranging from lifestyle changes, household moves, and health/aging. This reasoning was followed in rank by perceived decreases in general resource availability and thirdly by perceived competition from other users. This relative ranking of reasoning holds true for all households in general and when considering just those reporting recent use (between 2004 and 2008).

No households reporting recent use stated the intent to abandon areas because of competition with others. However, 17 percent of these households did report that they intended to use 37 individual areas less in the future due to competition. The majority of these areas are adjacent to the communities of Cordova, Whittier, Tatitlek, and Chenega—likely due in part to the fact that these are some of the most heavily used areas by harvesters. In these instances, locations of water-based competition centered around halibut and rockfish, while land-based competition focused on deer and berries.

Two additional areas were identified, including one which is in the highest category for commercial sport fish use as identified by Poe and Greenwood in chapter 6, in lower Port Bainbridge and at the south end of the Narrows. Given the intense pressure from commercial sport fish operators in the vicinity of Chenega Bay and Tatitlek, competition in these areas is unsurprising; however, the resources specified near Chenega were crabs and sea lion. Those identified near Tatitlek included halibut but also harbor seal. Some surprising areas of reported competition likely to result in less use include the Naked Island complex for deer (which may be a result of low deer numbers and poor hunter success as documented by ADF&G sport harvest data), an area of southwest Knight Island; and the lower end of Montague Strait between Latouche, Green, and western Montague Islands. In the case of these latter two units, resources involved included crab and sea lion.

Intent to use less or abandon does not necessarily equate with nonuse in the future, but these stated intentions parallel reported declines in overall resource use when comparing total use history to recent use from 2004 to 2008. The most common reported singular reason (outside of "other") for intended decreases in use was a perception of general resource decline. Some further investigation of those perceptions would prove useful to land and resource managers in the Sound.

There does appear to be a positive correlation between those areas of more intense harvest and areas of higher density recreation, and this varies by season. We know from contemporary evaluations (chapter 7) that there is substantially more total use in Prince William Sound during summer months (mid-June through August), so this could simply explain the result. A confounding effect associated with this analysis is that the recreation density surfaces used cannot separate out harvesting versus nonharvesting recreation. They were created using reported use locations of a variety of recreation groups, including a significant number of fishers and also locations of any vessel not readily identifiable as supporting commercial fishing or larger tour vessels. As a result, the uses being compared may, in part, be one and the same and thus may suffer from some autocorrelation effect.

Furthermore, these recreation density surfaces can only portray relative densities of use disconnected from any behaviors of groups involved. Many studies of use displacement in wildland settings (e.g., Hall and Cole 2007) highlight the behaviors of offending groups being a significant variable in predicting displacement. It is worth noting that the 15 households reporting they intended to use areas less as a result of competition perceived that their primary competitors were other local hunters, followed by commercial sport harvesters. Thus, recreation activities that are obviously not harvesting likely have less effect in terms of displacing harvesters from the region, suggesting that behavior of groups in these scenarios is a key component related to displacement potential. User group–specific evaluations, such as those explored in chapter 11, may provide further insights on potential for displacement of harvesters.

Certainly our efforts have significant limitations in terms of the degree of participation across communities. It would be difficult to argue that these results are statistically representative of any one community, particularly in the case of Tatitlek, where overall participation was limited compared to the other three communities. This effort should be regarded as offering a baseline of overall household use, spatial extent, and relative intensity of harvest effort across four seasons. The fact that this information was compiled spatially should allow for ease of integration with future analyses of human use and proposed management actions and should help to structure future monitoring of household resource harvest in Prince William Sound.

The Sound supports significant subsistence and sport harvest activities, as well as a thriving commercial industry for fish and game. Clearly, some level of displacement is already occurring around communities for local harvesters (i.e., residents of the Sound), but effects on other harvesters are not known. Similarly, the salient aspects of competition between harvesters, such as motivations, expectations, and general behavioral norms for each group, are not known. An effort to investigate these competition dynamics and other sociological factors with data on sport and commercial harvest would lead to a better understanding of harvest activities overall.

Land and resource managers in the region should work to establish a dialogue with households from the four communities in this study. These individuals have long-term knowledge of the resources in the region in general, as well as significant investments in their sustainable use into the future. Efforts to engage with those households should include informal, periodic consultation with those households demonstrating the greatest intensity and extent of harvest. This should be paired with systematic efforts to repeat the inquiry described herein in terms of spatial summaries of harvest effort and perceptions about resource availability and competition from others. Ideally, such monitoring would be entered into as a partnership with each of the four communities involved.

LESSONS LEARNED

With increased emphasis placed on providing opportunities for outdoor recreation, both for the individual and for economic growth in rural areas, it is vital that managers help maintain the traditional harvesting practices of resource-dependent, small communities. In many cases, the very nature of these communities and their harvest practices is a key draw for visitors. For the manager, this often means ensuring that local residents have abundant access to game resources without undue competition from visitors often eager to engage in similar pursuits. Consider the dynamics between subsistence and trophy hunters, which require balancing the needs of cultural heritage versus abundant game for commercial guide operations and commercial harvest. Then consider that many of these same wild landscapes also attract wildlife watchers, and things become even more complicated.

This work was a critical first check on ensuring traditional use practices in Prince William Sound are not being displaced by increasing levels of recreation. We created a baseline and a method for engaging harvesting households around questions of spatial use patterns and user conflict/displacement of local harvesting communities in the Sound. Though current dynamics don't indicate widespread overlap resulting in displacement, this effort affords repeated evaluation in order to detect this potential future circumstance.

Though highly sought game species and commercial activities have been quantified spatially (chapter 6), this was a novel approach for managers in Prince William Sound. A key to its utility was our attempt to summarize household harvest data across spatial units with boundaries common to a variety of managers. This approach allows the comparison of household harvest with large-scale economic activities like commercial fishing or outfitter and guide use. Similarly, results can be compared to those species whose total harvest is tracked by Alaska Department of Fish and Game—currently, bears, deer, and mountain goats. By approaching data collection in a fashion allowing for integrated analysis, Chugach National Forest managers lost some resolution as a result but ultimately boosted the broader utility of our data. Ideally, that tradeoff will result in better future management collaboration.

As land and resource managers make efforts to work beyond jurisdictional boundaries, a key divide often separates land managers from their game managing counterparts. In the case of Alaska, the state does not consider the implications of overall user experience for hunters/fishers and sees its focus as managing for sustainable harvested populations. On the other hand, federal agencies managing recreation are mandated to evaluate user experience and often shy away from playing a role in game management. This divide is similarly reproduced at the smaller scale within agencies, where recreation and wildlife/fish managers fail to address harvest user–related challenges in an integrated way.

This would be a nonissue, assuming that animals harvested directly equate to positive (or at least satisfied) hunter experiences. However, given the dearth of user experience evaluations for hunters, this assumption is troubling, and it's easy to imagine that overall experience is as multifaceted for hunters as it has been shown to be for a variety of recreation user groups. Certainly, hunters bring different objectives to the table; consider the issue of managing for trophy versus "meat" harvesters. Managers' continued acceptance of "game-in-the bag equals satisfaction" as the axiom driving engagement with their stakeholders will likely serve to increase potential intrauser conflict and result in uneven consideration of this important group (or perhaps more accurately, *groups*!) by recreation managers during land management planning.

Investigation of stakeholder perceptions is an important tool for managers across jurisdictions. This study revealed that a primary reason stated by users for the potential decrease in harvest use was a perception of general resource decline within a formerly harvested area. The perceptions about this declining resource could be correlated with existing harvest or other trend data in an effort to capitalize on what could be very useful local ecological knowledge. In our work, perceptions about resource abundance could be evaluated relative to harvest efforts for those species that are currently tracked closely, offering managers insight into the ability of hunters to perceive reduced harvest opportunities. An extension of this same line of inquiry could be taken to species whose populations are otherwise monitored by managers (like deer via pellet surveys conducted by ADF&G in the Sound) to see if perceptions about abundance correlate with reality.

Even as land managers aim to provide for a variety of recreation experiences, there appears to be profound lack of understanding of the behavioral and social norms of harvesting stakeholders—particularly hunters. We know from this work, as well as the work described in chapters 7 and 8, that this group is reluctant to report on their activities beyond numbers harvested and approximate effort and location (often these are difficult to come by). This likely results in harm to both managers and hunters. If the expectations and user experiences of hunters are poorly understood, they are less likely to be addressed in planning and outreach efforts conducted by managers. This is compounded into managers being seen as favoring other user groups and potentially allowing them to more easily make decisions that unknowingly and negatively impact hunters.

The face-to-face interviews conducted during our work *should not* be seen as a community engagement endeavor, but the results could be used to frame workshops that engage harvesters over spatially and temporally explicit results. The results of those discussions would aid both game and land managers with interpretation and validation of results and could be organized to foster greater understanding of the expectations and desired experiences of hunters. Traditionally, harvesting user groups are engaged by formal citizens' advisory boards, but these systems are likely not representative of all harvesters and probably don't facilitate data-driven, two-way conversations in the manner of a level of sustainable activity–type process (as described in chapter 8). Rigorously collected traditional ecological knowledge on species abundance, as provided by harvesters, or broad insights about social norms held by hunters, might assist managers. In recent years, attention to user experience and a citizen-science approach have helped managers work closely with and better understand outdoor recreation stakeholders. This approach should similarly be applied to hunters and fishers.

Communities around the world that rely heavily on the harvest of wild resources are uniquely positioned to detect changes in resource availability and seasonality, invasive species, and disease in the resources. The positioning of communities in proximity to or immersed within wildland settings makes them ideally located to detect changes to intact systems, both because proximity makes changes more obvious and because the lifestyle of these people is *dependent* on the products of those landscapes. As researchers and managers look to detect and track large-scale changes in species and resources due to climate change and other perturbations, they should invest in efforts that involve resident harvesters in the collection of structured feedback. Such efforts are most likely to succeed when residents trust the motivations of researchers, offering further incentive to managers to meaningfully engage with this group of stakeholders for the long term.

STAKEHOLDER ESSAY

SUBSISTENCE, HOW IT WORKS

PATIENCE ANDERSEN FAULKNER

THE TERM SUBSISTENCE HAS TAKEN many a turn in the modern world. Many folks participate in the "subsistence way of life" but don't take the time or have a need to define what it is. As a resident of the Last Frontier—Alaska—and an Alaska Native, it is a bit easier for me to do. Folks who live in the larger communities participate in some of the parts of subsistence but are not dependent upon the end product for their existence. They do realize after a time of only consuming that it is the interaction and sharing of the history, the skills, and the lessons taught that are missing. The issue of the entire state populace's need for the subsistence end product is continually debated as our population centers change.

The act of subsistence living is easy to separate into three parts of a continuous circle: the planning, the gathering/hunting, and the sharing. Each of these parts crosses over into the others and can also be distinct actions or activities.

The three parts begin with the planning. This comes from previous knowledge about the end product—be it fish, seal, berries, or birds. Over tea and pilot bread, the planning begins: determining what product is available and whether it is the right time to hunt or harvest. There are times that the hunted or gathered product is right in one's backyard and available with only a short walkabout, while at times the product has moved away and motorized transport is necessary. Those present can be parents and children, elder relatives, and neighbors. Excitement about the future activities is punctuated by recollection of a past success or disaster. Advice is offered about when it is the ideal time to harvest, the best and worst equipment to bring, telltale signs of the weather, the amount to harvest to sustain the wild stock, and environmental concerns about the area and the product.

The next part is the gathering or hunting, and those present for the planning may not always be the ones to participate. Often younger folks will be the worker bees at gathering. They will heed the advice of the Planner—advice that is relayed and debated. Observations on what is not present from previous hunts are shared and analyzed. If there is abundance, that too is analyzed. Perhaps it was an easy winter with slow transitions into the cold or the ash from previous volcanic eruptions has fertilized the berry fields. These conditions of normal or different are noted, to be woven into a planning session for future events.

This is how the storytelling of "the summer of . . ." or "the year we had no blueberries" falls into place.

Sharing is the final part, and it is the activity that is very broad. This is where the Elders are taken care of for their nutritional needs. They receive their product in a form that helps them and won't cause more labor than they can expend. The neighbors also receive some of the product, as a good neighbor also is an asset to all. While sharing the product, tea is the sit-down medium in which to hear the stories of how it was done and what further adventures are planned. Wistful memories and lessons are shared: remembering the first hunt and the ones in between, providing for an unfortunate family, community gatherings where the food was prepared, as well as remembering those who no longer are with the tribe or those who have moved away, and sending them some of the product.

The circle continues. Where the history of the hunting and gathering has been and where it will go. The lessons which were learned from the Elders—some told and retold throughout the times—and the advice to the younger ones on how it has traditionally been done and the changes in equipment for harvesting and getting around.

Subsistence is the using of what Nature provides and appreciating the abundance that we use. Overharvesting is discouraged through the lessons shared during the planning and harvesting. The best methods of preparation are taught as the product is in the hand. There is satisfaction in providing for those within the circle, and it warms the inner heart as well as encourages the hunters/gatherers to continue to work for the community.

With the *Exxon Valdez* oil spill, this circle was disrupted. Following the spill, I worked with the class action attorneys in advocating claims for the impacts from the *Exxon Valdez*. One of the 53 classes included the Alaska Native subsistence claims. The tribal governments were rejected in this process even as they stepped up and called in all favors for foods and cultural support from across the state (they later received pass-through funds from individual claimants to recognize their contribution in protecting the tribal member's rights). The affirmed statements requested that documentation be provided for the individual claims and that legal process required that a dollar amount be placed on the lost foods. This was overwhelming for most folks. When a citizen living in most of the rest of the United States places a value on a caught king salmon, they look to the newspaper ads at grocery stores for price per pound or think of the cost of an airplane ticket and travel expenses to catch one.

There were a few tribal members who actually had calendars of how much fish was caught and canned, how many meats or berries were harvested. I suppose they had a bit more time on their hands, but most tribal members accepted what they caught/harvested and immediately processed the food stuff and didn't keep records. The craft of processing—canning fish, making jam, freezing meat—was enough work, and keeping a journal was not productive. Even if it were possible to quantify, one doesn't look at the cost, as it would not make financial sense to place a value on the experience—it is priceless! When we gather our food harvest together, we aren't concerned whether we have a "plastic or paper" bag to hold it. We don't parade the trophy around to all of Main Street. We need to move on to the processing, the sharing, and the rest of our day. We have stories to tell, lessons to share, and food to prepare for dinner.

The oil industry responded with their best intentions of substituting canned sardines for herring. While the nutrition was needed, the product could not truly satisfy our need. It ignored the circle of subsistence. It disrupted the community life and continues to do so. Folks who fell into the worker bee age category worked on the oil cleanup and subsequent response programs. That left a gap for providing hands-on items for the reminiscing of the Elders and the visual presentation for the youth. The oil industry can and did provide donations to support cultural events, but community members also scattered with spill cleanup–related activities. We were busy salvaging the disruption from the oil spill and working aggressively to prevent the damages from the disaster. With the loss of mentors within a community, the onus is placed on fewer leaders. The burden of the *Exxon Valdez* oil spill disruption has enlightened communities as to how fragile our cultures are when an event so large can bring us to our knees.

We are growing new leaders with strong cultural experience. This disruption caused by the spill has been aggressively addressed through culture camps in most of the affected oiled communities. The culture camps recruit the Elders as well as worker bees to lead and encourage the youth in learning about this essential element of their lives. The skills are done through participation with an end product to bring home, to share with those left behind. The youth programs are thriving even with a few glitches. Dance groups perform and make their own regalia, sing songs in the language of the community, share their talents with other communities, and embrace the lessons they learn throughout their travels. We have a grand future as pride is grown within each child through the activities. We must support the Elders, encourage the youth, and embrace the uniqueness of our culture.

In reflecting upon subsistence, I recall that when times have been slim pickings for me, my friends have always shared their bounty. So, no matter how much or little I harvest, it is my obligation to do the same. We share the foodstuffs, the lessons, the skills of our culture freely; it keeps our culture alive and makes our lives so much richer. We help the Elders and we teach the youth.

CHAPTER 10

EVALUATING THE INTERACTIONS BETWEEN BLACK BEAR HUNTING AND KAYAKING ACTIVITIES IN WESTERN PRINCE WILLIAM SOUND, ALASKA

SPENCER LACE AND RANDY GIMBLETT

SOUND BITES

SIMULATION MODELING OF harvest data is a useful tool for understanding overlap between hunters and other groups of recreationists.

Bear hunters and kayakers have limited overlap with one another in western Prince William Sound as a whole, though key pockets of overlap and perceptions about incompatible behaviors like bearbaiting practices could exacerbate the potential for conflict.

Successful management of hunters in the Sound depends on managers having a better understanding of the desired experiences of this large and important user group.

INTRODUCTION

Black bear harvest levels have increased rapidly since 1995 in Prince William Sound. Bear hunters' eagerness to go afield after a long winter, the lure of fresh meat in the freezer, a love of the Sound, and relative ease of access via the Whittier Tunnel have all led a widely recognized high-quality spring hunt for black bears. Harvest in this area is managed by the Alaska Department of Fish and Game (ADF&G) as Unit 6D, which includes approximately four million acres of coastal temperate rain forest. A status report for 6D found a 100 percent increase in bear harvest between 1995 and 2001. In 2002, this area was open from September 1 to June 30, and the majority of harvest (~80 percent) occurred during May and June (ADF&G harvest data, 1989 through 2002). Harvest during this time reached a record of 435 bears, which was approximately 25 percent more than any other game management unit in Alaska. The season changed in 2006 to an opening from September 1 to June 10. A recent inventory has revealed this trend has continued. The Alaska Department of Fish and Game (ADF&G 2008) has reported that "hunter harvest reported during the past five regulatory years (2002–2007) averaged 449 bears, with the harvest in 2006–2007 at an all-time high. The majority of the harvest was male (75–85 percent), and most bears were taken in Unit 6D (82–86 percent)." The 2009 season was delayed by 10 days, resulting in the season now running September 10 through June 10, due to the continued high harvest (450 bears) and in order to reduce female harvest. According to ADF&G (2008), "Male black bears in Unit 6D tended to move down to beaches after emerging from winter dens to feed on new sedges and grasses, making them more vulnerable to harvest during this period. Females tended to remain away from beaches, instead favoring south-facing slopes and avalanche chutes that green up early in the season."

In the early spring harvest period (May), bear hunters likely become the most prevalent recreation use group on the shoreline of Prince William Sound. This harvest level represents an unknown number of individual hunting parties using the shoreline of the Sound. A hunt success of ~50 percent was documented by unguided hunters out of the port of Valdez during the late 1960s (McIlroy 1970). Harvest reporting required since 2009 identifies 948 hunters reporting 49 percent success rate in 2009 and 893 hunters reporting 51 percent success in 2010. According to David Crowley, Alaska Department of Fish and Game, with the opening of the Whittier Tunnel in June of 2000, it is thought that the majority of current bear hunter use enters the Sound through Whittier and likely occurs generally in the western and northern parts of the Sound (D. Crowley, personal communication).

The Chugach National Forest hosts a vast majority of these hunting parties and yet knows almost nothing specific related to their use patterns. This timing briefly overlaps with an increasingly popular early kayaking and pleasure boating season, causing concern. During the spring hunting harvest season, black bears concentrate their activity along shorelines (ADF&G 1982; McIlroy 1970), often with beach or estuary characteristics that are also attractive to kayak campers and other shoreline recreation uses. Beginning in the early 2000s, the Forest Service began to receive some individual reports of user conflicts in the western Sound between bear hunting groups and other nonharvesting use of the shoreline during late May and early June. Conflicts are reported to be exacerbated by the practice of bearbaiting, where hunters are permitted to establish stations baited with attractants for bears (often in the form of oily human foods). Approximately 60 bears or fewer are killed over bait annually due to this practice, which is approximately 12 percent of the bears taken. Many bait stations are established on beaches that may also be used by nonharvest users. The

Alaska Department of Fish and Game issues 130 to 140 bait station permits per season, allowing up to two per hunter, but reports indicate that many hunters only put up one. Though actual reports of conflict at bait stations have been limited, the perceived potential for individual conflict, especially with kayakers, is a concern to both Forest Service and ADF&G managers. To reduce potential conflict, ADF&G has closed bear baiting in two popular recreation areas in the western Sound, Blackstone Bay (AA 11), and Harriman Fjord/Barry Arm (AA 02). In addition, Alaska hunting regulations prohibit baiting within one mile of any dwelling structure or developed campsite. Anecdotal information suggests that this practice continues to be a problem in other areas in the Sound (e.g., Valdez Arm, as reported in chapter 8).

During 2004, the Chugach National Forest permitted nine hunting guide services that were authorized for a total of 790 user days in the western Sound alone. In addition, an unknown number of water taxi services specialize in supporting bear hunting groups throughout the Sound, but these numbers are not tracked by the Forest Service. The Chugach National Forest also permits another 24 nonharvest-oriented commercial operators for shoreline activities in the western Sound. With more harvest and nonharvest commercial operations being permitted, increasing private recreation, and new Forest Service efforts to manage shoreline camping areas, it is critical to have a comprehensive understanding of both the spatial and temporal patterns of activities related to the spring black bear harvest.

This chapter presents the results from an agent-based simulation to investigate the spatial and temporal distribution and overlap between hunters and kayakers during spring black bear harvest. This is a novel application of a scientific tool to explore this issue and hopefully offer managers some initial guidance as they attempt to understand the degree of importance of this potential issue of user conflict. It benefits from previous research efforts that were undertaken in the Sound to evaluate spring black bear hunting (Lace et al. 2008), and a recent evaluation of dispersed recreation use (see chapter 6).

METHODS

SPATIAL AGENT-BASED SIMULATION MODEL OF BEAR HARVEST ACTIVITIES

The simulation framework referred to as the Recreation Behavior Simulator (RBSim) (Gimblett et al. 1997; Gimblett 2002; Itami and Gimblett 2001; Itami et al. 2003; Itami 2002) was used to evaluate the spatial and temporal patterns of spring bear hunting activities in the western Sound. RBSim is a computer simulation tool, integrated with a geographic information system (GIS), which is designed as a general management evaluation tool in any human/landscape interaction setting where humans

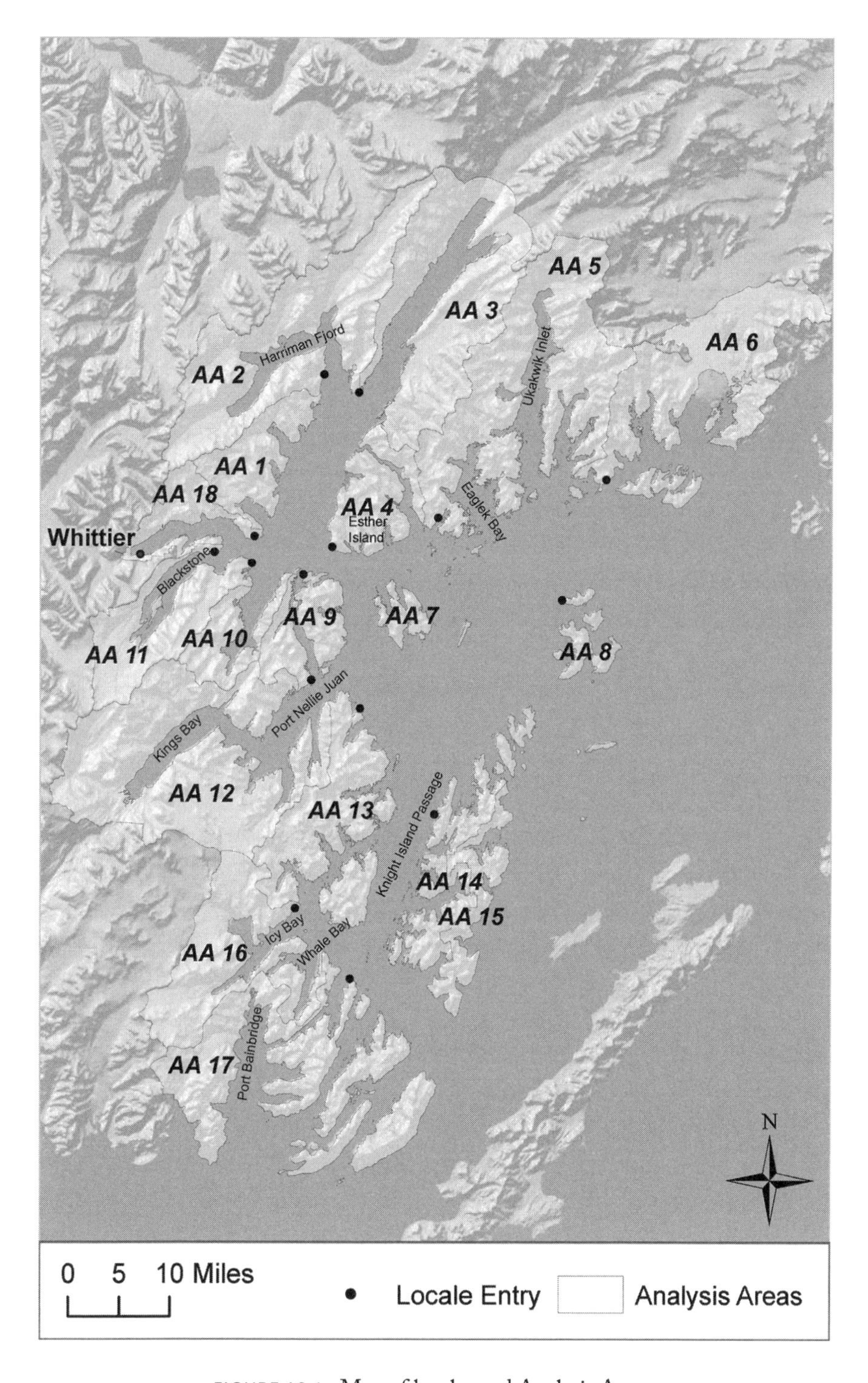

FIGURE 10.1. Map of locales and Analysis Areas.

travel on a linear network. RBSim has been applied in assessment of the dynamic spatial-temporal patterns associated with human/landscape interaction studies (Gimblett and Skov-Petersen 2008). Lace (2005) and Lace et al. (2008) specifically explored the use of a spatial agent-based simulation model in conjunction with a database containing location information for bear harvest sites in the area to analyze complex, spatially dynamic patterns of black bear hunting in the western Sound. While agent-based simulations have been successfully used in many situations, this study has used a rule-based approach, using a minimal amount of available data on hunter movement patterns and behavior. In contrast with some other studies, the rule-based nature of this project introduces several complications and some additional uncertainty. Trip data commonly compiled includes origin and destination locales, as well as other itinerary information such as party size, trip duration, and travel mode. With fewer inputs of trip data, the rules act in a strong way to determine the results, as did expert opinion and analysts who generated the rules. Not surprisingly, it would be possible to script the agent behavior to reflect a chosen scenario. Even so, with solid evidence of behavior and choice of rules, the scenario itself allows for emergent results.

The Lace et al. (2008) study evaluates through simulation the existing patterns of hunter use over the season, identifies peak visitation periods, evaluates the durations and destinations of hunter choice, assesses commercial and noncommercial hunting activities, and provides some guidance in describing the density of harvest per management unit. The spatial outputs from the simulation provide a means to evaluate hunter distribution across the Sound. The results of this study were used to inform management as to peak hunter use days, their cumulative visitor use days of assumed routes over a nine-year period and in specific areas: Eaglek/Unakwik (6,000 to 10,000 hunter days) and Kings Bay/Port Nellie Juan, West Knight Island Passage, Icy and Whale Bays, and Port Bainbridge (each with 2,000 to 6,000 hunter days) are the most heavily used in the study area. These areas consistently have the highest number of total visits and trends in bear habitat and overnight facilities. Eaglek/Unakwik (AA 05) receives the most visitor use days and has the highest number of bears taken. Similar to the findings regarding commercial use vs. private use (see chapter 7), Lace (2005) assumed that private boats account for approximately 75 percent of the bear hunter travel modes entering into the Analysis Areas, with water taxis accounting for an additional 15 percent. While water taxi services specialize in supporting bear hunting groups throughout the Sound, they are nowhere near as high as private boats, but do account for a significant number of trips entering the western Sound. While there is a significant amount of commercial use in the Sound, Analysis Areas Eaglek/Unakwik, Icy and Whale Bays, and Port Bainbridge are not only the sites most frequently visited by commercial activity, but also the most frequented overall. These areas account for the highest percentage of bear harvest, the longest duration of hunters' stay, and the most overnight activity use.

FOCUS GROUP EVALUATIONS OF IDEAL, EXPECTED, AND MAXIMUM TOLERABLE USE SCENARIOS

While the simulation by Lace et al. (2008) explored the spatial distribution of bear hunting activity in western Prince William Sound with limited trip data, many assumptions had to be made and behavioral rules were defined by expert analysts. This is appropriate for prototyping and testing a model, and deriving and verifying rules from those who physically use the landscape is a more accurate and defensible method that reduces uncertainty and improves the reliability of the model. A recent study using the limits of sustainability (LSA) framework was undertaken using focus group sessions in Cordova, Anchorage, and Valdez with a total of $n = 62$ participants from three user types (kayakers, recreational motor- and sailboaters, and hunters) to characterize and obtain a better understanding of the behavioral rules for each recreation group. Each focus group was presented with representative human use scenarios for three different subregions within the Sound (Blackstone Bay, Unakwik Inlet, and Sheep and Simpson Bays) known to have varying levels of existing use based on prior studies. They were asked to provide perspectives on the use levels presented as they related to their *ideal*, *expected*, and *maximum tolerable* scenarios of use for three different types of users.

Results suggest that hunters have the highest requirements for solitude and the lowest tolerance for competition (see chapter 8 for full details). Generally, hunters prefer "one boat per bay" and this "unwritten law" is generally acknowledged and respected among subsistence hunters in local communities like Cordova. However, conflict with other hunters occurs when this concept is not recognized—especially by commercial hunting guides. The key issues relating to the hunter groups had to do with the management of the hunting resource. Anchorage hunters felt that wildlife stocks were overhunted and that the practice of bearbaiting was undesirable. Hunting pressures, primarily from sports hunters as opposed to subsistence hunters, are considered to be intolerable during the peak hunting season, which causes local hunters to either move from traditional hunting areas or to hunt during the shoulder seasons. Hunters are generally in the Sound in the spring and fall and therefore do not overlap with the main summer kayaking season. When there is overlap, hunters see little problem with kayakers camping onshore, as they are quiet and don't interrupt game hunting. They therefore see little conflicts with kayakers and other general recreational boats but are concerned specifically about competition with other hunters.

Kayakers, on the other hand, follow hunters in the high value they place on solitude or low levels of use. This group values "low-impact" camping, self-sufficiency, camaraderie with other kayakers, and contact with nature. They are generally tolerant of other kayakers because of the quiet mode of transportation and strongly shared values of low-impact recreation. However, they are less tolerant of small motorized boats commonly used by hunters because of the noise, speed, wakes, and the impact boat wakes have on quiet and solitude. Also, if hunters camp onshore, they compete for campsites. Valdez kayakers were particularly opposed to the practice of bearbaiting undertaken by the bear hunting community because of the visual impact of litter from baked goods and the perceived danger posed by bears attracted to shoreline campsites. It is fairly clear that users perceive Blackstone Bay to be busy by kayaker standards and at or above capacity during peak periods. There is definitely concern among the kayaking community when it comes to interfacing with small motorized boats during hunting season.

SUMMARY

The studies described in chapters 7 and 8 revealed a complicated and differing set of needs and rules of behavior for each of the recreation groups studied. Due to the professed sensitivity of crowding for the hunter user groups and the perceived impacts of hunters on other groups, it is clear that more needs to be understood about hunters' patterns of use and a greater investment made in the management of this group. Hunting is certainly a key experience sought by individuals using the uplands of the Sound, and managers should actively work to facilitate this activity. The knowledge gaps that managers have relative to hunting in the Sound as well as the sensitivity of this group to changes in use levels and competition for resources make them, above all, most likely to experience declines in quality of experience in Prince William Sound.

Likewise, even though Valdez and Cordova kayakers perceive that use levels are less than what they consider to be *maximum tolerable* for small motorized boats during peak periods in Eaglek/Unakwik (AA 05), Lace et al. (2008) have shown this area to have the highest number of hunter days (6,000 to 10,000 days) and most bear taken during black bear season. Further study is warranted, given the growing number of visitors to the Sound, the sensitivity of the kayaking community to competition for campsites, the frequency of interaction with small motorized boats during hunting season, the practice of bearbaiting undertaken by the bear hunting community, and the potential danger posed by bears attracted to shoreline campsites.

EXAMINING THE INTERACTION BETWEEN BEAR HUNTING ACTIVITIES AND KAYAKERS

Understanding the spatial-temporal distribution and overlap between hunters and kayakers during black bear harvest is a first step towards filling this knowledge gap and is essential for management of the Sound. This analysis evaluates these interactions using a combination of GIS with existing standardized bear harvest data sets and a contemporary spatial analysis of user experience in the Sound. This spatial analysis is driven by bear harvest and user experience data sets from two separate studies and can begin to inform decision-making leading to proactive management of hunter/kayaker interactions in the Sound.

This analysis uses bear seal data for the years 1995 through 2004. Lace et al. (2008) developed a simulation model and summarized bears harvested for each of the Analysis Areas (see chapter 6 for full details on the creation of these units). Harvest locations were converted and summarized by Analysis Areas. To examine the interaction with kayakers from the 2009 study who reported visiting campsite locations and/or spent time on the shore, the total count of kayakers for each of the Analysis Areas was determined by associating each vessel point collected with a single Analysis Area. We recognize that these two data sets come from different studies and time periods, but feel they are a valid first attempt to understand potential bear hunter/kayaker interactions. Table 10.1 provides a summary of the number of bears that have been killed and kayakers who have visited or spent time in the same Analysis Area.

A Pearson's correlation analysis was performed on the data shown in table 10.1 to determine if there was a statistical relationship between kayak use and bear hunting in identified Analysis Areas. Results of this correlation analysis indicate that the overall correlation between kayak visits and bear hunting activity by Analysis Area

TABLE 10.1. Association of number of kayakers and bears per Analysis Area

ANALYSIS AREA (AA)	AA NO.	NO. OF REPORTS OF KAYAK USE	NO. OF BEARS HARVESTED
West Port Wells	1	56	62
Harriman Fjord/Barry Arm	2	47	31
College Fjord	3	17	66
Esther Island and Passage	4	86	115
Eaglek/Unakwik	5	54	206
Columbia Bay	6	56	35
Culross Island/Passage	9	121	65
Cochrane Bay	10	42	36
Blackstone Bay	11	75	47
Kings Bay/Port Nellie Juan	12	87	132
West Knight Island Passage	13	62	149
Icy and Whale Bays	16	10	90
Port Bainbridge	17	17	122
Passage Canal	18	291	28

TABLE 10.2. Pearson's rank correlation of bear harvest versus kayak interactions

	AA NO.	SCORE 1	SCORE 2	RANK 2	DI	DI2
Eaglek/Unakwik	5	5.29	20.18	14	8	64
West Knight Island Passage	13	6.07	14.59	13	4	16
Kings Bay/Port Nellie Juan	12	8.52	12.93	12	0	0
Port Bainbridge	17	1.67	11.95	11	8.5	72.25
Esther Island and Passage	4	8.42	11.26	10	1	1
Icy and Whale Bays	16	0.98	8.81	9	8	64
College Fjord	3	1.67	6.46	8	5.5	30.25
Culross Island/Passage	9	11.85	6.37	7	6	36
West Port Wells	1	5.48	6.07	6	1.5	2.25
Blackstone Bay	11	7.35	4.6	5	5	25
Cochrane Bay	10	4.11	3.53	4	0	0
Columbia Bay	6	5.48	3.43	3	4.5	20.25
Harriman Fjord/Barry Arm	2	4.6	3.04	2	3	9
Passage Canal	18	28.5	2.74	1	13	169

shows a weak positive association between the two activities when summarized at the zonal level of subregion polygons ($r = 0.07$). This weak correlation suggests that this is not a widespread problem in the western Sound (see figure 10.2).

However, this analysis does suggest that there are specific areas more susceptible to bear/kayaker interactions than others, though they are isolated.

There are popular hot spots for kayaking, including Harriman Fjord/Barry Arm, West Port Wells, Blackstone Bay, and Passage Canal, where lower numbers of bears have been hunted, and likewise there are hot spots for bear hunting activities including Eaglek/Unakwik, Kings Bay/Port Nellie Juan, and West Knight Island Passage, which are bear attraction sites. To determine which of the Analysis Areas rank highly for potential interactions between kayakers and bear hunting, we ran a Spearman's rank correlation analysis on normalized values of kayakers and bear hunters across all Analysis Areas. Results of this analysis again show a very weak overall correlation ($r = 0.119$) between these two groups. Results do suggest, however, that the top three areas with high potential for kayaker/bear interactions are Eaglek/Unakwik; Kings Bay/Port Nellie Juan; and Knight Island Passage, Esther Island, and Port Bainbridge.

Table 10.3 is an assessment from the Prince William Sound human use hot spots analysis (see chapter 6) revealing similar predictions between the interaction between beach use (kayakers) and bear hunting harvest. Both evaluations suggest competition for spring bear hunting in the Eaglek/Unakwik area, Port Nellie Juan, and Esther

TABLE 10.3. Analysis areas identified in the hot spot assessment (chapter 6) where both beach use and hunting occur simultaneously, suggesting potential for overlap and possible conflict between user groups

ANALYSIS AREA	GENERAL AREA
Eaglek/Unakwik	Cascade Bay
Eaglek/Unakwik	Olsen Island
Knight Island	Bay of Isles
Knight Island	Snug Harbor
Knight Island	Thumb Bay
Esther Island/Passage	Esther Bay
Esther Island/Passage	Esther Bright
Esther Island/Passage	Esther Passage
Esther Island/Passage	Granite Bay/Esther Island
Esther Island/Passage	Waterfall Cove
Kings Bay/Port Nellie Juan	McClure Bay
Kings Bay/Port Nellie Juan	Mink Island

Passage. A comparison of all results with the findings of LSA workshops suggests that while a use conflict is perceived by the kayak communities, it is not yet a dominant issue. While there are isolated incidences of bear hunter/kayak interactions, there is not enough evidence at this time to develop policy to address this issue. However, the practice of bearbaiting was of grave concern to kayakers and may serve to exacerbate a perceived problem of use conflict.

DISCUSSION

Hunting is certainly a key recreation experience sought by individuals and should be facilitated by land and resource managers in the Sound. Currently, managers' lack of knowledge about quality of experience dynamics associated with hunting and the sensitivity of this group to competition suggests that hunters may see declines in quality of experience that could become a problem in Prince William Sound. More robust information about kayaker motivations, expectations, and social norms is available and suggests a potential for conflict between these two groups.

This study illustrates that while there is a significant amount of bear hunter use in the Sound, Analysis Areas Eaglek/Unakwik (6,000 to 10,000 hunter days) and Icy and Whale Bays and Port Bainbridge (2,000 to 6,000 hunter days) are the most heavily used in the study area and account for a high percentage of the bears

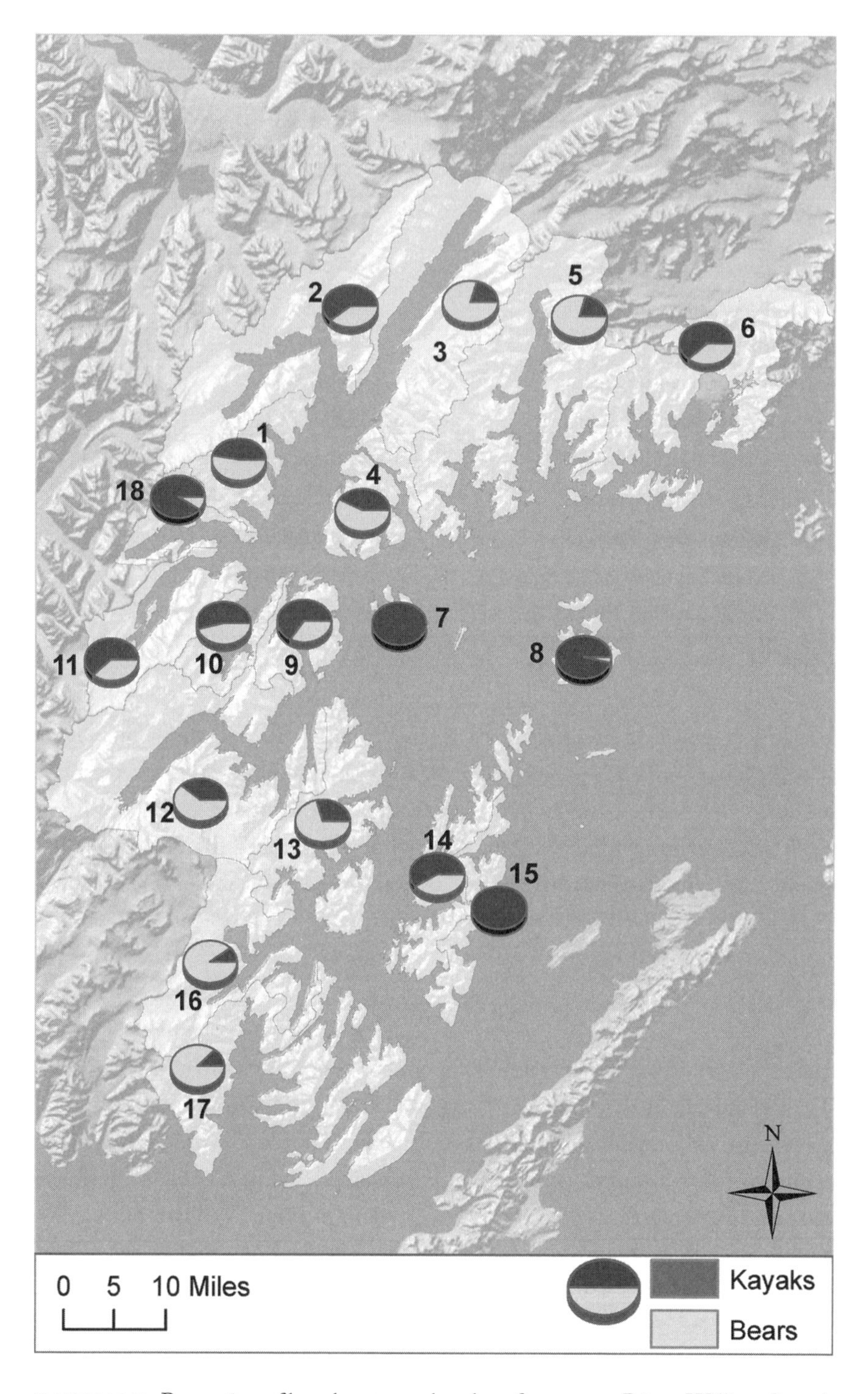

FIGURE 10.2. Proportion of bear hunters to kayakers for western Prince William Sound.

harvested, the longest duration of stay, and the locations where the most overnight activity occurs from hunters. The spatial analysis reveals that these bear hunting hot spots are not the most popular kayaking destinations. In fact, our analysis shows that the correlation between bear hunting activities and kayakers is weak and likely happens in isolated cases. Areas of potential concern include Eaglek/Unakwik; Kings Bay/Port Nellie Juan; and Knight Island Passage, Esther Island, and Port Bainbridge.

A limitation of this analysis is our inability to completely predict hunter effort in the Sound. The data used to predict hunter distribution comes only from those parties that successfully harvested bears. Assuming a 50 percent harvest success rate, there is a substantial amount of hunter effort that is not recognized in this analysis. Though it is reasonable to assume that unsuccessful hunters may focus on the same areas as successful hunters, there is also the possibility that their low success rate is a function of hunting in less productive locations which may be used by early season kayakers. Observations offered by experienced hunters who participated in focus group evaluations (chapter 8) suggest that there is a substantial number of bear hunters in the western Sound who have less total experience hunting in the region, and these individuals may end up hunting less productive locations frequented by kayakers. Without survey-based information relative to the empirical distribution of hunter effort, this possibility cannot be discounted by managers.

Prince William Sound hunters interviewed in chapter 8 had concerns about other hunter competition but perceived kayakers as only minimally interfering with their hunting activities. Kayakers from this same study were concerned about the competition for campsites during hunting season. Kayakers were also concerned about the practice of bearbaiting undertaken by some members of the bear hunting community because of the visual impact of litter from human or pet foods and potential danger posed by bears attracted to shoreline campsites in search of food. There was definite concern among the kayaking community when it comes to interfacing with small motorized boats associated with hunting seasons. Our analysis shows that while this might be a perceived concern, it is not a substantial issue that is happening throughout the Sound.

LESSONS LEARNED

While managers have been made aware of individual incidences of bear hunter/kayaker interactions, this does not seem to be a phenomenon that is widespread across the Sound. Therefore, making region-wide specific guidelines to address this issue is probably not warranted. However, it would behoove managers to invest in outreach efforts explaining the use patterns and behaviors of these two groups to each other. By making kayakers more aware of hunting hot spots and hunters more aware of beaches used for kayak camping, managers could facilitate voluntary separation between these groups. Anecdotal evidence suggests that the practice of bearbaiting is of grave concern to kayakers and could be a significant driver for perceptions about conflict with bear hunters. Communication efforts aimed at hunters and kayakers about this practice should be a high priority for land and resource managers in the Sound. By facilitating even a generalized understanding of each group's use patterns and behaviors in the Sound through maps, brochures, web content, and so on, managers could make a substantial contribution toward mitigating user conflicts.

This study demonstrates the potential of the use of tools like simulation modeling and GIS summaries of harvest data to explore questions about overlap between user groups. This can be a very useful first step for managers attempting to understand the scale and magnitude of potential issues associated with user conflict. Using these tools, we were able to identify a few key locations where managers may wish to target further inquiry, but we stress that overlap between user groups does not necessarily equal conflict.

Rather, these key locations might best be used to frame a study that could better understand the social dimensions of the problem and assess the importance of this potential conflict relative to other human use issues in the Sound. Key to the success of such an effort is understanding the dynamics *of both* user groups. Though the motivations of kayakers are relatively well understood (e.g., chapter 7), substantial information gaps relative to hunter motivations, quality of experience dynamics, intragroup competition, and social norms remain poorly understood by managers. In addition to designing a study, this first exploratory step could be used to target communication and stakeholder engagement efforts aimed at better understanding hunters as a large and important user group in the Sound.

CHAPTER 11

MODELED DISTRIBUTION OF HUMAN USE IN 2015 AND POTENTIAL WILDLIFE DISTURBANCE IN WESTERN PRINCE WILLIAM SOUND

LOWELL H. SURING, KAREN A. MURPHY,
SHAY HOWLIN, AND KARIN PRESTON

THE SOUND PROVIDES ESSENTIAL habitat for thousands of seabirds, marine mammals, and five species of salmon, as well as habitat for upland birds and mammals. The wealth of abundant wildlife and fish and impressive scenery has drawn people to the area for thousands of years. The 1989 *Exxon Valdez* oil spill was the most notable human-caused impact to the Sound's ecosystem (chapter 4). However, the opening of the Whittier Tunnel access road in the summer of 2000 brought new challenges to the Sound and to the species recovering from the spill as a result of increasing human use and associated development (Brooks and Haynes 2001).

This improved access was expected to result in increased human use (ADOT 1995), with possible consequences for wildlife and fish populations in the Sound. Concerns in other areas have led to many studies investigating whether recreation and other human activities caused disturbance to local wildlife. The results of these studies vary widely among species, season, and the intensity and form of human activity; however, the majority of studies document some disturbance effect on wildlife. Unfortunately, the consequences of such disturbances are poorly understood, with much less known about long-term effects on populations.

To demonstrate how human use in the western Sound may affect injured species, we examined the relationship of existing use and predicted future human use patterns relative to the general distribution patterns of nesting pigeon guillemots and harbor seal concentration sites. Our exploratory effort, originally published in its entirety as Murphy et al. (2004), is summarized here as the first spatially explicit and future-simulating tool used by managers in the Sound to help understand the potential relationships between human activity and wildlife resources. Harbor seals and pigeon guillemots were selected to represent two classes of animal species injured as a result of the spill, and geographic information system (GIS) layers depicting the distribution of these species were acquired from resource managers in the region.

ArcInfo GIS raster layers, or "grids," were created to represent the monthly distributions of kayak and motorboat use patterns from 1998 (from Murphy et al. 2004). A series of spatial and statistical modeling techniques (described in detail in Murphy et al. 2004) was used to evaluate landscape factors influencing use patterns of kayakers and/or recreational boaters in the western Sound (table 11.1). Landscape features of importance for kayakers used in predictive models included distances to campsites, shore, tidewater glaciers, upland recreation opportunities, and Whittier. Landscape features of importance used for recreational boaters included distances to upland recreation opportunities, upland glaciers, safe anchorage sites, sport fishing opportunities, and Whittier. The resulting models incorporating these factors were then used to create grids that projected likely use patterns by kayaks and recreational boats in the future.

A predicted annual rate of increase in kayak activity of 7.5 percent was assumed based on growth patterns observed by Twardock and Monz (2000) between 1987 and 1998. An increase in motorized recreation boat users was assumed to be 3.5 percent based on analyses by the Alaska Department of Transportation and Public Facilities and Federal Highway Administration (ADOT 1995). These rates were assumed to continue yearly and would simulate increased boat traffic in the western Sound through 2015. Additional increases of 100 percent were included for both kayaks and boats for years after 2000 once the tunnel had opened based on these same Department of Transportation and Federal Highway Administration impact studies.

In order to predict intensity of future overlap between kayaks and boats with wildlife resources, GIS techniques were used to place 1,000-m buffers around mapped locations of identified pigeon guillemot nesting areas and harbor seal concentration areas. Use projections for the months of May through August were summarized for potential overlap on pigeon guillemots, and June through August for harbor seals, to account for greatest sensitivity during breeding seasons.

In 2004, we projected that monthly use in 2015 near pigeon guillemot nesting sites ranged from no change to increases of more than four times 1998 use levels for kayaks and from no change to increases approximating 20 times for motorized recreational boats. Predicted monthly human use in 2015 near haulout sites for harbor seals ranged from no change to increases approximating 12 times by kayakers and from no change

TABLE 11.1. Landscape features of importance for kayakers and recreational boaters in Prince William Sound, used to predict likely use patterns

CHARACTERISTIC	NUMBER OF SITES	FORM OF VARIABLE AND MODELS CONSIDERED	
		DISTANCE	DENSITY
Safe anchorage	199	RB[b]	RB
Known sites suitable for camping	30	K[c], RB	K, RB
Known sites preferred for sport fishing	18	RB[b]	RB
Location of upland glaciers	14	K, RB[b]	K, RB
Location of fish hatcheries	5	RB	
Known sites preferred for black bear or Sitka black-tailed deer hunting	32	RB	RB
Location of lodges open to the public	3	RB	
Sites known to have outstanding or remarkable scenery	22	K, RB	K, RB
Location of tidewater glaciers	13	K[c], RB	K, RB
Sites adjacent to upland areas providing recreation opportunities	85	K[c], RB[b]	K, RB
Areas with a diversity of wildlife resources		K, RB	K, RB
Sites known to provide consistent wildlife viewing opportunities	18	K[c], RB	K, RB
Locations of anchor buoys		RB	RB
Locations of recreation cabins available to the public	6	K, RB	K, RB
Distance to shore	—	K, RB	
Distance to harbor	—	K[c], RB[b]	

[a] Considered for motorized recreation boat (RB) models and/or kayak (K) models.

[b] Selected for models used to estimate use by motorized recreation boats.

[c] Selected for models used to estimate use by kayaks.

to increases approximating 45 times by motorized recreational boats. Figures 11.1 and 11.2 show the relationship for pigeon guillemots, and figures 11.3 and 11.4 show the projected changes for harbor seals.

Natural resource management has become increasingly multijurisdictional and interdisciplinary over the last quarter century and, as a result, is extremely complex. Diverse groups representing a wide spectrum of interests are becoming increasingly

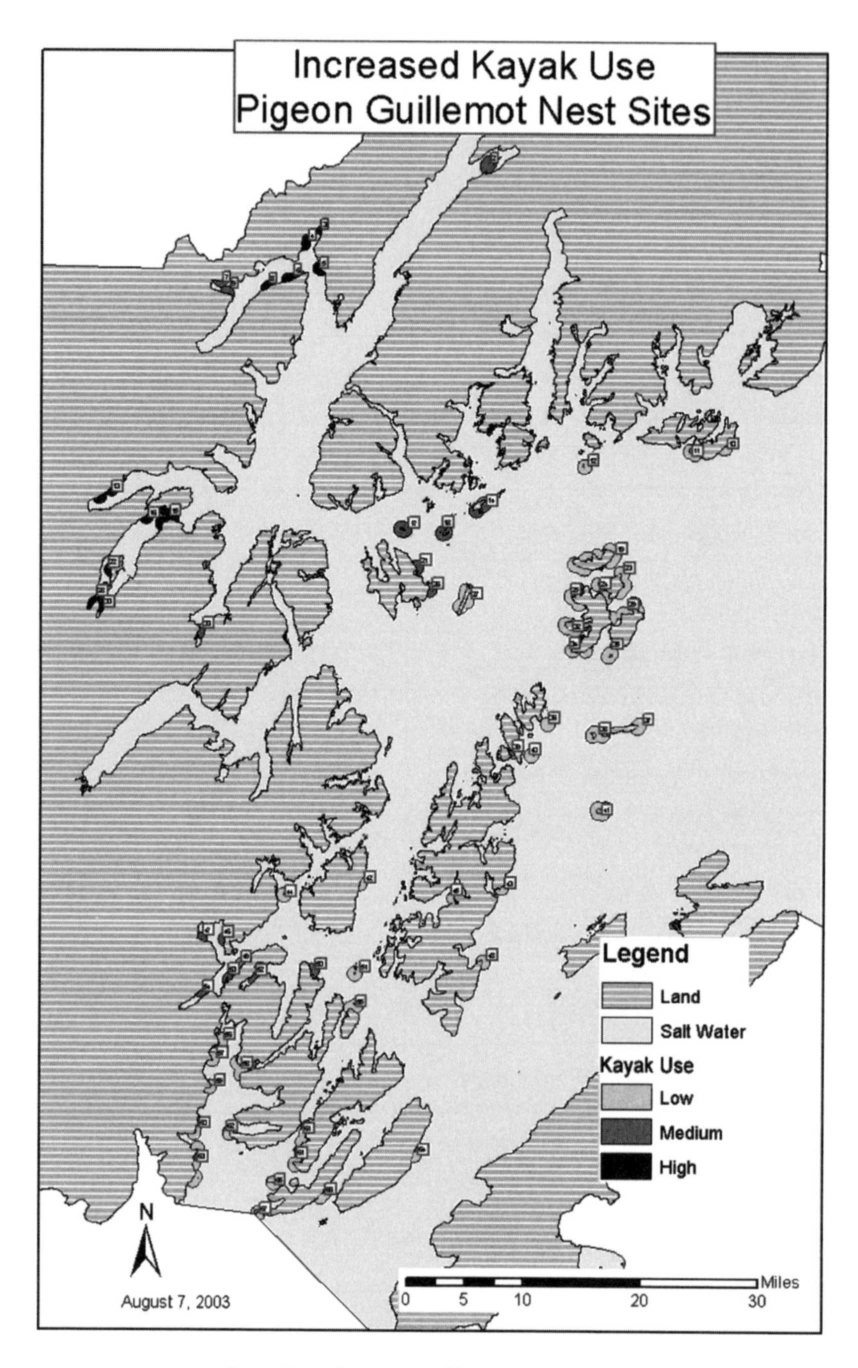

FIGURE 11.1. Location of pigeon guillemot nest sites and magnitude and location of predicted kayak use levels in 2015 within 1,000 m of nest sites in western Prince William Sound, Alaska.

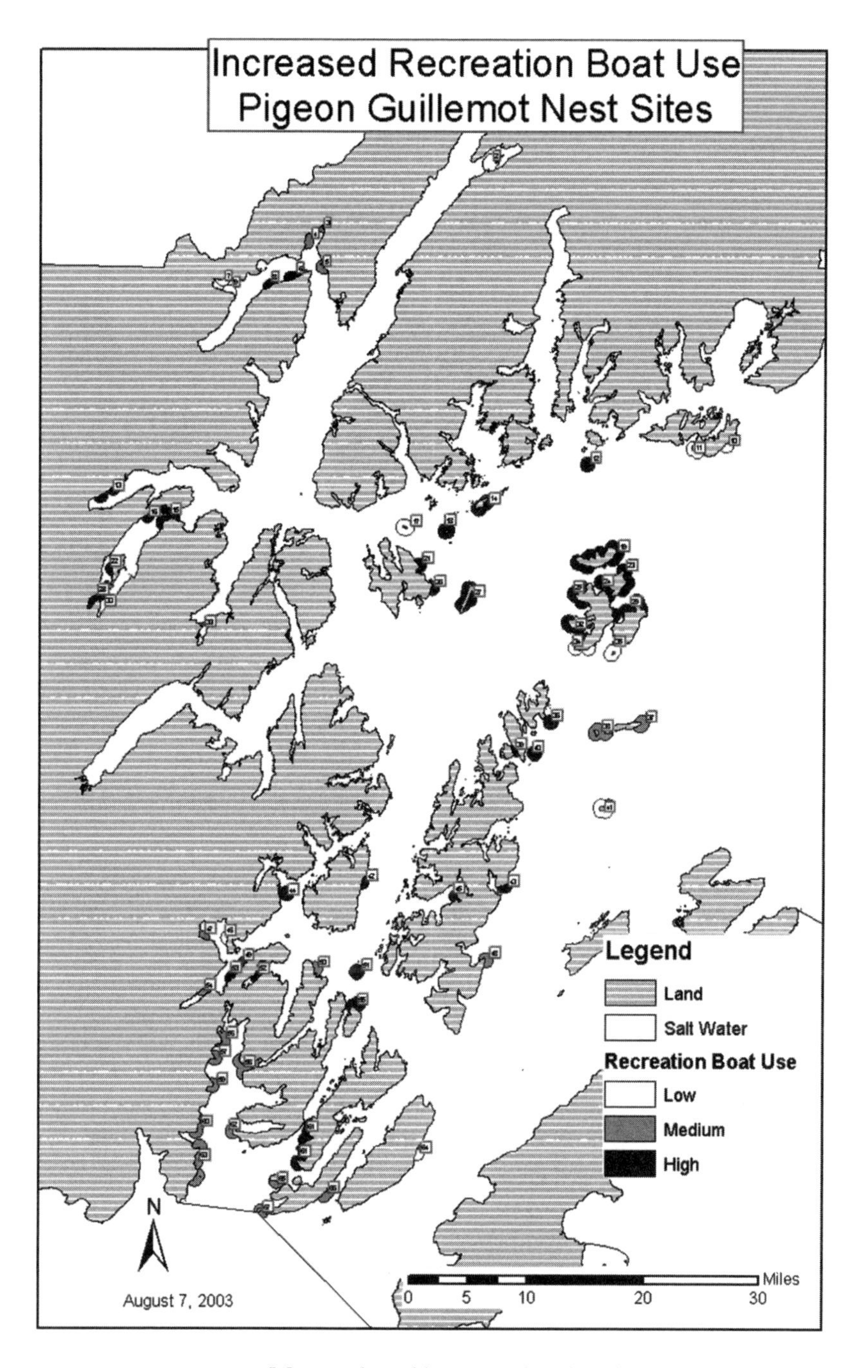

FIGURE 11.2. Magnitude and location of predicted motorized recreation boat use levels in 2015 within 1,000 m of pigeon guillemot nest sites in western Prince William Sound, Alaska.

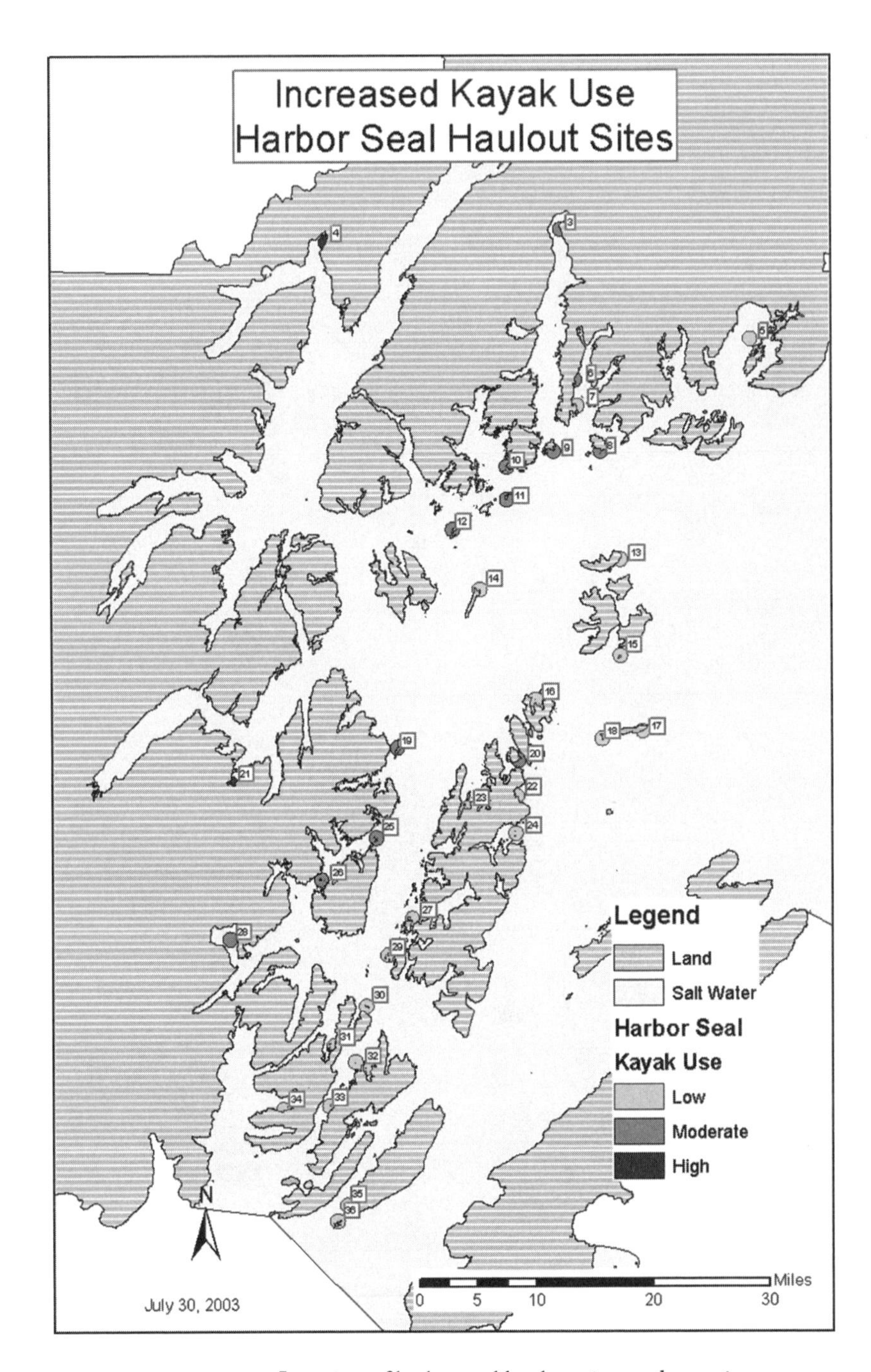

FIGURE 11.3. Location of harbor seal haulout sites and magnitude and location of predicted increase in kayak use in 2015 in the vicinity of haulout sites in western Prince William Sound.

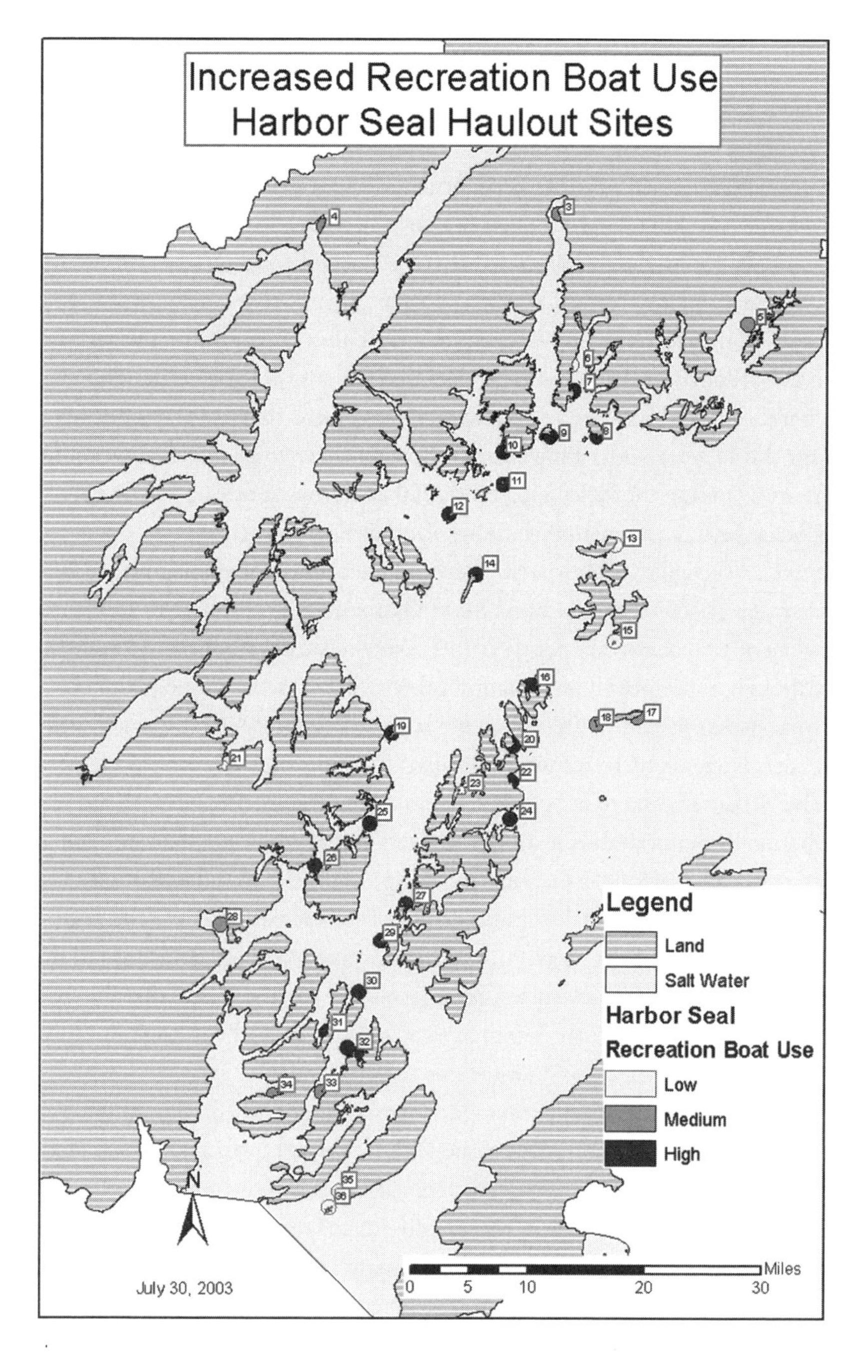

FIGURE 11.4. Magnitude and location of predicted increase in motorized recreation boat use in 2015 in the vicinity of harbor seal haulout sites in western Prince William Sound.

involved in influencing state and federal agencies' management of natural lands and wildlife. Adding to the complexity, many businesses market Alaska's scenery and wildlife to their customers. The challenge for managers is to provide opportunities for both commercial and recreational use of the environment without causing irreparable harm to wildlife resources. This is particularly important for species injured by the *Exxon Valdez* oil spill (chapter 4) that may not be resilient to further changes in their environment. The tradeoffs associated with decisions regarding levels of use by various user groups in the western Sound are often difficult to quantify. Management to encourage higher levels of use will help to ensure that large numbers of visitors have access to outdoor recreation resources and that vendors are able to develop and conduct traditional and innovative commercial operations that are economically viable (Brooks and Haynes 2001). However, choices may have to be made between limiting visitor use to ensure the well-being of injured and other species, and allowing higher levels of use (e.g., Lawson and Manning 2001) at some locations.

In 2004, we made a few specific recommendations for managing wildlife and human interactions in the Sound. Our study found that distance to campsites and upland recreation sites consistently helped explain the distribution of kayakers from May through September. Construction of new facilities should be planned in areas that offer access to other attractions for kayakers (e.g., glaciers) but are well away from areas with sensitive resources present. Existing sites that are in proximity to sensitive sites may need to be removed or closed during specific times of the year. The distribution of motorized recreation boat users was often explained by opportunities for sportfishing, black bear hunting in the spring, Sitka black-tailed deer hunting in late summer and fall, anchor buoys, and upland recreation sites. The response of motorized recreation boat users to these factors also offers opportunities to manage their distribution in areas with sensitive resources. Furthermore, the placement of anchor buoys may have great potential for directing motorboat users to attractions for this user group that are well away from sensitive areas.

Finally, we concluded that because land management jurisdiction in the Sound is so complex, public education may be one of the strongest tools available to managers. In 2004, we recommended that specific education materials be developed for distribution to recreational boaters in the Sound, identifying the situations and general habitats that should be avoided to minimize disturbance to sensitive wildlife. We also recommended education programs be developed and delivered to ecotourism guides and water taxi operators to ensure their operations do not result in increased disturbance. We felt these types of education efforts may be particularly effective since they can be concentrated at the origin of most use, in Whittier.

The opening of the Whittier Tunnel provided a strong impetus to plan for the future of the western Sound and changes in human use patterns. In order to make informed decisions about how to respond to or manage change in human use, it is important to have an understanding of current human use patterns, predicted future use patterns, and their potential impact on injured resources. This study was the first spatially explicit effort attempting to make such an assessment in the Sound region. Other studies exploring the specific dynamics of human use and wildlife will almost certainly be needed in order for managers to strategically identify those areas and species or habitats of greatest concern.

CHAPTER 12

CHARACTERIZING THE EFFECTS OF INCREASED RECREATION USE ON WILDLIFE WITHIN THE SOUND

LOWELL H. SURING AND AARON J. POE

SOUND BITES

THERE ARE RELATIVELY few areas of high-intensity recreational boating overlapping with species assemblages prone to disturbance by humans that were also affected by the *Exxon Valdez* oil spill.

Managers attempting to address recreation disturbance of species or habitats need spatially explicit characterizations of human use intensity equivalent in rigor to those traditionally used by ecologists to characterize species distribution or habitat use.

Efforts aiming to evaluate risk of disturbance for species and habitats are most useful to managers when they take a system-wide approach that returns spatially explicit results.

INTRODUCTION

Natural resource managers are becoming increasingly concerned about impacts of recreation and other human uses on wildlife (Knight and Gutzwiller 1995) because recreation has been identified as a major reason for the decline of species at risk (Losos et al. 1995; Czech et al. 2000) and because the number of people taking part in nature-based and wilderness recreation activities is increasing (Cordell et al. 2008). Wildlife viewing is one of the key recreational activities that take place on public lands, and this activity has grown in popularity in recent years (Cordell et al. 2008). Boyle and Samson (1985) and Knight and Cole (1995) have suggested that recreation may affect behavior and fitness of wildlife and have marked effects on individuals, populations, and communities. This is especially relevant in Prince William Sound, where wildlife and other sensitive resources are still recovering from the 1989 *Exxon Valdez* oil spill (EVOS) (EVOS Trustee Council 2010). To manage for coexistence between wildlife and human uses in Prince William Sound, particularly recreation, managers need to be aware of the potential consequences of recreation for wildlife. Competing uses (e.g., sightseeing and wildlife viewing) are likely to converge in desirable areas, resulting in potential detrimental effects upon each use and creating a larger cumulative impact on the associated resources.

For management of sustainable human use and sensitive resources (e.g., recovered and recovering EVOS-injured resources) in Prince William Sound, it is critical that the location, timing, and nature of human use intersecting with wildlife concentrations be well understood. Human use "hot spots," such as those described in chapter 6, are important areas in the Sound, where human use is more concentrated relative to surrounding areas. Human use is not evenly distributed in Prince William Sound (see chapters 6, 7, and 11). Particular locations are more desirable as water-based destinations for a variety of reasons including proximity to communities, presence of scenic glaciers and postglacial landscapes, availability of landing and camping areas, protected anchorages, sportfishing streams, cabins and other recreation infrastructure, and increased opportunities to harvest wildlife (Murphy et al. 2004, chapter 11). Our objective was to describe cumulative concentrations of 20 species and habitats previously injured by the oil spill, relative to concentrations of human use, and then to evaluate relative risk based on the literature, thus providing recommendations leading to management of sustainable human use and sensitive resources.

METHODS

Prince William Sound is a 160-km-wide embayment in Southcentral Alaska that extends 150 km inland from the Gulf of Alaska. Numerous bays, fjords, and inlets form over 2,900 km of mainland coastline encircling the Sound, and an additional 1,900 km of shoreline are associated with its many islands. The study area included the water and adjacent lands of Prince William Sound, bounded by Blying Sound and the Gulf of Alaska to the south and the Chugach Mountains to the north. We used the same General Areas delineated across Prince William Sound by Poe and Greenwood in chapter 6 as a midlevel spatial unit useful for management of recreation use and sensitive resources. General Areas were defined mainly by named geographic features, such as bays, as shown on standard USGS topographic quads. The resulting 568 General Areas varied widely in size, ranging from 1.2 to 80,900 ha (mean of 3,600 ha).

These units allowed us to investigate overlap at the tidal interface and nearshore habitats, which are the most heavily used in the Sound (see chapters 6 and 7) and have high values as habitat but are also particularly vulnerable to local and global human stressors, including climate change (Lubchenco et al. 1993; North Pacific Fishery Management Council 2002; Johnson et al. 2003). Therefore, a good understanding of how EVOS-injured resources in these areas may be affected by increasing human use is needed to help managers maintain these resources. The intent was to create analysis units large enough for ease of display and small enough to capture the spatial distribution of various landscape attributes, including features of human use and density of sensitive resources.

SENSITIVE RESOURCES

Several species of wildlife and habitat types are still recovering from the short- and long-term persistent effects of the 1989 *Exxon Valdez* oil spill (Peterson et al. 2003; EVOS Trustee Council 2010). The identified resources injured as a result of the oil spill provided a basis for evaluation in this analysis. All EVOS-injured resources were evaluated if they had comprehensive spatial databases describing distribution (table 12.1). Of the 14 species classified as threatened or endangered under the United States Endangered Species Act (Public Law 93-205) in Alaska (NOAA 2010b), only the Steller's sea lion occurred in Prince William Sound and had a comprehensive spatial database describing its distribution and abundance. Extensive survey and inventory work has been completed on the distribution and species composition of seabird colonies in Prince William Sound (USDI Fish and Wildlife Service 2004). These resources are particularly vulnerable to disturbance from human activities (Carney and Sydeman 1999). Considering these factors, they were also selected for analysis.

MODELED DENSITY INDEX

Spatial databases designed for use with a geographic information system (GIS) that described the distribution of individual breeding sites or colonies were available for all selected resources. These data represented the best available information at the time of this analysis. Predictions made on the basis of these data did not necessarily represent the total distribution of each species but rather represented a sampling of their distribution in Prince William Sound. The databases provided either point or linear locations of these sites. These data were used to develop an index that described the relative density in an area for each resource (i.e., density index). A kernel density interpolation that estimated distribution and density was created for each resource using these databases and ArcGIS Spatial Analyst 9.3 (ESRI 2008). We used a search radius of 7,000 m (the default value when the entire Sound was considered) and a 100 m x 100 m cell size (to correspond with a database of projected recreation density described in chapter 7). The kernel density algorithm provided an option for weighting the analysis for each species and habitat relative to both the species recovery status from the oil spill and its documented sensitivity to recreation activity.

TABLE 12.1. Sensitive resources identified in Prince William Sound, Alaska, with components of risk and associated weighting factors

SENSITIVE RESOURCE	CURRENT (2010) STATUS OF EVOS-INJURED RESOURCES[a]	THREAT FROM DISTURBANCE OR MORTALITY RESULTING FROM RECREATION OR COMMERCIAL ACTIVITIES		WEIGHTING FACTOR[c]
		LEVEL OF THREAT[b]	SOURCES	
Common loon (*Gavia immer*)	1	2	Caron and Robinson 1994; Heimberger et al. 1983; Kaplan 2003; Newbrey et al. 2005; Ream 1976; Ruggles 1994; Strong and Bissonette 1989; Titus and VanDruff 1981; Vermeer 1973.	2
Cormorant (double-crested) [*Phalacrocorax auritus*] and pelagic [*P. pelagicus*]	1	1	DesGranges and Reed 1981; Ellison and Cleary 1978; Henny et al. 1989; Kury and Gochfeld 1975.	1
Harlequin duck (*Histrionicus histrionicus*)	3	2[d]	Esler et al. 2000; Gilliland et al. 2002; Goudie 2006; Goudie and Jones 2004; MacCallum 2001; Rizzolo et al. 2005.	3
Bald eagle (*Haliaeetus leucocephalus*)	1	3	Anthony and Isaacs 1989; Becker 2002; Brown and Stevens 1997; Buehler et al. 1991a; Buehler et al. 1991b; Craig et al. 1988; Fletcher et al. 1999; Fraser et al. 1985; Grubb and King 1991; Grubb et al. 1992; Mathisen 1968; McGarigal et al. 1991; Montopoli and Anderson 1991; Rodgers and Schwikert 2003; Schirato and Parsen 2006; Stalmaster and Kaiser 1998; Stalmaster and Newman 1978; Steidl and Anthony 1996; Steidl and Anthony 2000; Watson 1993; Wood et al. 1989; Wood et al. 1998.	2
Black oystercatcher (*Haematopus bachmani*)	3	2	Andres 1999; Arimitsu et al. 2005; Golumbia et al. 2009; Irons et al. 2000; Lance et al. 2001; Lindberg et al. 1998; Morse et al. 2006; Murphy and Mabee 2000; Poe et al. 2009; Spiegel 2008.	2
Kittlitz's murrelet (*Brachyramphus brevirostris*)	4	2[d]	Agness 2006; Agness et al. 2008.	3
Marbled murrelet (*B. marmoratus*)	4	2	Bellefleur et al. 2009; Burger 1997; Burger 2002; Carter and Sealy 1984; Carter et al. 1995; Fry 1995; Hébert and Golightly 2006; Long and Ralph 1998; McShane et al. 2004; Speckman et al. 2004.	3
Pigeon guillemot (*Cepphus columba*)	5	2[d]	Ewins 1993; Nelson 1987.	3
Common murre (*Uria aalge*)	1	1	Beale and Monaghan 2004; McChesney et al. 2007; Parrish 1997; Rojek et al. 2007; Thayer et al. 1999.	1
Seabird colonies	0	3	Arimitsu et al. 2007; Carney and Sydeman 1999; Erwin 1989.	2
Sea otter (*Enhydra lutris*)	3	2	Curland 1997; Garshelis and Garshelis 1984; Lance et al. 2004; Sea Otter Recovery Team 2007.	2

continued

TABLE 12.1. *(continued)*

SENSITIVE RESOURCE	CURRENT (2010) STATUS OF EVOS-INJURED RESOURCES[a]	THREAT FROM DISTURBANCE OR MORTALITY RESULTING FROM RECREATION OR COMMERCIAL ACTIVITIES		WEIGHTING FACTOR[c]
		LEVEL OF THREAT[b]	SOURCES	
Steller's sea lion (*Eumetopias jubatus*)	0	2	Atkinson et al. 2008; Kucey and Trites 2006; Kucey 2005; Loughlin and Tagart 2006; Mathews 2000a; Mathews 2000b; NMFS 2008; Szaniszlo 2005.	2
Harbor seal (*Phoca vitulina*)	1	3	Henry and Hammill 2001; Johnson and Acevedo-Gutierrez 2007.	2
Dolly Varden (*Salvelinus malma*)	1	2	Cooke and Cowx 2006; Lewin et al. 2006.	2
Cutthroat trout (*Oncorhynchus clarkii*)	2	2	Cooke and Cowx 2006; Lewin et al. 2006.	2
Pacific herring (*Clupea pallasii*)	5	2	Jørgensen et al. 2005; Kvadsheim and Sevaldsen 2005; Misund et al. 1996; Schwarz and Greer 1984; Skaret et al. 2005; Vabø et al. 2002; Wilson and Dill 2002.	3
Salmon	1	2[d]	Cooke and Cowx 2006; Lewin et al. 2006.	2
Blue mussel (*Mytilus edulis*)	3	2[d]	Hamilton 2000.	2
Eelgrass (*Zostera marina*)	2[e]	3	Bell et al. 2001; Blaber et al. 2000; Francour et al. 1999; Hastings et al. 1995; Orth et al. 2006; Tallis et al. 2009; Walker et al. 1989.	3

[a]Status of resources injured from the effects of the *Exxon Valdez* oil spill (EVOS Trustee Council 2010). 0 = Not identified as a resource that was adversely affected by the *Exxon Valdez* oil spill. 1 = Recovered: recovery objectives have been met, and the current condition of the resource is not related to residual effects of the oil spill. 2 = Very likely recovered: while there has been limited scientific research on the recovery status of these resources in recent years, prior studies suggest that there had been substantial progress toward recovery in the decade following the spill. In addition, so much time has passed since any indications of some spill injury, including exposure to oil; it is unlikely that there are any residual effects of the spill. 3 = Recovering: recovering resources are demonstrating substantive progress toward recovery objectives, but are still adversely affected by residual impacts of the spill or are currently being exposed to lingering oil. The amount of progress and time needed to attain full recovery varies depending on the species. 4 = Recovery unknown: for resources in the unknown category, data on life history or the extent of injury from the spill is limited. Moreover, given the length of time since the spill, it is unclear if new or further research will provide information that will help in comprehensively assessing the original injury or determining the residual effects of the spill such that a better evaluation of recovery can occur. 5 = Not recovering: resources that are not recovering continue to show little or no clear improvement from injuries stemming from the oil spill. Recovery objectives have not been met.

[b]Threat of disturbance: 1 = The scientific literature is equivocal on whether disturbance or mortality from recreation or commercial activities is a threat to the resource. 2 = Empirical evidence reported in the literature indicated that the resource may respond negatively to disturbance, but population status may not be threatened because of disturbance or mortality from recreation or commercial activities or there is limited empirical evidence indicating that the resource may be threatened from disturbance or mortality from recreation or commercial activities. 3 = Substantial empirical evidence reported in the literature indicated that the resource may be threatened as a result of disturbance or mortality from recreation or commercial activities. Limited evidence has been published indicating that the resource is not threatened as a result of disturbance or mortality from recreation or commercial activities.

[c]See table 12.2 for calculation of weighting factors.

[d]Additional study is needed; life history characteristics indicate that the resource may be threatened by disturbance from recreation or commercial activities, but there is limited empirical evidence reported in the literature supporting this.

[e]Eelgrass did not have an EVOS-injured resource status assigned; the status for subtidal community was used.

As such, the species' or habitats' sensitivity was coded such that those still suffering a greater degree of injury from the spill and having greater vulnerability to disturbance resulted in a higher sensitivity, similar to the method of Haines et al. (2010) (table 12.2).

The current (2010) degree of injury from the EVOS was coded as follows:

- 0—Not identified as a resource that was adversely affected by the *Exxon Valdez* oil spill.
- 1—Recovered: recovery objectives have been met, and the current condition of the resource is not related to residual effects of the oil spill.
- 2—Very likely recovered: while there has been limited scientific research on the recovery status of these resources in recent years, prior studies suggest that there had been substantial progress toward recovery in the decade following the spill. In addition, so much time has passed since any indications of some spill injury, including exposure to oil; it is unlikely that there are any residual effects of the spill.

TABLE 12.2. Determination of weighting factors for sensitive resources in Prince William Sound, Alaska

CURRENT LEVEL OF RESOURCE INJURY FROM THE EVOS[b]	LEVEL OF THREAT FROM DISTURBANCE OR MORTALITY RESULTING FROM RECREATION OR COMMERCIAL ACTIVITIES[a]		
	1	2	3
0–1	1	2	2
2–3	2	2	3
4–5	2	3	3

[a]Values are from table 12.1.

[b]Status of resources injured from the effects of the *Exxon Valdez* oil spill (EVOS) (EVOS Trustee Council 2010): 0 = Not identified as a resource that was adversely affected by the *Exxon Valdez* oil spill. 1 = Recovered: recovery objectives have been met, and the current condition of the resource is not related to residual effects of the oil spill. 2 = Very likely recovered: while there has been limited scientific research on the recovery status of these resources in recent years, prior studies suggest that there had been substantial progress toward recovery in the decade following the spill. In addition, so much time has passed since any indications of some spill injury, including exposure to oil; it is unlikely that there are any residual effects of the spill. 3 = Recovering: recovering resources are demonstrating substantive progress toward recovery objectives but are still adversely affected by residual impacts of the spill or are currently being exposed to lingering oil. The amount of progress and time needed to attain full recovery varies depending on the species. 4 = Recovery unknown: for resources in the unknown category, data on life history or the extent of injury from the spill is limited. Moreover, given the length of time since the spill, it is unclear if new or further research will provide information that will help in comprehensively assessing the original injury or determining the residual effects of the spill such that a better evaluation of recovery can occur. 5 = Not recovering: resources that are not recovering continue to show little or no clear improvement from injuries stemming from the oil spill. Recovery objectives have not been met.

- 3—Recovering: recovering resources are demonstrating substantive progress toward recovery objectives, but are still adversely affected by residual impacts of the spill or are currently being exposed to lingering oil. The amount of progress and time needed to attain full recovery varies depending on the species.
- 4—Recovery unknown: for resources in the unknown category, data on life history or the extent of injury from the spill is limited. Moreover, given the length of time since the spill, it is unclear if new or further research will provide information that will help in comprehensively assessing the original injury or determining the residual effects of the spill such that a better evaluation of recovery can occur.
- 5—Not recovering: resources that are not recovering continue to show little or no clear improvement from injuries stemming from the oil spill. Recovery objectives have not been met.

The documented vulnerability to recreation disturbance was coded as follows:

- 1—The scientific literature is equivocal on whether disturbance or mortality from recreation or commercial activities is a threat to the resource.
- 2—Empirical evidence reported in the literature indicated that the resource may respond negatively to disturbance, but population status may not be threatened because of disturbance or mortality from recreation or commercial activities, or there is limited empirical evidence indicating that the resource may be threatened from disturbance or mortality from recreation or commercial activities.
- 3—Substantial empirical evidence reported in the literature indicated that the resource may be threatened as a result of disturbance or mortality from recreation or commercial activities. Limited evidence has been published indicating that the resource is not threatened as a result of disturbance or mortality from recreation or commercial activities.

Using Spatial Analyst (ESRI 2008), we categorized the resulting index numbers into four ordinal 25 percent quantile bins, and 0, for five classes (values 0 to 4) representing cells with progressively greater density and risk to disturbance (i.e., 0 to 1 = very low, > 1 to 2 = low, > 2 to 3 = medium, > 3 to 4 = high). This approach assumes that as density of sensitive species and habitats increases, the risk from potential disturbance increases. A cumulative density for all target species and habitats was calculated by combining cell-by-cell density index values for each sensitive resource using map algebra in ArcGIS Spatial Analyst 9.3 (ESRI 2008). The cumulative density index value was calculated for each General Area using zonal statistics in ArcGIS Spatial Analyst 9.3 (ESRI 2008).

DETERMINATION OF RECREATION INTENSITY

A database with the estimated density of recreational boat use characterized in 100 m × 100 m cells was previously developed from thousands of point locations of recreational vessels observed May through August of 2005, 2007, and 2008 (chapter 7). Using Spatial Analyst, we categorized these into four ordinal 25 percent quantile bins, and 0, for five classes (values 0 to 4) representing cells with progressively greater recreation density (i.e., 0 to 1 = very low, > 1 to 2 = low, > 2 to 3 = medium, > 3 to 4 = high). The mean recreation index value (ranging from 0 to 4) was calculated for each General Area using zonal statistics in ArcGIS Spatial Analyst 9.3 (ESRI 2008). A General Area was considered a *Primary Risk* if its cumulative mean density index for sensitive species and habitats was ≥ 30 (see figure 12.1) and its mean recreation index was ≥ 3. A General Area was considered a *Secondary Risk* if its cumulative mean density index for sensitive species and habitats was ≥ 13 to < 30 (see figure 12.1) and its mean recreation index was ≥ 3.

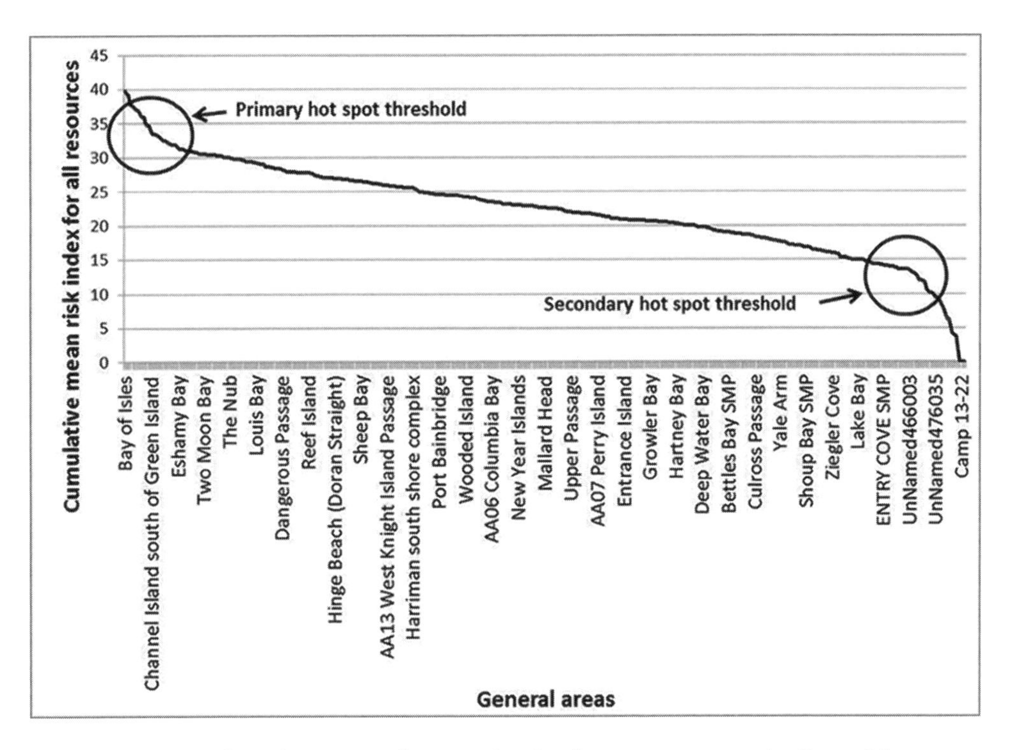

FIGURE 12.1. Cumulative mean density index for all sensitive resources by General Area and thresholds for primary and secondary hot spots in Prince William Sound, Alaska.

MANAGING RISK

Suggested risk management actions for regional managers based on classification of each General Area are given below.

For Secondary Risk:

1. Targeted outreach to appropriate user group(s) (e.g., Crawford et al. 2001).
2. Monitor disturbance events in General Areas identified as potential areas of concern (e.g., Cole 2004a).

For Primary Risk:

1. Monitor sensitive resource response to disturbance by recreationists (e.g., Steidl and Powell 2006).
2. Develop, implement, and monitor restrictions on use by recreationists in General Areas identified as potential areas of concern (e.g., Duffus and Dearden 1993).

RESULTS

OVERALL RECREATION USE

Projected centers of high-density recreation were concentrated in western and northern Prince William Sound (figure 12.2). Although the majority of the study area had General Areas with mean recreation index values that were very low or low (68 percent of the area), 32 percent of the area was moderate to high.

PRIMARY RISK AREAS

When General Areas with high total density index for all resources combined were compared with General Areas with high mean recreation density classes, 19 were identified as potential Primary Risk areas for management of interactions between sensitive resources and recreationists (figure 12.3). They were generally distributed across eastern and northern Prince William Sound and only made up 0.49 percent of the study area.

SECONDARY RISK AREAS

When General Areas with moderate total density index for all resources combined were compared with General Areas with high mean recreation density classes, 182 were identified as potential Secondary Risk areas for management of interactions between sensitive resources and recreationists (figure 12.4). They were concentrated across northwestern and northeastern Prince William Sound in the vicinities of Whittier and Valdez, with scattered areas elsewhere, and made up 13.63 percent of the study area.

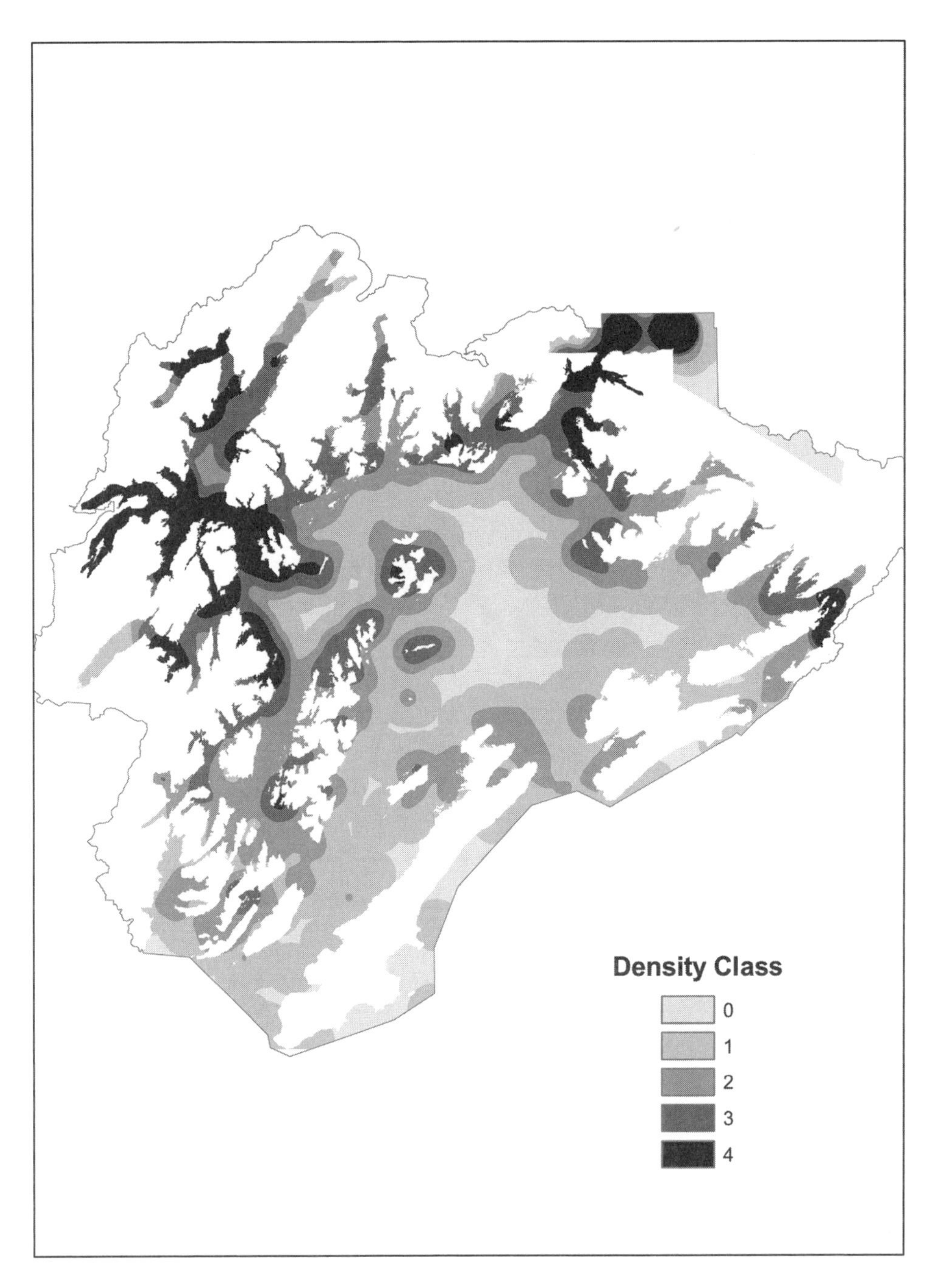

FIGURE 12.2. Distribution of density index classes for recreation in Prince William Sound, Alaska.

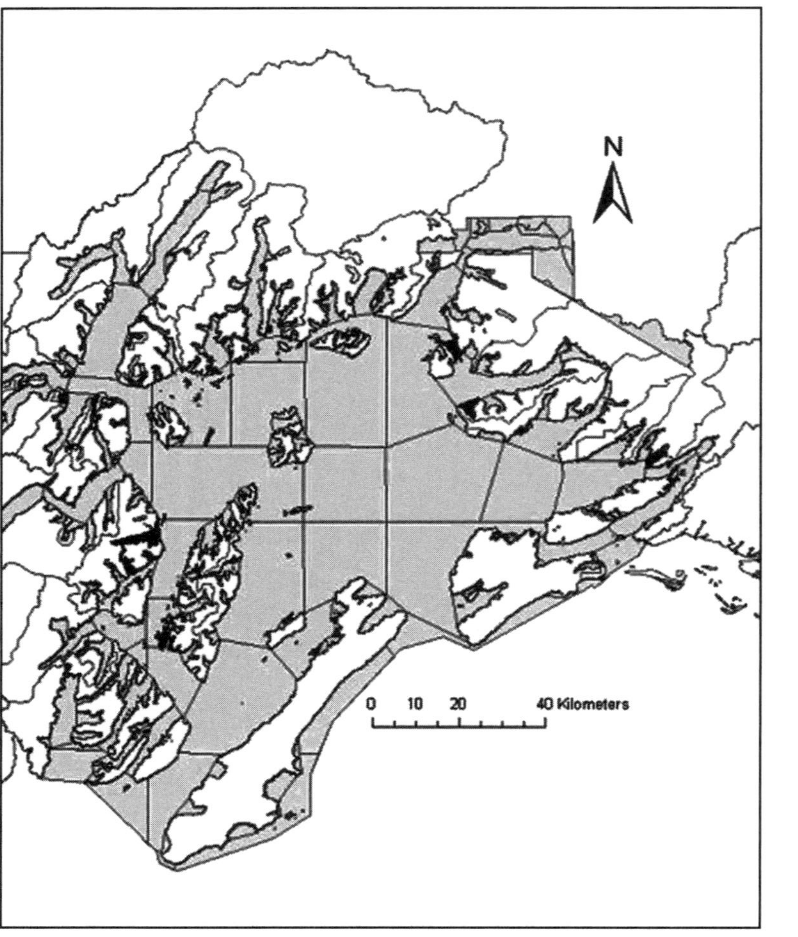

FIGURE 12.3 General Areas identified as potential primary hot spots for management of interactions between all sensitive resources and recreationists.

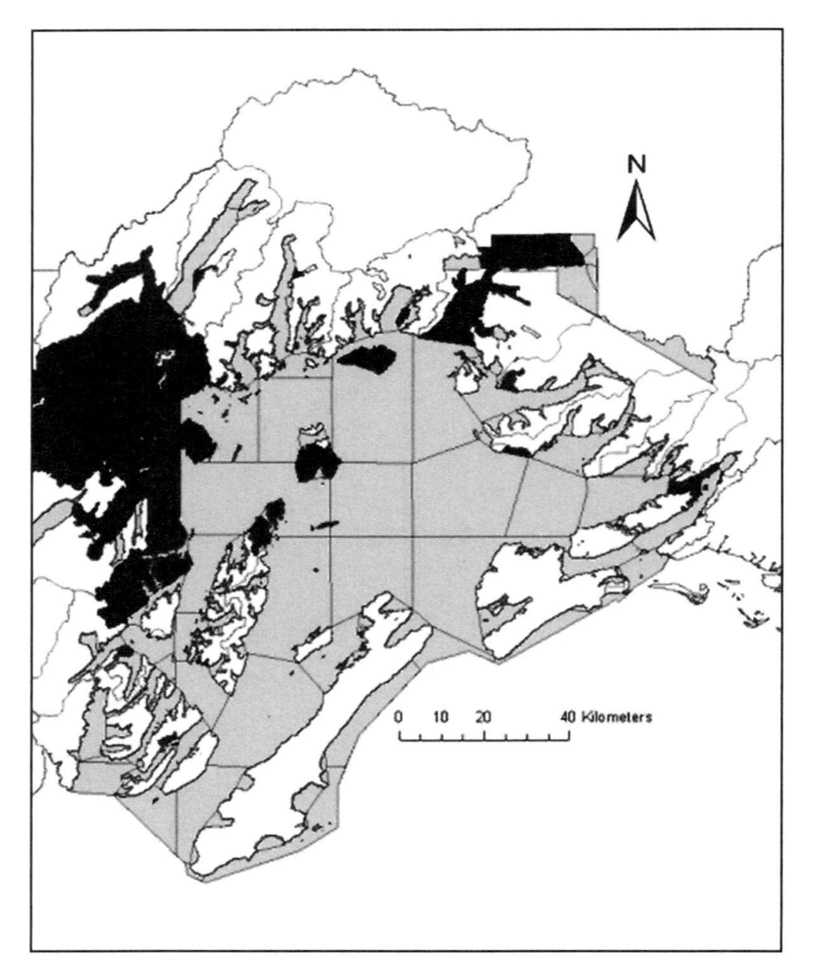

FIGURE 12.4. General Areas identified as potential secondary hot spots for management of interactions between all sensitive resources and recreationists.

DISCUSSION

The 19 General Areas that were identified as potential Primary Risk areas for management of interactions between sensitive resources and recreationists (figure 12.3) represent the most essential areas for development and implementation of innovative approaches to conserving sensitive resources. These are the areas with the most critical needs for onsite management of recreation uses and should be a priority for funding and staffing. These Primary Risk areas make up less than 1 percent of the area of Prince William Sound and offer the greatest opportunity to ensure that increasing recreation use does not further impact sensitive resources. As a first step, regional managers should invest in efforts to empirically measure the potential disturbance to those species and habitats identified as sensitive by this analysis (Management Category 3). In order to be useful, this analysis would have to demonstrate the mechanism for actual harm occurring from the recreation activity. Consideration should also be given to which active management tactics might best serve for redirecting detrimental recreation activities away from these areas (Management Category 4). Such action would almost certainly have to be implemented though a combined effort from a consortium of managers and would likely only come to fruition if detrimental effects of recreation could be adequately defined.

The 182 General Areas that were identified as potential Secondary Risk for management of interactions between sensitive resources and recreationists (figure 12.4) provide opportunities for identifying where ongoing management activities (e.g., education, monitoring of disturbance, enforcement) should be focused (Management Categories 1 and 2). There are also likely to be situations where it will be necessary to monitor sensitive resource response to disturbance by recreationists to develop and evaluate appropriate management approaches (Management Category 3). These sites, where recreationists overlap with sensitive resources, also offer opportunities to engage users in a dialogue to elevate awareness of these species and habitats and their role in the Prince William Sound ecosystem.

This analysis provided insights to where the increasing human use of Prince William Sound may potentially impact sensitive resources. The projected interactions may put these resources at risk to current threats to their well-being and may also impede their recovery from previous perturbations like the oil spill. These results offer an opportunity to focus development and implementation management actions in areas where the highest levels of human use are meeting the most sensitive species and habitats.

LESSONS LEARNED

When assessing relative risk of species and habitats to disturbance from human activities, this analysis underscores the importance of making the first step an inquiry that is system wide in scope. This effort attempts to look at species distribution data such that risk is not addressed for a single species but rather for groups of species interacting with recreation use. Similarly, it addresses the question of risk from a regional perspective, offering managers the ability to evaluate risk to species/habitats across Prince William Sound. This scale of analysis provides researchers with the information needed to plan future investigations of impacts to species/habitats along a gradient of potential disturbance. It also allows managers to make more strategic choices about investing in potential mitigation efforts.

A regional evaluation also forces managers to look beyond their individual jurisdictions for potential solutions to reducing impacts to sensitive species. By making obvious the need for regional collaboration on conservation objectives, this analysis promotes a key tenet of sustainable recreation management. Addressing the concern of increasing recreation impacts for a broad suite of species versus single species also allows managers to draw support from their individual agency mandates into an integrated analysis or management proposal. For example, the U.S. Forest Service, concerned about kayak camping impacts to black oystercatchers (chapter 15), joins forces with the U.S. Fish and Wildlife Service and Alaska Department of Fish and Game, concerned about potential disturbance of sea ducks in nearshore areas of the Sound (chapter 14).

This work, alongside other efforts in chapters 6, 7, and 9, demonstrates the utility of having spatially explicit data that defines relative intensities of human use. Too often, in the absence of these specifics, the ability to make critical assessments of relative risk is difficult. Many managers are left addressing the issue based only on literature findings, suggesting that overlap will equal some level of ubiquitous impact. The approach of rigorously defining human use intensity (as an ecologist might describe species density) affords a better result where thresholds of overlap (e.g., high, medium, low) can be developed to categorize the degree of the interaction. This allows managers to conduct a true risk assessment for sensitive species.

Another benefit of this approach is that it allows managers to look at how to best invest in outreach efforts to make recreation users aware of the presence of species and habitats that may be impacted by recreation activity. By having a specific spatial characterization of where recreationists and species have the highest levels of overlap, outreach campaigns can be targeted on those specific areas and species. For example, in the Sound, where the majority of recreation use occurs in the northwestern bays and fjords, managers can target outreach efforts in Whittier, the port most frequently used to access those areas.

STAKEHOLDER ESSAY

CAN PRINCE WILLIAM SOUND BE LOVED TO DEATH?

GERRY SANGER

IS IT POSSIBLE TO LOVE a place so much that what you do there imperils the very values that brought you there in the first place? What if there are so many like-minded people using the same place, that together you do just that, even if unawares? These are questions I have pondered for many years, as a working biologist, as a parent exploring the Sound by inflatable in the 1980s with my three young sons, and most recently, as a nature tour guide in Prince William Sound.

I was operating a Whittier water taxi and sightseeing business on the morning of March 24, 1989, when I was jolted by the morning news that shocked and angered Alaskans and people from around the world. That infamous event, the *Exxon Valdez* oil spill, soon became known by the acronym EVOS.

By the end of the 1990 season, the business was struggling, and when my fifth charter season ended the next year, it was clear that the business was failing for lack of a boat bigger than the four-passenger dory I was then using. A series of temporary EVOS-related jobs opened up in succession soon afterwards. The Whittier Oil Spill Response Office hired me for October 1991 to help with beach surveys looking for oil lingering from the spill, and later that winter, the Forest Service needed a wildlife planner on an interdisciplinary team that was rewriting the outdated management plan for Prince William Sound.

That yearlong Forest Service project was followed the next two years by jobs on more EVOS seabird and wildlife projects that greatly expanded my knowledge of the Sound's wildlife. I returned to chartering in summer 1995 with the old 21-footer and a new appreciation for the Sound and its wildlife resources that I could share with my charter clients.

I had also built up enough of a nest egg to have a bigger, six-passenger boat built that fall. With more publicity from guidebooks and a website, the company grew. By 2003, I was focusing the business more on nature tours than on hauling kayakers. I was well into my 60s then, and a slower pace of running just one tour a day allowed me to spend more time to "smell the roses" right along with my sightseeing clients. It was rewarding just being able to spend more time watching and photographing the Sound's wildlife.

The adjacent chapter considers the odds that increasing numbers of people using the Sound will adversely affect marine wildlife, and it offers suggestions for managers to better understand if impacts are actually happening. It examines how patterns of human use in the Sound overlap with the distribution of known wildlife that was harmed by the spill. A couple of key examples include harbor seals and pigeon guillemots. Harbor seals are charismatic, watchable wildlife, often seen from boats of all types when the seals are hauled out on isolated islands and rocky headlands or icebergs: areas visited infrequently by people on land due to their inaccessibility. Harbor seals have been officially declared recovered by the EVOS Trustee Council.

Pigeon guillemots (a seabird related to puffins), on the other hand, are little noticed by most visitors to the Sound. Regardless, they are a resource that was damaged by the *Exxon Valdez* oil spill and have officially remained in a "Not Recovering" status ever since. Their conservation situation is complicated by the fact that their population in the Sound was already declining when the spill hit. All these years after the spill, how can managers realistically balance the possibilities of an ongoing "natural" population decline, possible lingering effects of the spill, and possible harmful effects of boaters?

This can be a problem for many wildlife studies, where a large-scale synthesis approach overlooks possible specific problems at individual locations. An example is the large seabird colony in Passage Canal near Whittier, where thousands of black-legged kittiwakes, as well as several glaucous-winged gulls and pigeon guillemots also nest. This colony is visited daily in summer by hundreds of tourists in boats, large and small. Boats routinely approach this colony as close as 30 to 40 meters, and often as close as, literally, just a few meters. As the principal investigator of the most recent guillemot survey of the Sound in 1993, and from watching guillemots during my nature tours for many years since, I believe a distance of 40 to 50 meters to be a realistic minimum approach distance to guillemot colonies without the birds flushing or otherwise visibly changing their behavior. In any case, it is likely that all of the birds at this regularly visited colony have become habituated to the repeated close presence of boats to at least some degree, but it is unknown how that has affected their productivity.

For harbor seals, on the other hand, 100 yards is often advised by marine wildlife guidelines in the United States as a safe minimum approach distance by boats for marine mammals. In my view, harbor seals in the Sound need a much wider berth. Even at a distance of 200 meters, they will sometimes scramble into the water at the approach of a boat. That violates the technical definition of disturbing their "sheltering," as defined by the Marine Mammal Protection Act of 1972.

Some examples of management actions that make sense to me—including beefing up education efforts for boaters, as encouraged by this report—have merit and can be accomplished easily. Sorting out and enforcing legal jurisdictions among management agencies, however, could be a real challenge. The Forest Service can also restrict human presence on land at sensitive areas like breeding colonies: for example, Jackpot Island, an important guillemot nesting colony in southern Prince William Sound. Creating regulations, and especially enforcing them for boats on the water, is not nearly as clear cut, due to complex agency jurisdiction issues beyond the scope of this essay. Nonetheless, it is likely that most boaters would respect "advisory enforcement," especially if requested by uniformed agency personnel.

Regardless of the difficulties outlined above, I believe this study helps draw attention to the impact that growing numbers of humans could have on wildlife in Prince William Sound. Both managers and the public are being made more aware of potential problems and at the very least now have a yardstick to help direct future efforts at managing interactions between people and wildlife in the Sound. Continued monitoring of human use and wildlife by the Sound's management agencies should be encouraged. However, with the present political climate to reduce federal spending, an educated cadre of concerned boaters, outfitters, and guides will ultimately be the best means of assuring that the Sound will not be "loved to death."

CHAPTER 13

SPATIAL RELATIONSHIPS BETWEEN RIVER OTTER (*LONTRA CANADENSIS*) LATRINE SITES, SHORELINE FEATURES, AND HUMAN ACTIVITY, PRINCE WILLIAM SOUND, ALASKA

JESSICA B. FRAVER

SOUND BITES

RIVER OTTERS AND CAMPERS appear to prefer different shoreline habitat types and therefore may not have large-scale interactions with each other on the shores of the Sound.

Studies of disturbance need to go beyond simple overlap analyses and should attempt to address a range of variables to understand potential interactions between sensitive species and recreation activity.

Understanding potential recreation impacts to a harvested species that is wide ranging and capable of taking advantage of seasonal food resources results in complex questions for managers.

INTRODUCTION

Small-scale human activities, such as regulated harvest, commercial fishing, and recreation, are common along sheltered, remote Alaskan shorelines and may influence otter abundance and behavior, but these relationships are poorly understood (Kruuk 2006; Ben-David and Golden 2007a). River otters (*Lontra canadensis*) in the Sound spend most of their time foraging for intertidal, subtidal, and pelagic fish within 5 m of the shoreline and rarely extend farther than 20 m inland from the high tide mark (Larsen 1983; Larsen 1984; Woolington 1984; Blundell et al. 2002a). Similar to river otters, most human use in the Sound occurs on and near the shoreline (Twardock and Monz 2000).

Growing pressures from recreationists may increase the frequency and intensity of interactions between river otters and people. Yet, as with any evaluations of overlap between species and human activity, there are a number of complex questions that managers should try to answer before assumptions about impact are made. As in chapter 12, understanding the scope of the overlap at a broad spatial and seasonal scale is important—impacts from a few sites may not necessarily translate into widespread effects on well-distributed species. There can also be other stressors for species, including direct-harvest or ecosystem-level changes that complicate interpretations of disturbance effects. Finally, individual populations within species can also demonstrate varying levels of habituation to human activity (e.g., Golden 1996), and having a clear understanding of the specific type and magnitude of human activity species are exposed to is also important.

Previous research demonstrates both the sensitivity and resilience of river otters to recreational activity. Variables associated with recreation were not significant in models for predicting river otter occurrence in Spain (Barbosa et al. 2001). River otters in Idaho appear to tolerate human presence at sites with sufficient cover (Melquist and Hornocker 1983). In British Columbia, Canada, however, human pressures associated with the development and subsequent use of a new coastal hiking trail resulted in abandonment of latrine sites that were accessible from the trail (Giere and Eastman 2000).

Many species of mammals and birds being studied for disturbance effects are also directly harvested for food, trophies, or pelts. This likely has direct linkages to their individual disturbance responses to humans, but also complicates assessment of human impacts. Questions about the results of human disturbance of otters are complicated by the fact that trapping and hunting of otters is permitted in the Sound. Harvest impacts on populations of river otter increase and decrease with trapper/hunter interest, which is largely based on fluctuations in pelt prices (Melquist and Dronkert 1987; Chilelli et al. 1996; Golden 1996; Scognamillo 2005). To prevent overharvesting, managers must be able to predict sustainable yields and optimal harvest rates, but critical information about the impacts of harvest levels on otter populations is often lacking (Wolfe and Chapman 1987; Golden 1996; Kruuk 2006). During the years of this analysis the harvest season for river otter in Unit 6 was November 10 through March 31 with no bag limit.

Further complexity in interpreting human effects on river otters relates to commercial fishing activity. Dense concentrations of boats and shore use near hatcheries may deter river otters from foraging and terrestrial sites. Furthermore, nets and crustacean traps can contribute to river otter mortality (Mowbray et al. 1979; Melquist and Dronkert 1987; Polechla 1990). On the other hand, hatcheries and other fisheries enhancements may also be a rich source of food. For example, the potential for river otters to become a nuisance at hatcheries has been documented in some regions (Melquist et al. 2003; Sales-Luis et al. 2009). There is widespread commercial fishing activity in the Sound (see chapter 6), and there are five commercial salmon hatcheries operating in the Sound.

Coastal river otters favor prominent rocky points and tend to avoid large tidal flats and vertical cliffs, but they use sites in bays that include gravel beaches (Larsen 1983; Woolington 1984; Bowyer et al. 1995; Blundell et al. 2002b; Ben-David and Golden 2007b). In the Sound, little is known about the characteristics of these beach sites used by otters and whether or not human presence and disturbance at these locations impact otter use. As a first step to understanding potential disturbance, this study investigates the potential overlap in shoreline characteristics selected by campers and river otters. It also explores the relationships between their use of sites relative to commercial fishing activity, harvest rates, and shore use by recreationists at the localized site scale. It compares habitat features and impact levels of campsites used by river otters with other campsites. Finally, it specifies areas where lack of data merits further research and discusses data limitations for making this type of assessment of disturbance.

STUDY AREA

Prince William Sound is located in Southcentral Alaska at 61° north, 148° west. The Chugach and Kenai mountain ranges separate most of the Sound from Interior Alaska. Two large islands, Montague and Hinchinbrook, shelter the hundreds of bays and islands of the Sound from the Gulf of Alaska. The study area covers approximately 1,030 km (15 percent) of the Sound shoreline that had been previously surveyed for otter latrine sites by researchers and Alaska Department of Fish and Game (ADF&G).

The maritime climate of the Sound is characterized by heavy annual precipitation, much of which falls in the form of snow during long winters. Summers are generally cool and wet (Bowyer et al. 1995). Shorelines are exposed to large fluctuations in tide (+6 m to -1 m) and different levels of wave action. Terrestrial vegetation begins within 1 to 2 m of the high tide line and is dominated by old-growth hemlock (*Tsuga heterophylla*) and Sitka spruce (*Picea sitchensis*) forests (Larsen 1983; Bowyer et al. 1995).

Chugach National Forest manages most (approximately 85 percent) of the Sound's uplands, including the 855,000-hectare Nellie Juan Wilderness Study Area (Twardock and Monz 2000). The State of Alaska manages activity on the water and approximately 6 percent of the land in the Sound. Chenega, Tatitlek, Eyak, and Chugach Alaska Native corporations and private citizens own the remainder of the land. Most day use, backcountry camping, recreational boating, and commercial fishing occurs from mid-June through August.

METHODS

RIVER OTTER DATA

River otter latrine sites are specific shore locations used repeatedly by river otters for resting, grooming, drinking, feeding on large prey items, and intraspecific communication (Melquist and Hornocker 1983; Rostain et al. 2004; Ben-David et al. 2005). Managers and researchers commonly use indirect observations of river otters, such as latrine site surveys, to determine otter distributions and habitat preferences (Melquist and Hornocker 1983; Chabreck et al. 1985; Kruuk 1992; Bowyer et al. 1995; Swimley et al. 1998). Spatially referenced data on latrine site locations and biological and physical characteristics of the shoreline may be cost-effective data sources for determining habitat availability for river otter in Alaska.

Researchers have reported local differences in latrine site densities in Prince William Sound, which indicates variation in habitat availability for river otters (Ben-David and Golden 2007a). Variability in latrine site densities in an environment that is largely undisturbed by large-scale, onshore human activities provides an opportunity to study how the presence and proportions of specific shoreline attributes and exposure to different levels of small-scale human activities influence latrine site distributions.

ADF&G compiled latrine site locations from surveys conducted during various years between 1996 and 2008. I obtained geospatial data (i.e., Environmental Systems Research Institute shapefiles) from ADF&G in March 2009 that represented the locations of 647 latrine sites and surveyed shorelines. Researchers and ADF&G employees digitized the survey lines and some of the site locations from points recorded on field maps (1:63,000) and marked others directly with handheld global positioning system (GPS) units (Bowyer et al. 1994; Ben-David and Golden 2007b).

CAMPSITE ASSESSMENTS

From 2000 to 2008, Chugach National Forest rangers surveyed approximately 80 percent of the shoreline in the Sound for signs of human use and produced a dataset of 425 shoreline campsites (discussed in chapter 6). When Forest Service staff found or revisited a campsite, they marked the location with a handheld Garmin GPS unit. They assigned each campsite an impact level based on the overall size of the site, the amount of tree damage, and counts and conditions of tent pads, trails, trash, and fire rings (similar to methods described in chapter 17). Impact levels ranged from 0, a suspected or old campsite with regrowth of understory vegetation, to 3, a high-impact site (Neary et al. 2003). Most of the impact levels used in analyses were from surveys conducted after 2006. High-impact campsites were characterized by extensive trail systems linking large bare-soil sleeping and social areas, structures ranging from a small bench to a cabin, significant tree damage, piles of trash, and/or fire rings with deep ash deposits. In contrast, a low-impact campsite (level 1) might have a small tent pad with minimal disturbance to organic litter and surrounding vegetation and a few pieces of trash.

For this study, 165 shoreline campsite locations were surveyed from June through August 2008. An ADF&G protocol was adopted for assessing the habitat features of latrine sites to measure topographical, terrestrial vegetation, and intertidal substrate characteristics within a 10 m radius around the point of major entry to each campsite (Bowyer et al. 1995; Ben-David and Golden 2007a; table 13.1). An obvious trailhead leading to upland vegetation is a clear sign of latrine sites and campsites comprised at least partially of nondurable terrain (i.e., vegetation). All scat found within the full extent of a site was counted. A latrine site was considered active if it contained at least 10 scats (Ben-David and Golden 2007b).

Exposure was a subjective evaluation based on estimated length of open water (fetch) to which the site was exposed. Vegetated (above high tide direction) and intertidal (below high tide direction) slopes were measured with a clinometer and recorded in intervals of 5 degrees. Canopy and understory cover was estimated and recorded in intervals of 5 percent. The proportion of each substrate type that made up the intertidal slope was visually estimated and ranked (0 = < 25 percent, 1 = 25 percent, 2 = 50 percent, 3 = 75 percent, 4 = 100 percent). The proportion of canopy cover made up of old growth trees was similarly estimated and assigned a rank of 0 to 4. Burrows were natural cavities formed under the roots of large trees, decaying snags, or boulders. The burrow data was reclassified into a variable that indicated burrow potential (0 = no sites, 1 = 1 to 3 sites, 2 = 4+ sites). The presence and type of freshwater found within the extent of the campsite was also recorded.

SHOREZONE

The National Oceanic and Atmospheric Administration (NOAA) designed ShoreZone for the purpose of mapping and classifying geomorphic and biological characteristics of coastlines (Harney et al. 2008). ShoreZone protocol involved low-altitude (oblique) aerial surveys carried out during the lowest daylight tides of the year. Spatially referenced digital video and still photographs, as well as data on habitat features, of 45,000 km of Alaska coastline can be accessed online at https://alaskafisheries.noaa.gov/habitat/shorezone. Biologists and geologists conducted surveys of Prince William Sound during the summers of 2004 and 2007 and completed mapping and classification of habitat features in 2008. Intertidal and subtidal biota (called

TABLE 13.1. Habitat features sampled at field locations of campsites in Prince William Sound, Alaska, in the summer of 2008 (derived from Bowyer et al. 1995 and Ben-David and Golden 2007a)

HABITAT FEATURE	DESCRIPTION	VARIABLE DESCRIPTION
Exposure	Severity of wave action to which the site could be exposed	Exposed (0), moderate (1), or protected (2)
Vegetated slope	Slope above high tide	Integer recorded in 5-degree intervals
Tidal slope	Slope below high tide	Integer recorded in 5-degree intervals
Intertidal substrate: sand	Fine-grained material <0.5 cm in diameter	Ranked value: 0 (<25%) 1 (25%) 2 (50%) 3 (75%) 4 (100%)
Intertidal substrate: gravel	Rock material 0.5–10.0 cm in diameter	
Intertidal substrate: small rocks	Rock material 10.0–25.0 cm in diameter	
Intertidal substrate: large rocks	Rock material 25.0 cm–6.0 m in diameter	
Intertidal substrate: bedrock	Rock materials >6.0 m in diameter	
Percentage of canopy cover	Relative cover of trees taller than 3 m	Integer recorded in intervals of 5%
Percentage of understory	Relative cover of shrubs and tall grass	Integer recorded in intervals of 5%
Old-growth	Old-growth coniferous forest considered to be in a climax state	Ranked value: 0 (<25%)–4 (100%)
Burrow potential	Number of potential burrow sites	0 = no burrows, 1 = 1–3 burrows, 2 = 4+ burrows
Fresh water	Presence of streams, ponds, and pools	Indicator variable for each type

biobands), substrate type, and wave exposure were estimated for uniquely identified units of shoreline at a scale of better than 1:10,000 (Harney et al. 2008).

Each unit of shoreline was assigned one of 35 possible coastal classes. In order to compare specific shoreline types selected by otters and campers, an indicator variable was created for each of the 10 coastal classes that made up approximately 96 percent of the study area (table 13.2). Each coastal class that accounted for less than 1 percent of the surveyed shoreline was excluded from analysis. The coastal classes represented a unique combination of slope, width, and substrate type (Harney et al. 2008). Substrate

TABLE 13.2. ShoreZone coastal classes of surveyed shorelines (1,030 km) in Prince William Sound, Alaska

COASTAL CLASS (CC)	PERCENTAGE OF SHORELINE	SUBSTRATE	SLOPE	WIDTH
Rock cliff	7.09	Rock	Steep	Narrow
Ramp with gravel beach, wide	1.93	Rock and gravel	Inclined	Wide
Cliff with gravel beach	11.03		Steep	Narrow
Ramp with gravel beach	6.09		Inclined	Narrow
Ramp with gravel/sand beach, wide	5.19	Rock, gravel, and sand	Inclined	Wide
Platform with gravel/sand beach, wide	2.23		Flat	Wide
Cliff with gravel/sand beach	10.83		Steep	Narrow
Ramp with gravel/sand beach	14.07		Inclined	Narrow
Sand and gravel flat or fan, wide	17.95	Sand and gravel	Flat	Wide
Sand and gravel beach, narrow	15.83		Inclined	Narrow
Estuary	4.23	Organics	n/a	n/a

Note: Coastal classes were specific combinations of substrate type, slope, and width. Steep slopes were greater than 20°, inclined slopes ranged from 5° to 20°, and flat slopes were less than 5°. *Narrow* and *wide* referred to tidal zones less than and greater than 30 m wide, respectively.

type was made up of rock and/or sediment (organics, gravel, sand, or sand and gravel). Coastal classes could be flat (< 5°), inclined (5 to 20°) or steep (> 20°). Each slope level was reclassified into an indicator variable but included only inclined coastal units in analyses, because flat and steep slopes were correlated with different substrate types. Narrow (< 30 m) intertidal zones were assigned a value of 1 and wide (> 30 m) intertidal zones a value of 0. The vast majority of the Sound fell into either protected or semiprotected exposure categories, so semiprotected (SP) was entered as an indicator variable in analyses. A value of zero for the SP variable could therefore be considered protected, since so few of the surveyed units were exposed or were very much protected.

Two of the bioband fields were selected for inclusion in analyses. Eelgrass (*Zostera marina*) and soft brown kelps (*Laminaria* spp.) are associated with high fish diversity and abundance and therefore foraging opportunity for otters (Dean et al. 2000; Ostrand and Gotthardt 2005; Ben-David et al. 2005). Bioband distributions were recorded as patchy, continuous, or absent. Continuous biobands were dense enough to be visible in over half of the length of the shoreline unit, while patchy biobands covered less than half of the unit length (Harney et al. 2008). Based on results of preliminary analyses, the *Zostera* (ZOS) values were simplified to represent presence (patchy + continuous = 1) and absence (0).

HUMAN USE VARIABLES

The ADF&G Division of Wildlife Conservation developed the Uniform Coding Unit (UCU) for the purpose of geographically referencing harvest and management data. ADF&G supplied the spatial and associated tabular data for the present study, including the number of river otters harvested per year for 2005 through 2008. Harvest density was calculated for each UCU in the study area as the average number of river otters harvested per 100 km of shoreline (n = 23; the three Naked Island Group UCUs were combined into one). Latrine site and campsite densities per 100 km of surveyed shoreline and proportion of high use (level 3) campsites for surveyed portions of each UCU were also calculated. The two campsite values were multiplied to come up with an index of high-impact campsites.

The Chugach National Forest created spatial management units called General Areas (GA) in 2008 for the purpose of identifying and analyzing hot spots of human use, and provided client data from permitted companies that offer guided services in the Sound for the years 2005 through 2007 (See chapter 6). Daily client counts for camping and day use for the spring and summer seasons (April 1 through August 31) were combined to get an estimate of average shore use over the three years per General Area (n = 317). The density of shore use per 100 km of shoreline for each General Area was calculated. Each GA in the study area was associated with a UCU to determine the shore use density for each UCU.

Commercial fishing data (salmon only) for the years 2005 through 2007 were also provided by Chugach National Forest (chapter 6). ADF&G records permit and catch information for 44 geographical units, called Statistical Areas, in the Sound. The total number of daily permits that were associated with each Statistical Area was tabulated and used to calculate the average number of permits for the three years. The density of commercial fishing permits per 100 km^2 was also calculated. Where applicable, the values were combined for two statistical areas to calculate commercial fishing density per UCU.

STATISTICAL ANALYSES

A forward stepwise regression (p = 0.25 to enter; p = 0.1 to leave) was used to choose model parameters and the Akaike information criterion (AIC) to compare intercept only, global, forward selected, and reduced (removed parameters with a p-value > 0.1) models developed to explain latrine site and campsite occurrences compared to random and latrine site density (Shtatland et al. 2001 and Anderson et al. 2000, respectively). Independent variables for multicollinearity were tested prior to running the forward regression procedure and removed one of any pair of variables with $r > 0.4$ (Bowyer et al. 1995). All possible interactions between the remaining coastal class and bioband variable(s) were included in the selection process. Various tools in ArcMap 9.3 (ESRI 2008) were used to calculate shoreline lengths and associate feature classes. All regression analyses were performed in JMP 7.0.2 (a business unit of SAS Institute Inc.).

ArcMap 9.3 was used to generate a set of random points ($n = 1{,}920$, three times the total number of latrine sites) within a 5 m buffer around the surveyed shoreline that was used for both latrine site (0) vs. random (1) and campsite (0) vs. random (1) logistic regression analyses. Logistic regression was used to model the habitat features measured in the field (table 13.1) that best differentiated between campsites with (0) and without (1) otter latrine sites (> 10 scat). Canopy cover was moderately correlated with burrow potential ($r = 0.35$) and highly correlated with old growth ($r = 0.64$), and was therefore excluded from analysis.

Latrine site densities included all latrine sites located within an arbitrary zone of error (100 m) from the ShoreZone shoreline ($n = 637$). Proportions of rocky substrate type (rock; rock and gravel; and rock, sand, and gravel), continuous patches of soft brown kelp (SBR_C), eelgrass presence (ZOS), semiprotected shoreline (SP), and narrow intertidal zone (Narrow) were calculated for the surveyed portions of each UCU. SP, rock, and SBR_C were removed from analyses due to multicollinearity. Harvest density was highly correlated with commercial fish ($r = 0.63$) and was not selected as a model parameter. For these reasons, I excluded harvest density from analyses. The high-use campsite variable was removed from the analysis because it was not selected as a model parameter and resulted in a reduced sample size because three UCUs had not been surveyed for campsites. Interaction terms between ZOS and other shoreline characteristics did not improve the model and were not included in the final selection process for multiple regression analyses. Densities were log transformed and proportions were arcsine square root transformed. All estimates are reported as transformed.

RESULTS

SPATIAL RELATIONSHIPS

Both latrine sites and campsites occurred more often along semiprotected coastlines compared to random ($p = 0.002$ and $p < 0.0001$, respectively; table 13.3). Other shoreline characteristics, however, show stark contrasts at campsites and latrine sites, indicating that campers and river otters select very different shore types along surveyed Sound coastlines (table 13.3). River otter latrine sites were associated with narrow, rocky coastal classes, the presence of eelgrass, and continuous distributions of soft brown kelp. Campsites, on the other hand, were negatively associated with rocky

TABLE 13.3. Parameters entered in forward logistic regression analyses to select ShoreZone shoreline characteristics that differentiate latrine sites and campsites from random locations along 1,030 km of surveyed coastline in Prince William Sound, Alaska

	RIVER OTTER LATRINE SITE				CAMPSITE			
PARAMETER	ESTIMATE	SE	X^2	P	ESTIMATE	SE	X^2	P
Intercept	-1.91	0.14	177.47	<0.0001*	-3.45	0.19	316.71	<0.0001*
SP	0.25	0.11	5.44	0.02*	0.84	0.21	15.35	<0.0001*
SBR_P	—	—	—	—	—	—	—	—
SBR_C	0.33	0.11	8.68	0.003*	—	—	—	—
ZOS	0.55	0.11	24.81	<0.0001*	—	—	—	—
CC3	0.88	0.19	21.66	<0.0001*	-0.85	0.61	1.96	0.162
CC6	—	—	—	—	—	—	—	—
CC8	1.0	0.17	34.18	<0.0001*	-2.17	1.02	4.55	0.033*
CC9	0.49	0.21	5.59	0.02*	-1.9	1.02	3.44	0.064
CC11	-0.44	0.28	2.36	0.13	0.75	0.37	4.08	0.043*
CC12	—	—	—	—	—	—	—	—
CC13	0.72	0.16	20.22	<0.0001*	—	—	—	—
CC14	0.47	0.15	9.52	0.002*	—	—	—	—
CC24	-0.90	0.2	19.67	<0.0001*	0.86	0.24	13.21	0.0003*
CC25	—	—	—	—	—	—	—	—
CC31	-1.9	0.6	10.32	0.001*	0.78	0.46	2.87	0.09
SBR_C x CC11	0.82	0.52	2.46	0.12	n/a	n/a	n/a	n/a
SBR_C x CC24	0.65	0.37	3.07	0.08	n/a	n/a	n/a	n/a
ZOS x CC8	0.68	0.28	5.84	0.02*	n/a	n/a	n/a	n/a

Note: *SP* = semiprotected shorelines, *SBR_P/C* = patchy/continuous soft brown kelp distributions, *CC* = coastal class, *ZOS* = the presence of eelgrass, *n/a* = not entered in the forward selection process. Estimates, standard errors, Wald chi-square statistics, and p-values are included for parameters selected for each model. Interactions that were selected for the final model are included.

*$p < 0.05$, not selected for inclusion in the final model (i.e., lowest AIC)

shorelines, were found more often at wide, small substrate beaches, and were not influenced by the biobands.

Latrine site density was positively associated with the proportions of narrow intertidal zones ($p = 0.001$) and commercial fish density ($p = 0.031$; table 13.4). The presence of eelgrass was a weak predictor of latrine site density ($p = 0.11$). This reduced model did a better job of explaining variation in latrine site density (*AIC* = -20.13) than the other models, including habitat-only models with narrow intertidal zones and the presence of eelgrass (*AIC* = -16.37) and only narrow intertidal zones (*AIC* = -16.11; table 13.5).

TABLE 13.4. Parameters entered in multiple regression analyses to investigate the influence of commercial fishing and shore use by humans on latrine site density after accounting for shoreline substrate type

PARAMETER	ESTIMATE	SE	t-RATIO	P
Intercept	1.68	0.51	3.29	0.004*
Eelgrass	0.6	0.35	1.7	0.11
Narrow	2.17	0.56	3.87	0.001*
Rock and gravel	—	—	—	—
Rock, sand, and gravel	—	—	—	—
Commercial fish	1.4	0.6	2.33	0.031*
Shore use	—	—	—	—

Note: Latrine site densities were calculated for each Uniform Coding Unit that contained portions of the 1,030 km of surveyed coastline in Prince William Sound, Alaska. Estimates, standard errors, t-Ratio statistics, and p-values are included for parameters selected for the final model (i.e., lowest AIC).

*$p < 0.05$, not selected for inclusion in the final model

TABLE 13.5. Parameter codes, number of parameters (k), and Akaike information criterion (AIC) of multiple regression models compared to determine the best model for explaining latrine site density for Uniform Coding Units that contained portions of the 1,030 km of surveyed coastline in Prince William Sound, Alaska

MODEL	PARAMETERS[a]	k	AIC[b]
Intercept-only	I	1	-7.06
Global	N, Z, RG, RSG, CF, SU	7	-17.93
Forward	N, Z, CF, SU	5	-19.24
Reduced	N, Z, CF	4	-20.13
Habitat-only	N, Z	3	-16.37
Habitat-only reduced	N	2	-16.11

Note: The reduced model had the lowest AIC.

[a] *I* = intercept; *N* = proportion of narrow (intertidal zone < 30m); *Z* = proportion of eelgrass (*Zostera*) presence; *RG* = proportion of rock and gravel; *RSG* = proportion of rock, sand, and gravel; *CF* = commercial salmon fishing density (average number of permits/100 km^2); *SU* = shore use density (number of guide service clients/100 km of shoreline).

[b] $AIC = n * ln(RSS/n) + 2k$, where n = number of observations ($n = 23$) and RSS = residual sum of squares (Anderson et al. 2000).

FIELD ASSESSMENTS

Approximately 40 percent of the 165 campsites surveyed contained river otter scat. Twenty-four campsites shared shore locations with active latrine sites (> 10 scats). Only two of these latrine sites were included in the dataset provided by ADF&G. Campsites with latrine sites had higher burrow potential ($p < 0.0001$), more understory cover ($p = 0.002$), and were more likely to have ponds ($p = 0.014$) compared with other campsites (table 13.6). Positive associations with large rock ($p = 0.095$) and bedrock ($p = 0.144$) were weak sources of variation between campsites with and without latrine sites. Campsites with latrine sites had lower impact levels compared with other campsites, but this relationship was also weak ($p = 0.164$).

TABLE 13.6. Estimates, standard errors (SE), Wald chi-square statistics (X^2), and P-values (P) of selected parameters for a logistic regression model distinguishing between habitat features of campsites with latrine sites (> 10 scat) from campsites without latrine sites

PARAMETER	ESTIMATE	SE	X^2	P
Intercept	-4.25	1.03	17.04	<0.0001*
Exposure	—	—	—	—
Vegetated Slope	—	—	—	—
Tidal Slope	—	—	—	—
Burrow Potential	1.77	0.46	14.79	0.0001*
Impact	-0.38	0.27	1.94	0.1639
Ponds	1.7	0.69	6.01	0.0142*
Pools	—	—	—	—
Streams	—	—	—	—
Large rock	2.19	1.32	2.78	0.0954
Bedrock	1.38	0.95	2.13	0.1444
Gravel	—	—	—	—
Sand	—	—	—	—
Small rock	—	—	—	—
Old growth	—	—	—	—
Understory	2.26	0.74	9.45	0.0021*

*$p < 0.05$, not selected for inclusion in the final model (i.e., lowest AIC)

The number of scats found at campsites ranged from 0 to 136. Two of the campsites with otter scats had over 50 scats and eight (12 percent) had over 20 scats. By comparison, 60 percent of the over 200 latrine site locations provided by ADF&G that were surveyed in 2008 had over 20 scats and nearly 20 percent had over 100 scats.

DISCUSSION

River otters selected shoreline locations that had rocky substrates with narrow intertidal zones, the presence of eelgrass, and continuous distributions of soft brown kelp; they avoided flat sand and gravel beaches and estuaries. These findings corroborate results from previous studies of habitat selection by coastal river otters in Alaska (Larsen 1983; Woolington 1984; Bowyer et al. 1995; Ben-David et al. 2005) and indicate that ShoreZone may be a valuable resource for modeling and predicting latrine site occurrence.

A lack of overlap between most of the shoreline characteristics that distinguish latrine site and campsite locations from random suggests that river otters and campers in the Sound primarily look for different features when choosing shoreline sites. In contrast to shore types selected by river otters, campers avoided rocky shorelines and were attracted to flat, wide, sand-and-gravel beaches. The presence of eelgrass and soft brown kelp did not influence campsite occurrence.

Both latrine sites and campsites were associated with semiprotected shorelines. Shores with greater wave exposure tend to have steep, rocky intertidal zones and greater fish abundances than protected areas, which may explain river otter preference for semiprotected coastlines (Villegas et al. 2007). The protected heads of small bays are not common destinations for campers due to travel patterns and the sparseness of inclined small-sediment beaches with sufficient dry, flat surfaces above most high-tide levels.

It seems reasonable that inclined, rocky shorelines with sand and gravel beaches (coastal classes 9, 11, and 14; table 13.2) would be associated with both latrine sites and campsites, but width of the intertidal zone seemed to be a more important factor determining latrine site and campsite occurrences than the presence of rock with sediment. Narrow intertidal zones may be important to river otters for quick, consistent access to upland sites regardless of the tide level. Narrow beaches may have limited flat, durable, and nondurable space above high tide, which might deter campers but not otters.

Results of spatial analyses indicate limited potential for onshore conflict between river otters and campers in the Sound, but field data suggest that river otters use many sand, gravel, and small rock beaches that are also used by people. Scat counts at these beaches, however, were low compared with those at rocky sites. Otter scat was found at 40 percent of the 165 surveyed campsites, and 24 of these campsites overlapped with active latrine sites. Campsites with latrine sites had higher burrow potential and more understory vegetation, which supports the results of a previous study that demonstrated that otters have a high tolerance to human presence at shore sites with sufficient cover (Melquist and Hornocker 1983). Given the tendency for river otters to avoid gradually sloped intertidal zones with small sediment substrates, it was expected that the tidal slopes of campsites with latrine sites would be steeper and rockier compared to other campsites, which might be more evident with a larger sample size.

Campsites with latrine sites were also associated with ponds. Tracks, rub/scratch marks, and scat at the edges of these ponds indicate that otters may use these moderately sized bodies of standing water for drinking and/or removing salt from their fur. Ponds located at beach locations may be important resources for otters in some areas of Prince William Sound, and these sites often overlap with human destinations.

Campsites that shared locations with latrine sites may have lower levels of human impact compared with other campsites, but further investigation is needed to substantiate and explain this uncertain correlation. Damage to a shoreline site caused by heavy, repeated camping activity may have a greater influence on otter use of a site than mere human presence. Brief encounters with small groups of campers do not appear to deter use of an area by otters. Groups of otters occasionally used beaches that were occupied by people and tents (personal observation). Solitary individuals, including most females, may be more wary of human presence than groups.

Commercial salmon fishing was a strong predictor of latrine site density after accounting for proportions of narrow intertidal zones and eelgrass presence. Researchers have observed groups of otters using multiple salmon runs during spawning seasons (Bowyer et al. 2003). Diet analyses, however, suggest that salmon are not primary food items for otter in Prince William Sound and Southeast Alaska (Larsen 1984; Ben-David et al. 2005). Commercial salmon fishing may be an indicator of an unmeasured feature associated with foraging opportunity. However, three of the five salmon hatcheries (Main Bay, Armin F. Koernig, and Solomon Gulch) have the highest commercial fishing densities and have not been surveyed for latrine sites. In addition, drift nets have been identified as a drowning hazard for otter (Melquist and Dronkert 1987) and commercial drift net fishing in the Sound primarily occurs in only three locations, Main Bay (Main Bay Hatchery), the head of Unakwik Inlet (near Cannery Creek Hatchery), and the south end of Esther Island (Wally Noerenberg Hatchery), none of which has been surveyed for latrine sites.

Furthermore, the scale at which commercial fishing data are recorded may not be meaningful for looking at potential conflict with otters. Heavy commercial fishing activity that overlapped with surveyed UCUs was concentrated along shorelines that had not actually been surveyed for latrine sites. In Unakwik Inlet, for example, most commercial fishing boats were clustered near Cannery Creek Hatchery, which is located on the central-eastern shoreline. Latrine site surveys were carried out on the southwest corner of this inlet, which is just over 3 km wide.

Latrine site distributions may not be affected by the measured types and/or current levels of human activity in the Sound; this may also be the case for black oystercatchers in this region (see chapter 15). High impact camping, harvest rates, and shore use by permitted guide services did not influence latrine site density. Campsite densities are quite low compared with latrine site densities in the Sound, indicating that availability of desirable upland sites is more limited for campers than for otters. Camper density may also still be quite low relative to the number of shoreline campsites available. Shore use by permitted guide services is a limited representation of shore activity, because many residents and other experienced kayakers and boaters do not use guide services (likely only 10 percent when considered Sound wide). Furthermore, a number of private cabins and areas of regular occupation have not yet been mapped, and temporary cabins associated with set drift net fishing were not included in analyses of the present study.

This analysis did not detect a relationship between harvest density and latrine site density. Average harvest densities for 2005 through 2008 in the survey area ranged from 0 to 15 otters per 100 km of shoreline. Nearly 70 percent of the partially surveyed UCUs had a harvest density of less than 1 otter per 100 km of shoreline. Researchers estimate otter densities in the Sound to be around 28 to 80 individuals per 100 km of shoreline (Testa et al. 1994). Relatively low harvest rates across the survey area may explain the lack of association between harvest and latrine site density. One limitation of this analysis is that the latrine site densities were only based on partial (often less than 25 percent) surveys of the UCU shorelines. Moreover, none of the five UCUs with harvest densities over 10 animals per 100 km of shoreline had been surveyed for otter activity. UCU harvest densities vary from year to year, but trapping in some locations is consistently high relative to other locations. Several of these harvest hot spots correspond with high levels of commercial fishing activity.

According to Merav Ben-David, professor of wildlife ecology at the University of Wyoming, low levels of use and high abandonment rates of latrine sites, in particular passages, bays, and inlets of the Sound, may be related to human activities that were not measured in this study, including cabin construction; frequent, fast-moving boat traffic; and high levels of harvest (Ben-David, personal communication; Bowyer et al. 2003). Many areas with heavy human use in the Sound have not yet been surveyed for latrine sites and are ideal locations for future research designed to specifically address the impacts of small-scale human activity on latrine site distributions and use. Another limitation of this analysis is that the latrine site densities include the locations of potentially old or abandoned latrine sites from previous surveys, which is static and not the best measure of river otter response to human disturbances because both otters and people are highly mobile and populations are dynamic. Active latrine site densities and fecal deposition rates may be reliable measures of river otter abundance, but more research is needed to establish quantifiable relationships between these measures and estimations of population density based on DNA analyses (Hansen et al. 2007; Lanszki et al. 2008; Ben-David and Golden 2007b).

LESSONS LEARNED

Though river otters and shoreline recreationists both look for protected shorelines at large scales, smaller micro-habitat components relative to shoreline steepness and composition play an important role in assessing the likelihood of their interaction. As has been discussed elsewhere in this book (chapters 6, 7, and 17), humans using the uplands of the Sound often seek gentle-gradient, gravel beaches for specific access points. This is particularly true for those individuals intent on spending prolonged periods of time camping on the shoreline. These users seek broad stretches of gravel beach with flat areas above mean high tide—areas which seem less useful to otters.

Just as spatial scale is vital for exploring potential human disturbance or displacement impacts for this species, understanding the relative degree of intensity of use is of equal importance. Observations made in this study and by others have shown that otters seem to tolerate low amounts of transient use in proximity to latrine sites. Disturbance isn't likely until use becomes constant or the type of use changes (from incidental camping to permanently occupied cabins, for example). A similar dynamic is likely at play in the issue of nesting black oystercatchers as described in chapter 15, and in the issue of disturbance to Sound waterfowl as explored in chapter 14.

This study made an attempt to move beyond simple spatial association to look at the difference in microsite habitat variables (both above and below tide, courtesy of ShoreZone). This type of multidimensional analysis of wildlife disturbance questions is critical preparation before managers consider moving too quickly to solve human use "impacts" that may or may not exist simply because of spatial association. Furthermore, this study demonstrates that questions relative to displacement by human use are complicated, as illustrated by the case of a highly mobile species apparently capable of shifting behavior to take advantage of seasonal food sources (i.e., commercial fishing grounds and their support facilities). What may look like abandonment of favorable habitat by a species may in fact be the result of a shift in habitat quality that has little to do with the presence of humans.

RESEARCH NOTE

EXAMINING RECREATION AND WILDLIFE OVERLAP, AN APPLICATION FOR STRATEGIC CAMPSITE HARDENING

MARYANN SMITH FIDEL

A COMMON LANDSCAPE-LEVEL ANALYSIS is a spatial assessment of at-risk areas to minimize impacts to sensitive resources. As use increases in Prince William Sound and minimal development to contain campsite impacts becomes more commonplace, an analysis highlighting areas where recreation and sensitive wildlife overlap will prove useful to managers, allowing them to minimize impacts to sensitive wildlife resources. Several studies utilize geographic information systems, or GIS, to spatially examine overlap between recreation and sensitive resources. These studies provide management tools for visualizing, predicting, and evaluating human activity and potential impacts to wildlife and/or cultural resources. The power of these analyses is in their ability to reduce large landscapes into smaller areas for focused management attention.

Dispersed camping has potential to disrupt normal wildlife activity leading to deleterious effects. In the Sound, impacts from the *Exxon Valdez* oil spill still have lingering effects on wildlife. The recovery of species injured by the *Exxon Valdez* oil spill may be affected by increasing recreational use. Wildlife disturbance in the Sound can be minimized by locating recreational infrastructure away from sensitive wildlife habitats by concentrating locations of human use. This GIS assessment integrates spatial relationships of social, biological, and physical data. The goal of this research is to examine dispersed campsites' recreation use and impact in relation to the distribution of sensitive wildlife resources in order to inform strategic decisions about future campsite hardening.

Social data from surveys (see chapter 7) which identified recreation use points in the Sound were combined and analyzed using a technique known as *kernel density interpolation*. This process turns a distribution of hundreds of individual points into a map composed of thousands of cells, each with a predicted value for recreation use. This map surface is called a *raster* layer and allows for easy comparison to other data layers using GIS tools. Biological data layers included habitat for harlequin ducks, seabird colonies, black oystercatcher nests, bald eagle nests, harbor seal, sea lion, sea otter, blue mussels, and black bear. Based on research from the Sound and other coastal areas, these species are known to display disturbance responses to human activity. Each of these habitat layers was weighed equally, converted to raster, and combined. A point layer was created from this compilation of habitat layers, and a similar kernel density was completed to produce a surface representing concentration areas of all species.

The recreation density and biological layers were then combined to show areas with higher concentrations of species and recreation use—or areas where there is likely a high amount of overlap between wildlife and humans. The resulting map layer is displayed in the graduated color as figure 13.1. Red areas have high densities of recreation use and biological activity and indicate high potential for human wildlife interaction. Blue areas have less dense recreation use and biological activity and indicate relatively low potential for human and wildlife interaction.

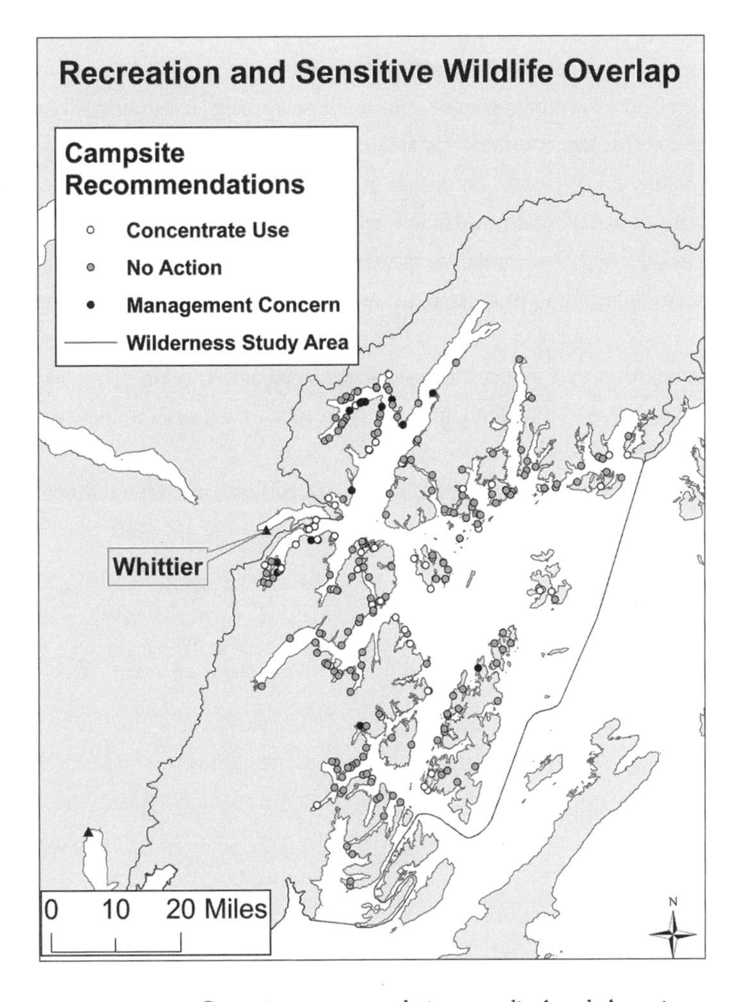

FIGURE 13.1. Campsite recommendations are displayed. Areas in black, the at-risk areas, have higher recreation densities and wildlife activities. Gray areas have low recreation densities and wildlife activity.

Physical data on campsite impacts included unpublished U.S. Forest Service data and Alaska Pacific University/National Outdoor Leadership School data (see chapter 17). These data were used to give each documented campsite an impact rating. Impact ratings were based on condition class, impacted area, and change in impacted area.

Campsite recommendations were based on the interaction of these three variables (recreation density, wildlife density, and campsite impacts). If a campsite was highly impacted with high recreation density and low biological activity, it was recommended for concentrating use and potentially campsite hardening as a means to accomplish this. If a campsite was highly impacted with a high recreation density and high biological activity, it fell into the category of "management concern." These campsites represent areas where human activity and wildlife are more likely to conflict. Disturbance to wildlife could be minimized by directing recreation elsewhere. Campsites of "no action" are those where minimal impacts have been documented, recreation density is low, and biological activity is also relatively low (see figure 13.1 for campsite recommendations).

When compiling an accurate picture of human use impacts, it is important to view these impacts at different scales. In a large system like the Sound, where use is widely distributed and doesn't follow specific paths (as it does in trail-accessed areas), it is important to identify "hot spots," or areas that may require specific management attention. Through this research, areas with higher potential for recreation and wildlife conflict are identified out of the large system. Integration of these data can assist land managers in minimizing human disturbance to sensitive wildlife resources by discouraging recreation in biologically sensitive areas and concentrating it in less biologically sensitive areas, using campsite hardening. This GIS analysis provides managers a good place to start when examining campsites for potential hardening projects.

CHAPTER 14

FLUSH RESPONSES OF TWO SPECIES OF SEA DUCKS TO KAYAKERS IN PRINCE WILLIAM SOUND, ALASKA

LAURA A. KENNEDY

SOUND BITES

EVIDENCE OF A POTENTIAL THREAT to both harlequin ducks and common mergansers goes beyond a simple overlap analysis with shoreline recreationists and includes demonstrated disturbance reactions to simulated kayak use.

Potential disturbance from summer shoreline recreation in the Sound is concentrated during the sensitive life cycle periods of brood rearing and molting for both species.

Understanding differences in habitat use as well as disturbance responses between and among species is vital for managers attempting to balance species response with sustainable human use.

INTRODUCTION

Distribution of sea ducks along the coastline is not uniform, and the preferred habitats of individual species determine the potential for human disturbance. Human presence has been shown to deter waterfowl from occupying preferred foraging, resting, and molting habitats (Korschgen et al. 1985; Belanger and Bedard 1989). As recreational pressures increase in natural areas such as Prince William Sound, it is important to identify factors that affect the shoreline distribution of sea ducks so that sea duck populations can be effectively managed and interactions with humans can be predicted and controlled. Understanding the dynamics around disturbance of wildlife species well enough to make effective management recommendations that won't artificially restrict human use results in complex research questions. This study looks at two different species of sea ducks common to the shorelines of the Sound to better understand the spatial association of their habitat with shoreline human use. It pairs that analysis with an evaluation of these sea ducks' specific behavioral responses to simulated disturbance from kayakers.

The harlequin duck (*Histrionicus histrionicus*) and the common merganser (*Mergus merganser*) are two of the most commonly occurring species of waterfowl that make use of the nearshore environment of Prince William Sound in summer, where and when interactions with human activities are most likely to occur (chapter 7). The types and severity of responses to human disturbance can vary greatly depending on the different life stages and social contexts. Giving managers specific information about the different types of human use in species' habitats as well as an understanding of the variation in disturbance response allows them to make informed decisions that balance species conservation with allowing human use.

Understanding the reactions of waterfowl to human presence requires information on the reactions of different species and the factors that affect those reactions. For example, harlequin ducks are known to be relatively tame and can be approached more closely than many species without eliciting a flush response (Robertson and Goudie 1999), as opposed to shy species, like common mergansers, which were described by Timken and Anderson (1969) as the first species of waterfowl to flush when approached by a human on foot or in a boat or automobile. In some sense, they may exemplify a good range in expected reactions from sea ducks to shoreline recreation in Prince William Sound. However, individual species may act differently during critical life stages, such as brood rearing and wing molt, because escape strategies are limited or negative impacts are potentially more severe during these stages (Korschgen and Dahlgren 1992).

Human disturbance during the brood-rearing season can break up broods or frighten parents into deserting their offspring, leaving ducklings vulnerable to predation and susceptible to death from starvation or exposure (Afton and Paulus 1992; Mikola et al. 1994). Wing molt is energetically costly and increases vulnerability to both predation and changes in environmental conditions; therefore, molting waterfowl minimize time spent in costly locomotive activities (Ringelman 1990; Flint et al. 2004).

Both species carry out two energetically expensive stages in Prince William Sound during the summer months, which are the peak season for shoreline recreation use (chapter 6 and chapter 7). Responding to human disturbance imposes direct costs on waterfowl, such as the energetic demands of flight, and time diverted from fitness-enhancing activities (Ydenberg and Dill 1986; Frid and Dill 2002; Blumstein 2003). This additional stress could pose a threat to both species, one that might be cumulative for harlequin ducks that declined in numbers following the 1989 *Exxon Valdez* oil spill and remain listed as not fully recovered (EVOS Trustee Council 2010).

STUDY AREA

This study was conducted from June to August, 2007 and 2008, in Prince William Sound, located in Southcentral Alaska and surrounded by the Chugach Mountains and Chugach National Forest. The eastern and western portions of the Sound were formed by separate glaciation events, resulting in drastic landscape and shoreline differences between the two sides. The result is a mosaic of sheltered and exposed shorelines, which potentially impacts the distribution of sea ducks across the Sound. Three main ports provide the majority of recreational and commercial fishing access to Prince William Sound: Whittier, Valdez, and Cordova.

Whittier is located about 60 miles from Anchorage, the most populous city in Alaska, and has seen a dramatic increase in use since the opening of the Anton Anderson Memorial Tunnel (or "Whitter Tunnel") to vehicular traffic in 2000. Previously, the only access to Whittier was expensive and by train through the tunnel. The Whittier Harbor Master reports that between 2000 and 2008, the number of boats launched from public boat ramps in Whittier increased by 152 percent (Whittier Harbor Master, unpublished data). Valdez and Cordova receive substantially less use than Whittier, but use has increased at all three ports as the Sound has become an increasingly popular destination for Alaskan residents and tourists (Wolfe et al. 2008).

METHODS

FLUSH BEHAVIOR DATA COLLECTION

Thirteen kayak-based surveys were conducted from June 28 to August 2, 2007, and June 12 to August 11, 2008. Each survey trip lasted seven or eight days and covered between 90 and 145 km of shoreline. Observations of harlequin ducks and common mergansers were recorded along continuous shoreline surveys. A survey started as soon as the research team launched their kayaks into the water and stopped if the researchers got out of the water for more than 15 minutes or if visibility became too poor to allow for accurate data collection. Use of kayaks enabled observers to stay close (5 to 30 m) to shore, to access areas too shallow for motorboats, and to approach ducks very quietly. This simulated predator stimulus also allowed for experimental evaluation of the effect of a very common recreational activity in the protected waters of Prince William Sound.

At the initial sighting of harlequin ducks or common mergansers, the number of individuals of each species and sex was recorded. Only individuals in obvious male alternate plumage were recorded as males. Individuals were recorded as females if the observer was confident an individual was a female rather than a male in eclipse plumage, as unknown when the observer was not confident of an individual's sex, and as juvenile when young were accompanied by a female or were in obvious juvenile plumage.

Flush distance was recorded as the distance (in meters) between the observer and the duck or center of group of ducks at the onset of a flush response by any member of the group. A range finder was used when possible; otherwise, flush distance was estimated. All observers used range finders to train at distance estimation over water to minimize error. Flush response was characterized on the behavior of the majority of individuals in a group, and was recorded as *run*, *dive*, *swim*, or *fly*. Molt status was characterized on a majority basis for each group. Individuals were determined to be in molt only when the observer saw a failed attempt at flight or could see that flight feathers were absent. Individuals were determined not to be in molt if they could fly. Molt status for all other observations was considered unknown. For the purpose of analysis, ducks with unknown molt status were assumed to be in molt because these groups always ran or swam in response to an approaching kayaker. Females with broods were never considered to be in molt, regardless of flush response, because female waterfowl do not undergo wing molt until they abandon their broods (Oring 1964).

HABITAT DATA COLLECTION

At the initial sighting of a single duck or group of ducks, a GPS point was taken, and the distance from the observer to the duck (a range finder was used whenever possible), angle (360°) from the nose of the kayak to the duck (or center of the group), and species were recorded. In 2008, I also characterized shoreline features at duck locations to physically describe the shoreline and to measure shoreline features related to foraging opportunity and predator avoidance of the two sea ducks. A 100-m segment immediately behind the duck(s) from the observer's perspective was approximated visually (i.e., 50 m to the left and right of duck along shoreline), stream presence or absence within this segment was noted, and the substrate type below the high tide line was characterized. Stream presence was limited to presence of moving water on the beach; dry stream channels were excluded. Substrate types were defined as sand (< 0.5 cm in diameter), gravel (0.5 to 10 cm in diameter), small rocks (10 to 25 cm in diameter), large rocks (25 cm to 6 m in diameter), bedrock (> 6 m in diameter), and estuary (absence of rock material, e.g. tidal flat). The 100-m segment was described in four 25 percent intervals. Any one substrate type had to cover at least 25 percent of the 100-m segment to be counted. At least one and no more than four substrate types were recorded for any 100 m segment. Stream presence or absence and shoreline characteristics were also recorded at 1,000-m intervals throughout every survey in 2008. These "nonduck observations" were collected to describe the general characteristics of coastline features along the survey routes.

The GPS point taken at every duck observation corresponded to the observer's location, not to the duck's location. I used the distance and angle recorded at every observation and the latitude and longitude of the original GPS point in a conditional formula to shift the GPS points to a more accurate representation of the duck's location:

Latitude:
If([direction]=0,[x_coord],If([direction]=180,[x_coord]If([direction]=90,
[x_coord]+[distance_min], If([direction]=270,[x_coord]-[distance_min],
[x_coord]+Sin([direction]*0.0174533)*[distance_min]))))

Longitude:
If([direction]=90,[y_coord],If([direction]=270,[y_coord],If([direction]=0,
[y_coord]+[distance_min], If([direction]=180,[y_coord]-[distance_min],
[y_coord]+Cos([direction]*0.0174533)*[distance_min]))))

Where direction = the angle from the nose of the kayak to the duck (0 to 360)
and distance_min = the distance from the observer to the duck

Of 425 total campsite locations known across the Sound, shoreline substrates and stream presence were classified at 145 campsites in 2008. The proportion of each substrate type (sand, gravel, small rocks, large rocks, and bedrock) was estimated in a 10-m radius around the main use area, which was evident by presence of tent pads, fire rings, or other signs of human use.

I used National Geographic TOPO! tagged image file format (TIFF) images in ArcMap to hand digitize all small islands and rocky outcroppings within 500 m of shore in the 13 surveyed areas of Prince William Sound. I used the ArcGIS point distance function on the resulting point feature class of rocky islets to determine the number of rocky islets within a 500-m radius of all harlequin duck and common merganser observations and of campsites.

I acquired an ArcGIS point feature class of bald eagle (*Haliaeetus leucocephalus*) nests from the Chugach National Forest. Bald eagle surveys were conducted in Prince William Sound in 1996 and 2001. Bald eagles hold the same territory for many years and often use the same nest year after year, up to 34 years (Buehler 2000), therefore I assumed the point feature class of nests, though eight years old, was still representative of active bald eagle territories. Bald eagles are known predators of harlequin duck (Robertson and Goudie 1999) and common merganser (Mallory and Metz 1999) broods and adults. I used the ArcGIS near analysis function on the bald eagle nest point feature class to determine the distance from each harlequin duck and common merganser observation and each campsite to the nearest bald eagle nest. I used the near analysis function in ArcGIS to find the distance from each harlequin duck and common merganser observation to the nearest campsite.

ShoreZone is an online resource that provides aerial imagery of the coastal environment integrated with classified geomorphic and biological features of the intertidal zone. Intertidal vegetation was classified as patchy or continuous based on observation of the occurrence and extent of species assemblages (Harney et al. 2008). Of the intertidal vegetation features available from ShoreZone, rockweed (*Fucus* spp.), salt-tolerant grasses (*Puccinellia* spp.), kelp (*Laminaria* spp.), and eelgrass (*Zostera* spp.) were the most relevant to my study, as they provide fish spawning habitat (Bustnes et al. 2000) and therefore are expected to have high densities of juvenile fish and fish roe, which may attract harlequin ducks and common mergansers. For analysis, I considered all patchy and continuous classifications of the biological features as presence and all observations without patchy or continuous classification as absence of the feature. Exposure was classified in ShoreZone as very protected, protected, semiprotected, semiexposed, exposed, and very exposed. Nearly all (93.5 percent) of the shoreline in Prince William Sound is classified as protected or semiprotected, because of the protection tendered by Montague and Hinchinbrook Islands from the Gulf of Alaska; therefore, my goal was to discriminate between protected and semiprotected shorelines. To do this, I grouped very protected and protected into one group labeled "protected" and semiprotected, semiexposed, exposed, and very exposed shorelines into a second group labeled "exposed." I used the ArcGIS linear referencing tool to associate all duck, random, and campsite locations to a unique identifier on the ShoreZone shoreline, which enabled me to label duck, random, and campsite point locations with the four vegetation features of interest and the classification of exposure.

STATISTICAL ANALYSES

Two-sample t-tests were used to compare flush distances and contingency tables to compare flush behaviors between species. Data from 2008 were used in analysis of flush distance and from 2007 and 2008 in analysis of flush behaviors. Flush distances between early and late season were compared because adult waterfowl may react differently to approaching kayakers as they habituate to human presence (Madsen 1985; Belanger and Bedard 1989; Korschgen and Dahlgren 1992) or may attempt to minimize energy expenditure in preparation for molt and migration (Korschgen and Dahlgren 1992). Also, females with young may respond to predation threats differently based on the size of their young and the young's ability to escape, which changes as the season progresses. Early versus late season were separated at July 10, the earliest date that a common merganser or harlequin duck was in molt. Group size was categorized into small and large for harlequin ducks and nonbreeding common mergansers based on the median group size of each species (harlequin duck median group size = 7; common merganser median group size = 3).

Analyses of flush distance may be more accurate when the approach distance, or distance a threat has been visible, is accounted for (Blumstein 2003). Approach distance at duck observations was collected, but it is unlikely that our estimates were accurate because ducks often were already reacting to our presence when they were detected. Flush distances were transformed to the log scale to meet the assumptions of normality and variance (Ramsey and Schafer 2002). Values reported are median ± SE. All statistical analyses were performed by JMP 7.0.2 (SAS Institute Inc. 2007).

Shoreline GIS data were transformed as necessary to meet assumptions of statistical tests (Ramsey and Schafer 2002), and all parameter estimates are reported as transformed. I used stepwise selection ($p = 0.25$ to enter; $p = 0.1$ to remove) to determine shoreline characteristics that best differentiated presence of each duck species from random sites. I measured and acquired shoreline characteristics related to the foraging behavior (i.e. shoreline substrate, rocky islets, stream presence, intertidal vegetation, exposure), potential predators (i.e., bald eagles and humans), and predator avoidance behavior (i.e., rocky islets, shoreline substrates) of each species (table 14.3).

To examine the potential for overlap between duck and human use of shorelines, I used stepwise selection ($p = 0.25$ to enter; $p = 0.1$ to remove) to create a model from all field-based and spatially acquired variables used to construct the duck models (table 14.4), excluding distance to the nearest campsite. All statistical analyses were performed by JMP 7.0.2, (SAS Institute Inc. 2007).

RESULTS

Of $n = 358$ observations of harlequin ducks and common mergansers made during the summers of 2007 and 2008, only 5.8 percent of ducks did not flush in response to approach by kayaks ($n = 21$). Common mergansers flushed at a greater distance (71.66 ± 0.82 m) than harlequin ducks (47.07 ± 0.86 m) ($t_{222} = 3.4, p = 0.0004$). The two species did not differ in their method of fleeing ($G_3 = 4.6, p = 0.21$; Pearson $X^2_3 = 4.706$, $p = 0.19$); both were most likely to fly as their first reaction to human disturbance ($p < 0.0001$; table 14.1).

Adult (non-breeding) common mergansers flushed at greater distances from kayakers (82.11 ± 0.08 m) than females with young (56.04 ± 0.11 m; $t_{152} = -2.8, p = 0.003$). Females with young showed a seasonal trend in flush response; their flush distance was greater late in the season (67.1 ± 0.14 m) versus early in the season (40.81 ± 0.20 m; $t_{55} = 2.1, p = 0.02$). Flush distance of nonbreeding mergansers did not differ between small and large groups, or by molt status regardless of group size (table 14.2). Harlequin ducks flushed at greater distances late in the season (51.44 ± 0.12 m) versus early in the season (32.81 ± 0.21 m; $t_{66} = 1.9, p = 0.04$), and large groups flushed at greater distances (57.56 ± 0.14 m) than small groups (38.12 ± 0.14 m; $t_{66} = -2.04, p = 0.02$). Molt status did not affect the flush distance of groups of adult harlequin ducks, regardless of group size (table 14.2).

Of the variables chosen to construct duck presence models (table 14.5), nine factors were selected by stepwise procedures to differentiate campsites from random sites (table 14.3). Of these nine factors, four overlapped with shoreline features important to harlequin ducks and common mergansers (table 14.3), including proportion of gravel and small rocks, stream presence, and number of rocky islets. Presence of harlequin ducks was related to the number of rocky islets along the shoreline ($p = 0.02$;

TABLE 14.1. Proportion of behaviors used by harlequin ducks (*Histrionicus histrionicus*) and common mergansers (*Mergus merganser*) to flee from kayakers in Prince William Sound, Alaska, during the summers of 2007 and 2008

	HARLEQUIN DUCK[a]		COMMON MERGANSER[b]	
BEHAVIOR	*n*	PROPORTION	*n*	PROPORTION
Dive	2	0.02	1	0.004
Fly	54	0.5	116	0.5
Run	25	0.2	68	0.3
Swim	28	0.3	44	0.2

[a] $G_3 = 60.6, p < 0.001$; Pearson ratio $X^2_3 = 49.86, p < 0.001$

[b] $G_3 = 155.9, p < 0.001$; Pearson ratio $X^2_3 = 120.64, p < 0.001$

TABLE 14.2. Compared flush distances (in meters) between harlequin ducks (*Histrioniucs histrionicus*) and common mergansers (*Mergus merganser*) in Prince William Sound, Alaska, in the summers of 2007 and 2008

GROUP	COMMON MERGANSER			HARLEQUIN DUCKS		
	MEDIAN ± SE	*t*	*p*	MEDIAN ± SE	*t*	*p*
Young present	56.04 ± 0.11 (yes) 82.11 ± 0.08 (no)	2.8	0.003	n/a n/a	n/a	n/a
Broods: season	40.81 ± 0.20 (early) 67.1 ± 0.14 (late)	2.1	0.02	n/a n/a	n/a	n/a
Adults: season	79.26 ± 0.10 (early) 85.51 ± 0.12 (late)	0.5	0.32	32.81 ± 0.21 (early) 51.44 ± 0.12 (late)	1.9	0.04
Adults: molt status	83.7 ± 0.21 (yes) 81.73 ± 0.08 (no)	0.1	0.46	49.9 ± 0.93 (yes) 43.1 ± 0.83 (no)	0.7	0.24
Adults: group size	80.78 ± 0.10 (small) 84.11 ± 0.13 (large)	0.2	0.4	38.12 ± 014 (small) 57.56 ±0.14 (large)	2.04	0.02
Large adult groups: molt status	86.21 ± 0.85 (yes) 83.13 ± 0.79 (no)	0.1	0.45	66.96 ±0.18 (yes) 46.14 ± 0.23 (no)	1.3	0.21

table 14.4). The odds of harlequin duck presence increased by 1.14 for each additional rocky islet present within 500 m of the shoreline (95 percent confidence intervals for multiplicative change from 1.09 to 1.19; table 14.4). Presence of common mergansers was associated with gravel ($p < 0.0001$) and small rock beaches ($p = 0.04$), rocky islets ($p = 0.002$), and presence of streams ($p < 0.0001$; table 14.4).

TABLE 14.3. Factors examined to differentiate harlequin duck and common merganser observations and campsites from random sites on shorelines, in Prince William Sound, Alaska, summers of 2007 and 2008

FACTOR	VARIABLE TYPE
Exposure	Categorical
Stream presence	Categorical
Salt tolerant grass presence	Categorical
Rockweed presence	Categorical
Kelp presence	Categorical
Eelgrass presence	Categorical
Proportion of sand	Continuous
Proportion of gravel	Continuous
Proportion of small rocks	Continuous
Proportion of large rocks	Continuous
Number of rocky islets within 500 m	Continuous
Distance to nearest campsite*	Continuous
Distance to nearest bald eagle nest	Continuous

*not included as factor to differentiate campsites from random

TABLE 14.4. Parameters selected by stepwise procedures, their standard errors, Wald chi-square, P-value, and odds ratios for the probability of harlequin duck and common merganser presence versus random observations in Prince William Sound, Alaska, in the summers of 2007 and 2008

MODEL	PARAMETER	ESTIMATE	SE	WALD X^2	p	ODDS RATIO
Harlequin duck	Number of rocky islets	0.28	0.18	5.67	0.02	1.3
	Proportion gravel	0.80	0.47	2.96	0.09	2.2
Common merganser	Stream presence	1.51	0.23	44.10	<0.0001	4.5
	Proportion gravel	2.06	0.31	45.70	<0.0001	7.9
	Proportion small rock	0.76	0.38	4.08	0.04	2.1
	Number of rocky islets	0.28	0.09	10.05	0.002	1.3

TABLE 14.5. Parameters selected by stepwise procedures, their standard errors, Wald chi-square, P-value, and odds ratios for the probability of campsites versus random sites in Prince William Sound, Alaska, in the summer of 2007 and 2008

PARAMETER	ESTIMATE	SE	WALD X^2	*p*	ODDS RATIO	OVERLAP WITH*	
						HARLEQUIN DUCK	COMMON MERGANSER
Rockweed presence	-0.92	0.33	7.9	0.005	0.4	No	No
Kelp presence	0.54	0.27	3.9	0.05	1.7	No	No
Eelgrass presence	-0.51	0.25	4.3	0.04	0.6	No	No
Stream presence	1.003	0.31	10.6	0.001	2.7	No	Yes
Proportion sand	8.15	1.02	63.9	<0.0001	3,456.5	No	No
Proportion gravel	6.60	0.60	123.2	<0.0001	738.2	No	Yes
Proportion small rocks	5.74	0.66	76.02	<0.0001	310.5	No	Yes
Number of rocky islets	0.48	0.13	13.6	0.0002	1.6	Yes	Yes
Nearest eagle nest	0.43	0.13	10.2	0.001	1.5	No	No

*Yes = factor overlapped between campsite model and presence of duck species indicated ($p < 0.05$);
No = factor did not overlap

DISCUSSION

The flush response of waterfowl to kayakers depended on the species, breeding status, season, and group size. Common mergansers flushed at greater distances than harlequin ducks, which suggests a potential link between inherent nervousness and life history strategy. Harlequin ducks are thought to exist at an energetic threshold (Goudie and Ankney 1986), which may inhibit them from being nervous, because nervous animals spend more time in flight than bolder animals, thus expending more energy (Ydenberg and Dill 1986).

Once a potential predator has been detected, animals begin weighing the costs and benefits of flight (Ydenberg and Dill 1986; Blumstein 2003). One cost is attracting the predator's attention by moving. Some species rely on crypsis (the ability to blend in with the environment for protection) rather than immediate flight and therefore flush at shorter distances (Ydenberg and Dill 1986), as was the case of female common mergansers with young. Young mergansers cannot fly and are vulnerable to predation and separation from their mother and siblings while running across the water

(Mikola et al. 1994; Keller 1991), and therefore may rely on being difficult to detect as they assess the predation threat. The shorter flush distances of females with young also could be a strategy to minimize energy expenditure of young, which need to eat between 40 and 80 percent of their body weight per day (Mallory and Metz 1999), and females, which after incubation are at their lowest body mass of their annual cycle (Mallory and Metz 1999). Females with young flushed at greater distances late in the season (after July 10) versus early in the season, which could be explained by the fact that the young were older, larger, and more mobile (although still flightless), thus less vulnerable to predation and separation. Larger young may rely less on crypsis and more on their ability to flee or, because of their increased size, may be less energetically stressed than early in the season. Harlequin ducks and nonbreeding common mergansers also flushed at greater distances late versus early in the season, although the relationship was weak for mergansers.

Many species show the opposite trend later in the season, which is thought to indicate habituation to human presence (Belanger and Bedard 1989; Peters and Otis 2006). Explanations of this seasonal trend are complicated by many life history and behavioral traits. Molt in harlequins is a protracted process beginning with nonbreeding males, then breeding males and nonbreeding females, followed by failed breeding females, and lastly breeding females; it begins in mid-July and may last until early October. The changing composition of flocks and flight status makes it unclear whether behavior is being driven by molt status, sex and age, or other factors. According to Dan Rosenberg, Alaska Department of Fish and Game, harlequins may also be forming pair bonds at this time of year, which could have an effect. As some birds have been in the Sound all summer, and others have not, it would be hard to attribute behavior to habituation, at least for harlequins (D. Rosenberg, personal communication).

Large groups of waterfowl tend to have a larger flush distance than small groups of waterfowl due to increased vigilance (Madsen 1985; Belanger and Bedard 1989; Korschgen et al. 1985; Korschgen and Dahlgren 1992) and dilution of risk (Ydenberg and Dill 1986). Harlequin ducks and common mergansers followed this expected trend, but the effect was very weak for mergansers. Wing molt is energetically expensive (Oring 1964; Ringelman 1990; Adams et al. 2000), therefore it was expected that nonmolting ducks would flush at greater distances than molting ducks when the cost of flight is greater than the risk of staying (Blumstein 2003). However, this trend was not found in this study as it related to either harlequin ducks or common mergansers. Molting waterfowl may be so vulnerable that their strategy is to flee as soon as a threat is perceived in order to escape at all.

Most animals face a tradeoff between maximizing energy intake and minimizing exposure to predators (Lima and Dill 1990; Blamires et al. 2007; Halstead et al. 2007).

Therefore, habitat features associated with the presence of a species likely provide both foraging opportunity and predator avoidance. My results were consistent with this idea; harlequin ducks and common mergansers were associated with habitat features that provide foraging and predator avoidance opportunities.

Harlequin ducks are small-bodied sea ducks with high energetic demands (Adams et al. 2000; Esler et al. 2000) and therefore are strongly associated with areas of high prey density (Robertson and Goudie 1999) where energy intake can be maximized, such as at rocky islets (Lindstrom et al. 1999; Esler et al. 2000). Common mergansers were associated with streams, which have a high abundance of juvenile salmon and salmon roe (Mallory and Metz 1999), their main prey in Alaska (Willson and Halupka 1995). Nest density of common mergansers has been correlated with the abundance of juvenile salmon (Wood 1987b; Willson and Halupka 1995), and mergansers may time reproduction and migration to coincide with salmon spawning and juvenile salmon migration from streams to the sea (Wood 1987a; Wood 1987b; Willson and Halupka 1995). Fish species richness and abundance are greater at shorelines with macrophytes (Newbrey et al. 2005), such as gravel beaches (Lindstrom et al. 1999). Therefore, gravel beaches, which were associated with presence of mergansers, may offer a high-density foraging opportunity.

In addition to foraging opportunity, rocky islets provide ducks a resting place that has good visibility for detecting the presence of land- and water-based predators, such as American mink (*Mustela vison*), American marten (*Martes americana*), bear (*Ursus* spp.) and river otter (*Lontra canadensis*) (Mallory and Metz 1999; Robertson and Goudie 1999). Occasionally we observed single or small groups of mergansers swim behind islets as part of a flush response. These same areas might serve as refuge from human use disturbance given that boats typically avoid rocky shallow areas with numerous islets. The algae- and barnacle-covered rocky islets may also provide crypsis (Endler 1978; Merilaita and Lind 2005) to harlequin ducks. Presence of common mergansers at gravel and small rock beaches may also indicate predator avoidance by crypsis; the white and gray belly of common mergansers coupled with the rusty markings on the breast of females and upperparts of juveniles (Mallory and Metz 1999) match their gravel and small rock background.

Four shoreline features—gravel, small rocks, streams, and rocky islets—were associated with campsites and one or both sea ducks, suggesting the potential for displacement of ducks at these shoreline features. Common mergansers may face a higher potential for conflict with campers because the overlap of shoreline features associated with their presence was complete with features associated with campsite locations. They are also less likely to habituate (Timken and Anderson 1969) to human presence than harlequin ducks (Robertson and Goudie 1999).

LESSONS LEARNED

The approaching threat used in this study was a pair of sea kayakers, which moved slowly and quietly across the water. Impacts on ducks may be more severe from larger groups of kayakers or faster, louder watercraft (Tuite et al. 1984; Korschgen et al. 1985; Keller 1991; Korschgen and Dahlgren 1992), such as power boats, fishing boats, and tour boats, all of which occur in variable numbers throughout the Sound. Of these, smaller power boats are most likely to have widespread distribution across nearshore waters of the Sound (chapter 7). The duration of the disturbance and distance from the disturbance may be more significant than size, speed, or noise, and independent of the type of vessel. For example, a fast-moving, loud powerboat at a farther distance may illicit a different overall response (other than just flush distance) than a slow-moving kayak at close range, but these differences require testing.

Our observations show that both species flushed with regularity at distances less than 100 m when approached by kayaking observers. This separation distance is currently used by the Chugach National Forest as a guideline to keep water-based recreation from disturbing concentrations of waterfowl and shorebirds (USDA Forest Service 2002). A contemporary pilot study exploring flush distance from motorized vessels found flocks regularly exhibiting flush response at distances of ~150 m for these same two species (see research note by Youngstrom, this chapter). Future studies should be undertaken using other nearshore watercraft in areas with varying levels of existing human use intensity to obtain a better idea of the short response to different recreation user types and to explore any potential for habituation to these disturbances.

As recreation use in the nearshore zone continues to increase, there is a realistic threat to both harlequin ducks and common mergansers that reside in these same locations. Observational data used in this study to compare campsite locations and shoreline features important to ducks found a significant overlap, suggesting the potential for conflict.

In chapter 15, researchers found that human use was associated with black oystercatcher nest territories at the scale of an area like Prince William Sound but that individual territories still had substantial separation between birds and camping humans. This may also be the case for these sea duck species whose individual shoreline roosts or feeding areas may not directly overlap campsites. Similar to black oystercatchers, these sea duck species show an affinity for rocky islets which likely affords them some refuge from nearby boat traffic. Understanding these dynamics relative to spatial association, as well as the potential for species to habituate to predictable patterns of recreation use, is key to assessing actual conservation threats from human disturbance.

While this study has shown a relationship between ducks and shoreline recreation and very clear disturbance responses, the direct impacts of recreation on waterfowl specific to energetic and reproductive costs are still unclear. There are tradeoffs that managers must make as they evaluate these relatively rigorous indicators of disturbance effect. The decision to step up efforts to attempt to further restrict use may be appropriate in this instance but could have unintended consequences for other species or result in negative responses from recreation users that may be counterproductive to other management efforts. Though the specific decisions may be less clear, in this instance managers are at least able to make those decisions with a couple of lines of evidence indicating a potential problem. This evidence may be enough to engage users in the region in a discussion about potential impacts and work with them to develop acceptable management guidelines.

RESEARCH NOTE

BIRDS AND BOATS

SADIE YOUNGSTROM

AS HUMAN ACTIVITY INCREASES in Prince William Sound, new impacts are introduced to the ecosystem. Wildlife populations are still recovering from the devastating 1989 *Exxon Valdez* oil spill. Information on ecology of nearshore birds was lacking prior to the oil spill and the need was recognized during planning of oil spill restoration. Many years of research have been dedicated to monitoring the effects of the oil spill on wildlife populations, especially on water birds (seabirds, shorebirds, and ducks) in the Sound. In the years since the oil spill, researchers have begun asking whether populations in the area might be affected by other factors that may be hampering species' ability to rebound from the devastation caused by the spill.

Most visitors access the Sound with watercraft ranging from kayaks to motorboats. As a result, marine species now experience previously unprecedented levels of vessel traffic in many parts of the Sound. Many studies have shown that human disturbance resulting from boat traffic has negative impacts on wildlife. Effects may vary from gross changes in distribution, in which sites are totally abandoned, to more localized small-scale alterations in feeding or roosting behavior in response to recreational boating activity. Disturbances displace waterfowl from high quality feeding grounds, increase energetic costs associated with flight, and may lower productivity of nesting or brooding waterfowl.

In late July and early August of 2008, as part of my senior research thesis at Alaska Pacific University, I worked alongside biologists from the Chugach National Forest to conduct a series of disturbance trials on molting common mergansers and harlequin ducks using sea kayaks and motorboats. I was specifically interested in evaluating differences in disturbance responses relative to vessel type for these two species common along much of the shoreline of the Sound.

The Forest Plan guiding the management of most of the region's uplands advises Chugach National Forest managers to attempt to prevent disturbance of molting sea ducks, specifically identifying harlequin ducks as being a concern as a result of the oil spill and due to the ducks' relatively high concentrations along the shores of the Sound. Harlequin ducks spend most of their molting period resting and preening, so ensuring that those molting congregations are not disturbed is important. Molting harlequin ducks are obligated to feed where they become flightless, indicating the potential for continued exposure. Chronic disturbance may compromise the ability of harlequin ducks to molt, as they alter their time budgets by continually avoiding sources of disturbance.

The study area included all water within the Sound, as well as land within 100 m of the shoreline, excluding waters on the Gulf side of Montague and Hinchinbrook Islands. I took advantage of an existing framework of survey transects established by the U.S. Fish and Wildlife Service to monitor trends in water bird populations in the Sound. Disturbance response observations took place on a random selection of those transects. Observations were made from a motorboat or sea kayak within 100 m of shore at a pace, course, and distance that assured complete coverage of the survey area and maximized the opportunity to observe ducks.

Statistical analysis found that both species were significantly disturbed by approaching watercraft. Kayaks were able to approach these birds at closer distances than motorboats (a median of 50 versus 160 m for harlequins and 76 versus 150 m for mergansers) before disturbance behavior was exhibited. However, kayaks caused a greater frequency of disturbance than motorboats (97 percent of the time versus 88 percent for motorboats), and this was probably due to the close proximity kayaks were able to approach before ducks felt the need to escape.

In 270 observations, I found that 110 groups of common mergansers exhibited greater sensitivity, flushing at farther distances of approach and ultimately fleeing farther than 160 groups of harlequin ducks, once disturbed. When comparing dominant flush behaviors of the ducks to watercraft type, some differences were found. Dominant flush responses for common mergansers, once disturbed by motor boats, were swimming and flying away from the approaching boat. When disturbed by kayaks, the birds were most likely to fly or dive in response. Harlequin ducks "ran" (rapid flight across the surface of the water without taking to the wing) or swam away from approaching motor boats once disturbed. When disturbed by kayaks they most often ran or dove away. The two most dominant avoidance responses are charted in percentage in the table below for the 1,848 harlequins and 591 individual mergansers that exhibited disturbance when approached by motorboats or kayaks.

Assuming energetic cost goes from greatest to least by running, diving, flight, swimming (without factoring in any sort of distance of flush), it would appear that harlequins respond with behavior that might result in more energy expended. Also, it appears that both species seem to have the option of swimming away from an approaching motorboat, but that is not the case when approached by kayak. This same general conclusion would be true if flight is more costly than diving, which it certainly could be.

Approach by both types of watercraft was observed to result in a significant disturbance for these two species of sea ducks in the Sound. Such disturbance could reduce time spent foraging, increase time spent in energetically costly behaviors, and ultimately decrease

The two most dominant response behaviors of disturbed individual harlequin ducks and common mergansers exposed to experimental boat approach by kayaks and motorboats

HARLEQUIN DUCK AND KAYAK		COMMON MERGANSER AND KAYAK	
Running	75%	Flying	63%
Diving	25%	Diving	37%
HARLEQUIN DUCK AND MOTORBOAT		COMMON MERGANSER AND MOTORBOAT	
Running	58%	Swimming	65%
Swimming	42%	Flying	35%

overall fitness or increase offspring mortality, as molting groups include large numbers of juveniles. Motorboats caused sea ducks disturbance from greater distances and caused greater flush distances than sea kayaks in this study. This disturbance may cause a reaction to escape out of the immediate area of disturbance and participate in high-energy activities like flying or "running" that allow escape.

This might mean that disturbance from motorboats has more potential than kayaks to displace ducks farther from foraging, molting, or resting habitats. However, likely due to closeness of approach, kayaks had a greater frequency of causing disturbance responses and initiated more responses that are likely to be energetically costly ("running" versus swimming away). Furthermore, considering that kayak travelers have more potential to interface with these shoreline species than powerboat traffic, which is typically focused in main channels in the Sound, the overall threat assessment becomes less clear. The complexity around disturbance responses by these two species and the dynamics of these two types of vessel traffic questions the utility of a "one-size-fits-all" approach when attempting to mitigate disturbance to wildlife species.

Continued research is needed to determine long-term effects of human disturbance on sea ducks in the Sound. As recreational activities increase in the Sound, it is essential to find a balance that will allow wildlife populations and humans to coexist. It appears that humans create at least short-term impacts on sea ducks. With educational programs targeted at different types of users, setback distances, or seasonal buffer zones for human activity, it should be possible for humans to decrease the amount and type of disturbance on sea ducks.

CHAPTER 15

BLACK OYSTERCATCHERS AND SHORELINE CAMPSITES IN PRINCE WILLIAM SOUND

AARON J. POE, MICHAEL I. GOLDSTEIN,
BRAD A. ANDRES, AND BRIDGET A. BROWN

SOUND BITES

MANAGERS OF REMOTE WILDLANDS often lack basic information about the overlap of sensitive species and human use and yet are often pushed to make assessments about potential disturbance impacts.

Because black oystercatchers favor similar beach habitat types used for shoreline camping yet have limited direct overlap with shoreline campsites, caution should be used in predicting disturbance based only on habitat association.

Targeted education efforts aimed at recreationists are likely the best way to prevent disturbance of black oystercatchers and other widespread sensitive species in remote regions.

INTRODUCTION

Black oystercatchers (*Haematopus bachmani*) are conspicuous, long-lived shorebirds adapted to life in rocky intertidal zones along the entire Pacific Coast of North America (Andres and Falxa 1995). Susceptible to both human and natural disturbance on the breeding grounds (Andres and Falxa 1995), this species is vulnerable to catastrophic environmental events such as oil spills, and invasive predators have been introduced by humans in parts of its range (Andres 1997). By the late 1990s, black oystercatchers were considered recovered from the effects of the *Exxon Valdez* oil spill, but a more recent assessment of long-term water bird monitoring efforts in the Sound suggested that black oystercatcher populations in oiled areas had not recovered to prespill levels (Irons et al. 2000). Alaska Shorebird Coordinator Richard B. Lanctot, of the U.S. Fish and Wildlife Service, Migratory Bird Management, in Anchorage reports that elevated liver cytochrome P450IA levels were documented in 2004, indicating continued exposure to oil (R. Lanctot, unpublished data), which resulted in the species being downgraded; it is still listed as recovering (EVOS Trustee Council 2010).

The Chugach National Forest designated black oystercatchers as a Management Indicator Species under its 2002 revised Forest Plan in part because of its history of impact from the oil spill but also because of concern about disturbance from shoreline recreation, particularly in the western Sound (USDA Forest Service 2002). Murphy et al. (2004) predicted an increase in the recreational use of shoreline habitats in the western Sound with the most popular months for shoreline recreation in the western Sound being June and July (Murphy et al. 2004). This same period is also a critical time in the life cycle of black oystercatchers when they are incubating eggs and rearing chicks.

Given the concerns about increased amounts of recreation along the shorelines of the western Sound, the Forest Service began a program to inventory the shoreline of the Sound for black oystercatchers. This program aimed to better understand the distribution of nesting black oystercatchers and document habitat associations. It also had the important management goal of assessing if current efforts by the Chugach National Forest to prevent disturbance from shoreline recreation were adequate. For example, the Forest Service had published a management guideline of keeping "concentrated recreation activities" greater than 330 feet away from known black oystercatcher nests. They enacted the guideline without understanding where nests were for most of the Sound or how they overlapped with shoreline campsites. They also had no specific information that supported this separation distance—opting to borrow it from the Bald Eagle Protection Act for active nests of that species.

The following study draws upon the first four years of data from that shoreline inventory work as well as parallel efforts by the Forest Service to map the locations of coastal campsites in the Sound. It represents the first analytical efforts taken by managers in the region to attempt to understand the relative risk of nest disturbance based on proximity to shoreline campsites. The approach described is iconic of wildland managers trying to gain some cursory understanding of the scope of interactions between wide ranging human use and a sensitive species whose distribution is poorly understood.

METHODS

STUDY AREA

Located in Southcentral Alaska, western Prince William Sound (~60° north, 147° west) is separated from Interior Alaska in the north and west by the steep slopes of the Chugach and Kenai Mountains. Western Prince William Sound has approximately

2,030 km of mainland shoreline and 2,630 km of island shoreline. Shorelines of this region are generally steep and rocky, but are punctuated by more gradually sloped beaches composed of gravel, cobble, and rocky debris, which is deposited by glaciers, avalanches, or streams. In addition to the many beaches, there are hundreds of small rocky islets, wave-cut platforms, and emergent glacial moraines.

Under existing forest management guidelines, most of the western half of the Sound is part of the Nellie Juan-College Fjord Wilderness Study Area (USDA Forest Service 2002). Developed recreation sites are not widely distributed, and there are no significant upland resource extraction activities, such as forestry or mining. With the exception of commercial fishing, the only widespread human activity in the western Sound is recreation, both private and commercial. Activities include boating, sportfishing, kayaking, wildlife viewing, hunting, and sightseeing, which are the bases for ecotourism and charter businesses located in Whittier. Much of this activity is shoreline associated and uses primitive backcountry campsites dispersed throughout western Prince William Sound (Murphy et al. 2004; Colt et al. 2002).

SHORELINE SURVEYS

We distributed our survey effort throughout Chugach National Forest–managed western Prince William Sound based on a suite of competing objectives, such as areas with high and low human activity (eventually summarized in Murphy et al. 2004), the inclusion of both mainland and islands, visitation to areas with little or no previous survey effort, and travel logistics. Thus, we did not select units randomly; instead, we delineated survey units based on physical boundaries (topographical breaks in shoreline segments such as bays, islands, and archipelagos). Because of the large area and the costs of access, we surveyed 18 units, ranging in size from 21 to 254 km in length, over four years from 2001 to 2004 (figure 15.1). We surveyed a total linear distance of 1,943 km of shoreline, approximately 64 percent of the western Sound shoreline managed by the forest. Because black oystercatchers have a high breeding site fidelity (Tessler et al. 2007), surveys over multiple years along different shorelines likely encountered different (i.e., independent) breeding pairs.

We surveyed each unit twice per season. During each survey, two trained observers visually surveyed the entire shoreline of each survey unit from small inflatable power boats. We conducted surveys at ≤5 km/hr and ≤50 m from shoreline between 0800 and 1900 h (Alaska Daylight Time) at various tidal stages, although we maximized our effort during high tides to increase our likelihood of finding birds near nest sites. When we detected black oystercatchers, we went ashore to collect habitat information and determine breeding status. Although we recognize that detection rates may differ by observer, habitat type, tidal cycle, and within and between breeding seasons, we minimized these effects as much as possible by using techniques that were successful in previous Prince William Sound work (Andres 1997, 1998, 1999). Through these efforts, we believe we were able to detect virtually all of the black oystercatchers occurring on the shoreline but acknowledge that the number of territories found during our surveys is likely a minimum.

Because most black oystercatchers initiate breeding in May and early June in the Sound (Andres and Falxa 1995), we completed our initial shoreline surveys to obtain territory and nest information within a 10-day period during late May and early June. For each bird located, we assigned status as territorial or nonterritorial using a combination of behavioral observation and nest search results. To reduce disturbance, we conducted these assessments for less than 30 minutes on each territory. We defined a reproductive pair by the presence of eggs or chicks or by reproductive behaviors such as courting, nest building, copulation, or territorial aggression, and we defined nonreproductive status as individual, pair, or more than two birds not engaged in reproductive behavior. Our approach provides a conservative estimate of the number of reproductively active pairs. We georeferenced all locations with a GPS and later developed a GIS overlay with individual territories and nest locations. We used locations of territories to estimate linear pair density (pairs/km) for each survey unit.

We assigned a shoreline type for each nest location based on the following five categories: salt marsh and tidal flat; wave-cut platform; exposed rocky shore; sheltered rocky shore; and gravel beaches. Nests were small enough that they did not overlap two categories. These shoreline types represent an aggregate of 10 base shoreline types defined for an Environmental Sensitivity Index GIS layer produced for Prince William Sound by the National Oceanic and Atmospheric Administration (NOAA 2000). We calculated the total available km of each shoreline type for all survey units.

SPATIAL ANALYSES

We pooled GPS locations from all territories and survey units across years and conducted a chi-square goodness-of-fit analysis to determine if black oystercatchers selected habitat types in proportion to their availability (Sokal and Rohlf 1995). We calculated a Jacob's D electivity index to determine preference or avoidance of particular shoreline types, and we used the following equation: where r = *the proportion of nests on that shoreline type* and p = *the proportion of shoreline type from all survey units combined* (Jacobs 1974). The Jacob's D shows habitat preference values relative to availability, from -1 (selection against) to +1 (selection for). We used an occupied territory to indicate shoreline class membership.

We used the Chugach National Forest Backcountry Ranger Program primitive campsite inventory for the western Sound to determine how humans selected habitat types for camping. This layer represents a combination of known sites identified during a five-year, complete shoreline inventory effort of western Prince William Sound (Chugach National Forest, unpublished data). We conducted a chi-square goodness-of-fit test to determine if humans selected habitat types in proportion to their availability (e.g., Johnson 1980) and calculated a Jacob's D electivity index to determine preference or avoidance of particular shoreline types using the following equation: $D = (r - p)/(r + p - 2rp)$, where r = *the proportion of campsites on that shoreline type* and p = *the proportion of shoreline type from all survey units combined* (Jacobs 1974).

To evaluate the association between campsites and nest territories at the landscape scale, we used a Kolmogorov-Smirnov (KS) distribution test for continuous distributions (Zar 1999). We measured the Euclidean distance from campsites to nests and from campsites to random points and summarized nearest distance values from all known campsites and 270 random shoreline locations from the raster layer. We computed a cumulative distribution function (CDF), plotted the resulting values, and then measured the maximum distance (Dmax) between the observed (distance-to-nest) and expected (distance-to-random) curve and compared that result to the critical value of ($D\alpha$) for the KS goodness-of-fit test for continuous distributions (Zar 1999).

RESULTS

Between 2001 and 2004, we identified 291 black oystercatchers and 94 unique breeding territories along 1,943 km of shoreline (table 15.1). Linear pair density ranged from 0.03 to 0.38 pairs per km, with Harriman Fjord and the Dutch Group having the highest density of nesting black oystercatchers (figure 15.1). Of the 94 territories evaluated, 50 percent were on gravel beaches, 21 percent were on sheltered rocky shores, 15 percent were on exposed rocky shores, 14 percent were on wave-cut platforms and rocky islets, and none were in either salt marsh or tide flats. Nest territories were not distributed in proportion to available habitat ($\chi^2 = 9.20$; critical value = 9.488; df = 4; $p = 0.059$). Using Jacob's D electivity index, black oystercatchers selected for wave-cut platforms and gravel beaches, and selected against salt marsh and tide flats, sheltered rocky shores, and exposed rocky shores (table 15.2).

We assessed the five shoreline types for the 186 primitive campsites within our study area. Campsites were not distributed in proportion to available shoreline type ($\chi^2 = 58.396$; critical value = 9.488; df = 4; $p < 0.0001$). The Jacob's D electivity index showed that camps were on gravel beaches and not on exposed or sheltered rocky shores or on wave-cut platforms (table 15.3). Black oystercatcher territories averaged 1,775 m away from shoreline campsites (SD = 1,426; range 60 to 5,865 m). Four of the territories occurred <100 m from campsites, but the majority (74 percent) occurred >500 m from campsites. Territories on gravel beaches ($n = 47$) averaged 1,596 m from campsites (SD =1,603; range 60 to 5,843 m). Four territories occurred <100 m from campsites; of the 24 territories <500 m from campsites, 17 (71 percent) were on gravel beaches. When evaluated at the landscape scale, nest sites were positively associated

TABLE 15.1. Summary of black oystercatchers detected in western Prince William Sound from 2001 to 2004

YEAR	SURVEY (KM)	TOTAL BIRDS	TERRITORIAL PAIRS	BIRDS FOUND IN CONFIRMED TERRITORIES (%)
2001	498	81	22	54
2002	421	53	18	68
2003	296	75	24	64
2004	728	82	30	73
Total	1,943	291	94	65

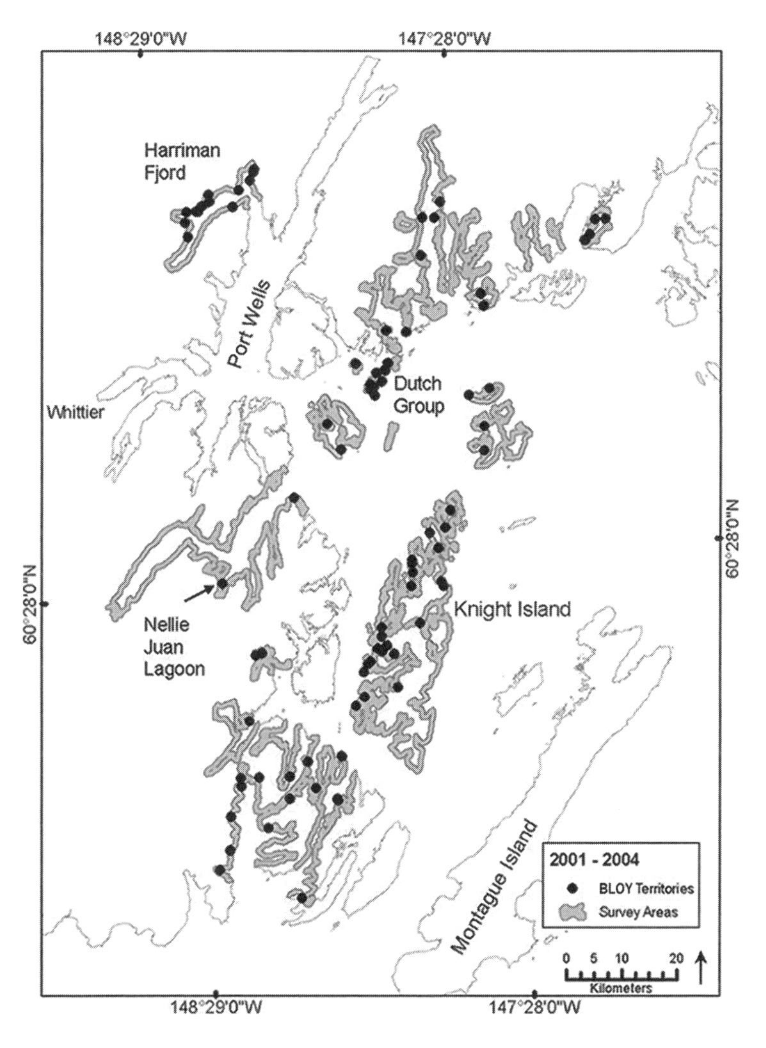

FIGURE 15.1. Ninety-four black oystercatcher territories in western Prince William Sound, detected between 2001 and 2004.

TABLE 15.2. Chi-square and Jacob's D electivity index for black oystercatcher territories relative to available shoreline type in western Prince William Sound, based on surveys completed from 2001 to 2004

SHORELINE TYPE	SHORELINE LENGTH (KM)	PROPORTION OF STUDY AREA (%)	NUMBER OF TERRITORIES EXPECTED	NUMBER OF TERRITORIES OBSERVED	JACOB'S D
Salt marsh and tide flat	46	2	2.3	0	-1
Wave-cut platform	164	8	8.0	13	0.24
Exposed rocky shore	309	16	15.1	14	-0.04
Sheltered rocky shore	583	30	28.5	20	-0.18
Gravel beach	841	43	40.7	47	0.08
Total	1,943	99	94.6	94	

TABLE 15.3. Chi-square and Jacob's D electivity index for primitive campsites relative to available shoreline type in western Prince William Sound, based on surveys completed from 2001 to 2004

SHORELINE TYPE	SHORELINE LENGTH (KM)	PROPORTION OF STUDY AREA (%)	NUMBER OF CAMPSITES EXPECTED	NUMBER OF CAMPSITES OBSERVED	JACOB'S D
Salt marsh and tide flat	46	2	4.4	4	-0.05
Wave-cut platform	164	8	15.7	11	-0.18
Exposed rocky shore	309	16	29.6	4	-0.76
Sheltered rocky shore	583	30	55.8	19	-0.49
Gravel beach	841	43	80.5	148	0.30
Total	1,943	99	186.0	186	

with shoreline campsites (Dmax = 0.179; critical value = 0.168; $p < 0.01$). Random points averaged 2,329 m from campsites (SD = 1,700; range 85 to 7,388 m).

DISCUSSION

We documented the distribution of black oystercatcher nest territories on nearly 2,000 km of shoreline in western Prince William Sound. We recognize the limitations of this information, particularly the trade-off between covering a greater extent of available shoreline and being able to compute detection rates by running survey segments multiple times. Our results provide a baseline inventory for the distribution of territories in the western Sound and identify the Harriman Fjord/Barry Arm complex, the Dutch Group archipelago, and the Nellie Juan Lagoon as important breeding areas for black oystercatchers in Prince William Sound, in addition to Montague and Green Islands identified earlier (Andres 1997, 1998).

Competing management objectives and logistical constraints (e.g. cost and access) limited our ability for 100 percent coverage of western Prince William Sound. We surveyed 64 percent of the shoreline managed by the Chugach National Forest, and subsequent monitoring efforts should consider these limitations. Not computing detection rates could have resulted in underestimating total territories, potentially complicating our ability to rigorously identify preference of shoreline types. Detection rates by habitat type likely vary most significantly when attempting to answer nest distribution questions, and, thus, we limited our analyses to territories. The relative preference identified from such a large survey area and the fact that it was characterized conservatively (five general shoreline categories) make our results representative of use patterns by this species in the Sound. The relative associations of territories with campsites will aid in prioritizing management questions and improving long-term species monitoring.

The selection of habitat by shorebirds in general (del Hoyo et al. 1996), and black oystercatchers in particular (Nysewander 1977; Hockey 1987; Andres 1998), is driven by direct and indirect human influences. Our results indicate that campers recreating in this backcountry are seeking the same general beach shoreline type as ~50 percent of nesting black oystercatchers detected in our study. However, the distances between campsites and nest territories are great, averaging 1.8 km, and the overall direct influences may not be a concern. Furthermore, it is not immediately clear that shoreline camping has direct disturbance potential in Southcentral Alaska, whereas it may in areas of higher human population. Kayak camping disturbance trials did not reach thresholds that lowered productivity of black oystercatchers nesting in Kenai Fjords National Park (Morse et al. 2006), but nest failure has been attributed to human disturbance in Oregon (E. Elliot-Smith, cited as personal communication in Tessler et al. 2007). However, the levels of human use in the Sound (e.g., chapters 6 and 7) are likely much lower than much more accessible areas in other parts of the Pacific Northwest.

LESSONS LEARNED

Black oystercatchers are one example of a species that is highly sought by wildlife watchers traveling through Prince William Sound. It's an iconic, raucous bird prone to dramatic aerial displays and predator defense. Though our study didn't demonstrate widespread, direct overlap between shoreline campsites, these behaviors often attract attention from shoreline users (particularly those interested in birds) and unfortunately can draw individuals into nesting territories, where they have the potential to disrupt breeding or worse, trample eggs or chicks. For a species that is dependent on both upland (U.S. Forest Service–managed) and tidal (State of Alaska–managed) environments, as well as receiving protection under the Migratory Bird Treaty Act and state wildlife plans, there are a number of land and resource managers with responsibility for ensuring limited impact and population viability of this species. Furthermore, each of these entities has differing approaches to addressing and permitting management activities along the shoreline. In the case of black oystercatchers, an interagency working group was established to foster communication toward a consistent management approach through a Conservation Action Plan developed by both federal and state entities (Tessler et al. 2007). For this species, upland managers have been able to leverage funding support from this collection of interested partners to specifically look at human disturbance relative to black oystercatchers as part of two empirical studies (Morse et al. 2006; Spiegel 2008).

Similar joint efforts *if* initiated by this working group, targeted at spreading awareness about the potential for disturbance by campers and beach walkers, would likely help mitigate disturbance of nesting sites. Through relatively simple education messages for iconic species like the black oystercatcher, managers can turn public interest into a vehicle for stakeholders to help spread the word about best practices for viewing and moving through their territories.

In Prince William Sound, where intensive management of human use is difficult given its huge extent and management prescription as a Wilderness Study Area, managers are limited in the actions they can take to prevent species disturbance. The Chugach National Forest's Forest Plan stipulates that concentrated human activity should be kept at least 100 m away from black oystercatcher nesting areas. However, such restriction cannot be effectively enforced, given the size and complexity of the region for anyone but their permitted commercial operators, who are likely less than 10 percent of the total use occurring in the Sound (see chapter 7). The Forest Service is also somewhat limited in the ability to develop facilities that could concentrate use (see chapter 16). As use increases, and given the compounding effects of a species still dealing with aftereffects of the *Exxon Valdez* oil spill (EVOS Trustee Council 2010), managers need to be proactive in their efforts to mitigate disturbance. We suggest an effort focused on communicating with Prince William Sound stakeholders about the sensitivities of this species during the breeding season (approximately May to mid-July) and best practices to avoid disturbance. Managers should support interagency groups like the Black Oystercatcher Working Group to develop a single consistent message targeted at beach users and leverage the capacity of that group to spread the message to Sound stakeholders.

STAKEHOLDER ESSAY

EXPERIENCING OYSTERCATCHERS

SARAH WARNOCK

AS A SCIENCE EDUCATOR, I've seen the impact that first-hand encounters with the natural world can have on people. The magnitude of the experience is reflected in their faces as they peer into the magical realism of a tide pool or hold a wild bird for the first time at a banding station. It's a feeling of personal discovery, of verification that Earth can be a pretty cool planet.

We know that exposure to the natural environment is directly correlated with attitudes about conservation. But as technology develops, as we become more and more an indoor society, there is a loss of opportunity to experience the natural world and a concomitant loss of support for the natural environment. The lack of public interest in and knowledge of environmental processes is coming at a particularly bad time; the complexity of the issues facing us today requires at least some knowledge of ecosystem interactions to fully comprehend their implications. In the recent past, support for environmental issues could be garnered through images—of polluted air and water, of cleared rain forest and urban encroachment. The major issues of today are much more difficult to explain; climate change and ocean ecosystem collapse aren't easily captured in a single image or sound bite.

Clearly we need to encourage people to connect with the natural world in order to build support for its protection. The type of future we pass on to our kids depends on the choices we make and whether we select behavioral changes that will slow or reverse current trends. But all the data in the world won't change behavior. Only an emotional connection will spur action. First-hand, direct involvement with the natural world is the best way to elicit a positive emotional response and support for the environment.

Therein lies the rub. Encouraging interaction with wild places can have detrimental effects on the environment. Sensitive species are negatively affected by visitors: wildlife-watching cruises can affect feeding patterns of whales; watercraft near seabird colonies can cause temporary nest abandonment leading to chick and egg predation; invasive plant species can be unknowingly trekked into pristine areas. Where is the balance between protection of a natural resource and the value of increasing public support for that resource through direct interaction?

The challenge for managers of our wildest places is to protect vulnerable species and habitats while encouraging recreational use of the environment. Public education may be one of the strongest tools available, and the story of black oystercatchers in Prince William Sound is a good example of how science and education can come together to allow for life-changing personal experiences in nature, without the detrimental effects of wildlife disturbance.

Jet black with a strikingly orange bill and eye-ring, black oystercatchers are large and charismatic shorebirds that inhabit rocky intertidal coastlines throughout western North America. As a group, oystercatchers are in decline around the world, due in part to human disturbance of their preferred nesting sites: flat, pebbly beaches which are also attractive to people. This is certainly true for the steep, rocky shorelines of the Sound, where gravel beaches are often a bottleneck point for human access. Even temporary disturbance by visitors can result in crushed eggs or the defensive behavior of distraught adults that draws the attention of nearby predators.

Excluding visitation to sensitive areas of the Sound is a daunting proposal to say the least, when considering its many miles of remote shoreline. By focusing on visitor education, managers can reduce disturbance while promoting responsible wildlife viewing and the personal experience that garners an overall ethic of stewardship benefitting the environment as a whole. In the end, managers and the species in their charge gain support from a populace more appreciative of natural systems.

This is not to say that education and outreach alone *solves* managers' challenges around wildlife disturbance—rather, it is a wise first step. Increased education and experience in the natural world will lend support for what is truly needed in nearly every large, intact ecosystem we have left: a mosaic of protection that includes inviolate zones in highly sensitive areas, in concert with temporally restricted areas and unrestricted access in places where the effect of human disturbance is low. If visitors (and society at large) gain increased understanding and greater appreciation for these species and habitats, then potential future beach closures or setback distances from nesting areas will seem not only reasonable but necessary. In the end, it will be our choice to protect wildlands, and that choice will depend on our collective experience with nature.

CHAPTER 16

CAMPERS' PERCEPTIONS OF WILDERNESS AND CAMPSITE HARDENING IN BLACKSTONE BAY, PRINCE WILLIAM SOUND, ALASKA

MARYANN SMITH FIDEL AND PAUL TWARDOCK

SOUND BITES

HARDENED CAMPSITES ARE man-made developments that may be perceived as contradictory to the concept of wilderness.

Recently, there have been proposals to harden more campsites in the Wilderness Study Area, primarily to prevent resource damage.

This research suggests that campsite hardening is a publicly acceptable management tool, but it is not without cost.

INTRODUCTION

The U.S. Forest Service is faced with a dual mandate when managing forest lands within the Nellie Juan-College Fjord Wilderness Study Area. The 1964 Wilderness Act requires agencies to preserve the natural integrity of the wilderness resource while providing for its continued recreation use. Often these two mandates conflict when recreation impacts the very resource on which it depends. Recreation in protected areas may threaten the biophysical condition of campsites (Cole 1981a) and social conditions (Stankey 1973). Campsite hardening is often employed to concentrate use and mitigate biophysical impacts of camping and provide recreation opportunities. Campsite hardening employs some type of development to increase the capacity of a campsite to withstand use. In Prince William Sound, the level of development varies from raised wooden tent platforms, outhouses, and bear-proof boxes to simple gravel pads in a wilderness setting.

Evidence suggests recreation use is increasing in Prince William Sound (Twardock and Monz 2000; Colt et al. 2002; ADOT 2009). Estimates range from 4.3 to 16.8 percent yearly increases in recreational use of the Sound. Research suggests campsite impacts are also increasing in the Sound (Monz 1998; Monz and Twardock 2004; chapter 17). Recent work (Twardock et al. 2010) found that campsites in the Sound were increasing in size and that campsite proliferation (user-created new campsites) was more common than campsite abandonment (campsite recovery). Permanent impacts such as tree damage were found to be increasing.

Campsite hardening is a tool used by land managers to mitigate landscape impacts from recreation. Evidence suggests campsite impacts occur within the first few uses of a natural site (Marion 1998; Hammitt and Cole 1998). This has led many managers to employ spatial containment strategies in an effort to reduce the overall area impacted (Leung and Marion 1999a). Campsite hardening is used in containment strategies and is usually accompanied by public education (pamphlets, maps) or use restrictions (designated camping policy). Hardening increases a campsite's resistance to additional impacts, so as the use increases, few additional on-site impacts occur. Hardened campsites can also provide amenities, such as an outhouse and a flat surface. Enhanced functionality of a campsite is intended to attract visitors. As visitors migrate to these durable hardened campsites, other campsites receive less pressure and may recover.

Currently there are seven Alaska State Parks–hardened and five Forest Service–hardened campsites in the Wilderness Study Area. David Sanders, a kayak ranger with the Forest Service, started hardening campsites in 2003 (D. Sanders, personal communication), and Jack Sinclair with the Alaska State Parks began campsite hardening in 1994 (J. Sinclair, personal communication). Camping is unrestricted, and campsites are filled on a first-come, first-served basis. A recent investigation in user experience was unable to document competition for campsites from a pool of survey respondents (see chapter 7), but anecdotal reports by the U.S. Forest Service have been received by the Chugach National Forest (USFS, unpublished data).

Recently there have been proposals to harden more campsites in Prince William Sound, primarily to prevent resource damage. Public scoping to enable campsite hardening projects in Forest Service–recommended wilderness areas occurred in the summer of 2010. Other land management agencies (Alaska Native and state) have also considered more campsite hardening projects. A Marine Trail System consisting of a string of hardened campsites between Valdez and Whittier has been discussed. A memorandum of understanding outlining a cooperative framework for development of the Valdez-Whittier trail has been signed by Alaska State Parks, Chugach

National Forest, National Park Service, National Wildlife Federation, and the Prince William Sound Economic District. Public meetings were held in Anchorage and the five Prince William Sound communities to gather public input about the trail system.

The 1964 Wilderness Act defines wilderness as a place "without permanent improvements," with "the imprint of man's work substantially unnoticeable." Hardened campsites are man-made developments that might be perceived as contradictory to the definition of wilderness. Some evidence suggests that hardened campsites are inconsistent with some visitors' perceptions of wilderness character (White et al. 2001; Cole and Hall 2009). Other research indicates that the functionality of a campsite, which is enhanced by campsite hardening, is the most important aspect of campsite evaluation (Heberlein and Dunwiddie 1979; Brunson and Shelby 1990). No research has examined the public's perceptions of hardened campsites in Prince William Sound, although about 5 percent of known dispersed campsites are hardened, and more are likely to be built in the future.

This research examines campers' perceptions and evaluation of campsite hardening in a coastal wilderness setting. Open-ended interviews were chosen to explore perceived functional and symbolic qualities of campsites, perceptions of crowding, coping, and opinions of campsite hardening in Blackstone Bay, Prince William Sound.

CAMPERS' PERCEPTIONS

Public opinion of campsite hardening is likely formed by two themes. White et al. (2001) suggest two potential dimensions of campsite evaluation exist: symbolic and functional. The first has to do with what visitors feel is consistent with the character of an area. The second relates to the amenities and functionality that a campsite offers.

SYMBOLIC

The symbolic dimension has to do with how campers perceive wilderness. The theory of cognitive dissonance (Festinger 1957) states that when a situation or setting fits into our understanding of the thing, an individual is satisfied; if not, an individual feels stress. Setting attributes may be consistent or inconsistent with the idea of wilderness. Setting attributes of many campsites include natural beauty, privacy, human impacts, litter, and/or degree of development. Campers may symbolically perceive ecological impacts and campsite hardening as inconsistent with the idea of wilderness, especially if these impacts are perceived as anthropogenic changes. Farrell et al. (2001) found that obvious human-caused impacts such as litter and graffiti carved in trees received negative reactions, while less obvious impacts which may or may not be recognized as human caused, such as vegetation loss due to trampling, did not receive negative reactions. Hardened campsites are anthropogenic modifications in an otherwise natural setting. If the pristine state of wilderness is highly valued by visitors, and campsite hardening is seen to conflict with that value, it is likely that hardened campsites will be perceived negatively by some users. Although in Prince William Sound there is a diversity of expectations and perceptions among users (see chapter 8), others may not perceive wilderness and hardened campsites as being in conflict.

FUNCTIONAL

The functional dimension of a campsite refers to the amenities needed in order to camp at a site. Amenities may include a flat spot to pitch a tent (or tents), access to water, and a place to hang food.

Hardening enhances the functionality of campsites, and the functional aspect of campsite evaluation is of primary importance (Brunson and Shelby 1990). This would suggest that hardened campsites are publicly desirable. Daniels and Marion (2006) found visitors' satisfaction increased after managers hardened campsites and restricted use to designated areas.

Functionally, campers may view some human-caused campsite impacts as desirable amenities (Farrell et al. 2001). For example, trampled surfaces devoid of vegetation may be viewed as a comfortable place to pitch a tent, and social trails may be perceived as convenient routes. Heberlein and Dunwiddie (1979) found through systematic observation that more parties selected and stayed longer at worn and littered campsites. Highly used campsites devoid of vegetation are functionally better in that they offer a flat spot for a tent. Symbolically, these campsites displayed evidence of past use and could not be considered pristine wilderness campsites. This suggests functional amenities are more important to campsite selection than the symbolic quality of the campsite.

In an open ended question—"What makes wilderness experiences different from other experiences?"—Cole and Hall (2009) found that 10 percent of respondents mentioned the lack of development and 14 percent said challenge. In the same survey, directional signage and well-constructed bridges were overwhelmingly considered to add to the wilderness experience. The authors conclude that either the added convenience of directional signage and bridges outweighed the deleterious effect of man-made objects in the wilderness or they were not considered to detract from the challenge. It seems that *in the abstract*, development is inconsistent with the symbolic nature of wilderness, but in reality, the convenience of minimal development adds to an experience.

It is important to note that the previously mentioned studies all occurred in higher use density areas than Blackstone Bay. White et al. (2001) suggests visitors may react more negatively to anthropogenic changes in areas of lower use densities, since wilderness character may be highly valued in these areas.

COPING MECHANISMS

Coping mechanisms occur in outdoor recreation as a response to conflict and crowding. They can manifest themselves behaviorally or cognitively. Displacement is a form of behavioral coping. When visitors are not satisfied, they simply go elsewhere. Cognitive coping includes rationalization and product shift.

Rationalization is based on the fact that visitors often put considerable effort, time, and money into their backcountry experience, and this is a voluntary investment. Since the perceived cost is high, high satisfaction is rationalized. Product shift occurs when expectations about an experience are not met. Visitors are likely to redefine the experience as something different than they initially thought in order to rectify inconsistencies between the expected and the actual experience. Both coping strategies are rooted in the theory of cognitive dissonance (Festinger 1957), which states that people will rectify inconsistencies in perceptions in order to reduce stress. Johnson and Dawson (2004) noted that working definitions of rationalization and product shift are not detailed enough to separate, so they are dealt with together.

Evidence of coping mechanisms (displacement, product shift, and rationalization) is prevalent in qualitative outdoor recreation research. In three Adirondack wilderness areas, Johnson and Dawson (2004) found coping in 53 percent of respondents. Manning and Valliere (2001) found that 94 percent of residents near an increasingly popular park employed some type of coping. Propst et al. (2008) found that although all respondents were either "satisfied" or "very satisfied" with their trip, 63 percent utilized some type of coping mechanism.

METHODS

STUDY SITE

Blackstone Bay's proximity to Whittier (about 10 miles away), the presence of two spectacular tidewater glaciers, and abundant wildlife make this area one of the most visited bays in the Sound (Wolfe et al. 2006; see chapters 7 and 19). Use is increasing in Blackstone Bay. Guide and outfitter use increased from 75 "visitor nights" in 1987 to 500 in 1998 (Colt 2002). Blackstone Bay is one of the areas in the Sound that is managed and monitored closely by the Forest Service for the important recreation experience it provides and its relatively high level of use, which peaks during summer and holiday weekends (see chapter 19).

Campsites in this area have three levels of development (figure 16.1). Decision Point State Park is a more heavily used campsite, which many pass on their way to Blackstone Bay and other areas. It is a hardened campsite at the mouth of Passage Canal. This is one of the most developed campsites in the Sound. There are

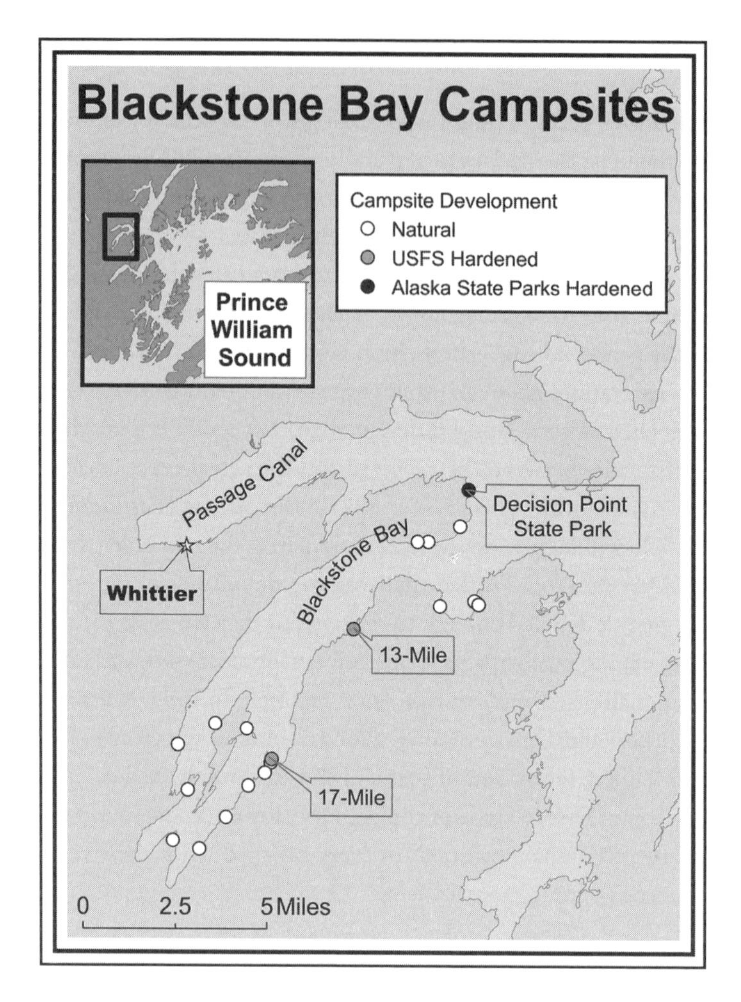

FIGURE 16.1. Map of natural and hardened campsites in and around Blackstone Bay.

an outhouse, boardwalk, four wooden tent platforms, a bear-resistant food safe, a bear hang, and several constructed gravel sites near the shoreline. Thirteen-Mile and Seventeen-Mile are Forest Service–hardened campsites, named for their relative distance from Whittier. An effort was made to keep these campsites as natural appearing as possible. Development is limited to three to four gravel pads and retaining walls made out of natural materials. There are 16 documented, naturally occurring campsites in Blackstone Bay. These campsites have formed naturally without any type of development. The size and popularity of these dispersed campsites vary.

Interviews were conducted in Blackstone Bay and at Decision Point State Park. These popular areas were chosen because they offered both hardened campsites and natural campsites, and the increased likelihood of finding visitors to survey.

INTERVIEWS

On-site, open-ended interviews were chosen to gather data. By conducting interviews on-site, campsite evaluations are less abstract. White et al. (2001) noted that on-site, open-ended or observational studies tend to contradict closed-ended surveys. Beaulieu and Schreyer (1985) conducted 325 on-site interviews in wilderness areas in Utah and Wyoming, and compared the results to a similar mail-back questionnaire from 255 participants. They found statistically significant differences between factors involved in hypothetical choice situations and actual on-site decisions. The authors conclude that on-site interviews more accurately reflect factors involved in real choices.

Qualitative research is not intended to be extrapolated to include the general population. It is an exploration of potential ideas and themes present within a group about a specific subject (Patton 2002). Twenty-two interviews were conducted ($n = 22$). This is a small sample of total users of Prince William Sound and does not represent a random sample of users.

Sampling was opportunistic. Campers were interviewed if they were seen at a set-up camp. Day users were not interviewed, because individuals who are not utilizing the site for camping are unlikely to be able to accurately evaluate its qualities. When approaching a group, a brief explanation of the research was given and the group was asked for a volunteer. A quiet spot nearby was selected, and an interview was conducted. Interviews were kept brief in an effort to preserve the visitors' wilderness experience. Most interviews were less than five minutes, as the questions asked were fairly simple. All groups approached participated, for a 100 percent response rate. All groups were kayakers, except for one in a Zodiac (small inflatable motor boat). All groups were independent travelers, without a guide, but many had rental kayaks.

Interviews occurred during the 2008 and 2009 summer seasons. Thirty-one days, including weekends, holiday weekends, and weekdays, were spent interviewing in Blackstone Bay.

Exactly half (11) of the interviews took place on unmodified, natural, dispersed campsites. Six interviews were conducted at Decision Point State Park. Three interviews took place at Forest Service–hardened site Thirteen-Mile and two were conducted at Forest Service–hardened site Seventeen-Mile for a total of 11 interviews conducted at hardened campsites.

Questions were intended to elicit information concerning campers' perceptions and evaluations of the campsite in an open-ended format prior to discussing campsite hardening. Visitors were asked what they liked and disliked about the campsite they were camped on. They were asked what they thought about the number of other campers in the bay in order to gather information regarding visitor perceptions of crowding. Then, if camped on a hardened campsite, visitors were asked what they thought about hardened campsites.

The interviews were recorded using a digital recorder and then transcribed and imported into NVivo7 software. Content analysis was used to categorize transcripts into themes. First-order coding was completed within a week of returning from a trip. Code categories were developed as they were discovered in the interviews. The interviews were then revisited and analytic coding completed the process (Richards and Morse 2007). Interviews were, once again, read through to insure the correctness of the coding.

RESULTS

Major themes from the interviews included functional and symbolic factors (to campsite evaluation), expectations of crowding, coping (displacement, rationalization, and product shift), and opinions regarding hardened campsites.

FUNCTIONAL AND SYMBOLIC FACTORS

Functional factors were coded to include anything affecting the functionality or usability of the campsite. Symbolic factors were coded as anything affecting the experience. When advantages and disadvantages of campsites were discussed, functional factors were mentioned more frequently (92 times) than symbolic factors (36 times). Functional factors also had more categories than the symbolic factors, suggesting more complexity in visitors' perceptions. In order from most frequently mentioned to least, functional factors included: flat spot (15), staying dry (13), wind cover (7), bear box (6), outhouse (6), view (6), access to water (5), boat concerns (4), bugs (4), easy access (4), fire (4), logistics (3), bear concerns (2), bear hang (2), boardwalk (1), and tent platforms (1). Again, from most frequently mentioned to least, symbolic factors included scenery (18), wildlife (8), campsite impacts (4), social factors (4), and litter (4).

EXPECTATIONS AND CROWDING

Perceptions of crowding were explored in the third question. Only one visitor reported feeling crowded at the time of the interview: "There are a lot of people." Many people indicated an acceptable level of crowding at the time of the interview but referred to crowding in a different area or at a different time. Visitors more frequently did not feel crowded (mentioned 29 times). Crowding was mentioned 18 times. There was much variability in attitudes in this category. For some, getting away from people was a main goal of the trip: "I'm trying to get away from people, generally speaking." Others felt a safety net associated with having others around, which was mentioned 3 times: "People are around so you know that you are secure if anything happens. You've got a safety net." Some felt glad that others were also out enjoying the wilderness: "It's good that people are getting out and they are recognizing the value of having wilderness and having places to use so . . . we can't just hog it all for ourselves."

Expectations of crowding were mentioned on 8 occasions, although no questions were directly asked about expectations. Visitors had either visited the area before, heard about it from a friend, or just expected crowding based on the bay's proximity to Whittier and the presence of two tidewater glaciers.

COPING

Coping was expressed in the form of displacement, rationalization, and product shift. Many visitors had a hard time being critical of campsites and their experience in general. Overall, visitor satisfaction was high. Four respondents could not think of anything wrong with their campsite, even when prodded with additional questions like, "Is there anything that could maybe just be a little better or is less than ideal?" Although satisfaction was high, 55 percent of interviews expressed some type of coping mechanism.

DISPLACEMENT

Displacement was mentioned in seven interviews (32 percent). Three respondents reported that their campsite choice was, in part, a result of other campsites being occupied: "We went past a few sites on the way in and for the most part they were all filled up." Some spoke of displacement in the abstract. They would in the future go somewhere else if the area became too crowded, or they had in the past gone elsewhere because of crowding (two interviewees): "Since they put the tunnel in, since you can drive through and that increased the number of people quite a bit I just sort of boycotted it for a while and went to other places." One spoke of being displaced because of resource conditions: "I did notice that at some of the sites it is very muddy and so that is what deterred us from staying there, is that it was just really muddy and wet and trampled and so it wasn't a site that was very desirable." Interestingly, one visitor spoke of being displaced by hardened campsites: "I usually try to avoid areas like this that are all built up. I generally don't like it too much."

PRODUCT SHIFT AND RATIONALIZATION

Evidence of product shift or rationalization was found in six interviews (27 percent). Product shift involves the negative perception of an experience attribute followed by a statement defining the area so it is consistent with perceptions. For example, "I don't know I'm not a huge fan of the, like, that's a tent spot, and that's a tent spot. That's not really my thing, but it's also the nature of the bay, I think." And, "You think you are in a beautiful remote wilderness, but because of the accessibility to Whittier, there are quite a few boats driving by and the big tour boats are loud, a daily thing, so you can count on big boats driving by right past your camp, which is not that cool, but it is what it is."

Rationalization was expressed when visitors said something negative about the campsite or their experience and then followed it up with a "but it's okay" type of statement. They are rationalizing their experience attributes as positive. For example: "I would love to go back in the early mid-eighties, when there were only a few people out here and everybody you met you knew, that would be great, but that's not the reality."

PERCEPTIONS OF HARDENED CAMPSITES

Support for hardened campsites was driven by four factors: an understanding that hardened campsites are intended to protect the resource, they are limited to high-use areas, they offer functional amenities, and they are natural appearing. Opposition to campsite hardening centered on the perceived loss to wilderness character.

Eight visitors expressed an understanding that campsite hardening was initiated to protect resources from damage caused by recreation. For example, "I think that in this spot it is totally justified to have a structure for people to be, so it doesn't deplete more of the surrounding forest. People kind of scatter and make more, sort of muck holes. This keeps people contained."

Five visitors understood the need for hardened campsites in high-use areas but expressed an appreciation that they were not everywhere. For example, "I think this would not be applicable out in places that don't see as much kayak traffic or I think that it would be sort of overkill in places that don't see the kind of use that this place does, so I think that if I was way out in the Sound away from any towns like Cordova, Valdez, or Whittier I would be kind of overwhelmed with this sort of platforms and structures and stuff." And, "I guess it would be harder to swallow if you were farther out and looking for that remote wilderness and you found this, as opposed to being so close to Whittier at such a popular site running into this is a little easier to take right here."

When asked what they liked about their campsites, three visitors camped at natural sites mentioned the lack of development: "It's kind of nice because it's not developed." Alternatively, two visitors, both camped at Willard Island, expressed a desire for some type of campsite hardening. These two parties were camped in crowded conditions (two parties camped nearby) at the most highly used natural campsite in the bay.

Nine out of 11 visitors camped at hardened campsites mentioned an appreciation for the added functionality that hardened campsites offer. For example, "It's convenient, you know." And, "It's a no brainer. You pull up and it's not like you have to look for a place to set your tent. You know, it's dry, it's flat, so it's good."

The appearance of a campsite was brought up three times. Natural-looking campsites seem more consistent with what visitors feel is symbolically appropriate in wilderness: "I don't like to see it (the campsite) where, like, it is so obvious. Obviously subtlety is nice."

Forest Service–hardened campsites strive to be as natural appearing as possible, and to some visitors they looked like naturally occurring campsites. One visitor did not realize that Seventeen-Mile was a man-made spot, and two other visitors suspected that the Forest Service–hardened campsites were man made and asked for verification from the interviewer.

Two visitors felt overwhelmingly negative about hardened campsites, and the conflict of development in wilderness was mentioned or alluded to by 7 out of the 11 visitors camped at hardened campsites. For example, "I was out here before there was any hardening at all at this site, it was just a wilderness site, like every other one, and, again, you could come here and feel like you were the first person here, it was very easy to believe that and you can't do that anymore, so that's a loss." And, "I don't like to see that sort of thing, because it is not really in the spirit of wilderness."

DISCUSSION

FUNCTIONAL AND SYMBOLIC FACTORS

The fact that functional concerns were mentioned more frequently and contained more complexity implies that these factors were more important to campsite selection and evaluation than symbolic factors. This is consistent with other research (Brunson and Shelby 1990; White et al. 2001). It intuitively makes sense that when people are searching for a campsite, their first considerations would be the following: *Does the campsite work? Is it large enough for the group? Is there a flat place to put a tent?* Although these considerations are usually primary, the importance of symbolic factors should not be dismissed. These were frequently discussed and usually with much enthusiasm.

EXPECTATIONS AND CROWDING

Visitors more frequently reported lower encounter levels than expected. This is consistent with other research. In 2006, 71 percent of visitors at the Forest Service Backcountry Informational Yurt in Whittier who were returning from Blackstone Bay reported levels of crowding were less than or equal to expectations (USFS, unpublished data). This was a similar finding to that of recent user experience work in chapter 7. Generally speaking, visitors expect some level of crowding in Blackstone Bay, and actual conditions don't seem to be exceeding expectations of crowding.

Expectations play an important role in trip satisfaction. When the number of people encountered exceeds expected encounter levels, satisfaction is likely to decline, but when encounters are less than expected, satisfaction remains high (Shelby et al. 1983; Tseng et al. 2009). This may partially explain why visitors more frequently reported not feeling crowded (96 percent), although this is one of the highest-use bays in Prince William Sound. Similar results were found throughout the Sound in the Evaluation of Prince William Sound User Experience study, where 95 percent reported crowding had no effect on their experience (chapter 7).

Although most visitors did not report perceptions of crowding, there was evidence of displacement, a result of crowded conditions.

COPING

Evidence of coping emerged during this process, although no questions were specifically asked pertaining to coping mechanisms. It is possible that coping mechanisms would have become apparent had questions been designed toward that end. However, it is significant that visitors discussed some type of coping even without specific questions to prompt them. Overall, 55 percent of interviews contained some evidence of a coping strategy being used. The pervasiveness of the use of coping strategies is corroborated by similar research (Manning and Valliere 2001; Johnson and Dawson 2004; Propst et al. 2008).

Manning and Valliere (2001) add a few cautions based on evidence that coping mechanisms are prevalent in outdoor recreation research. The authors warn that measures of visitor satisfaction may be misleading. One cause occurs when displacement removes visitors who are more sensitive to crowding from the respondent pool in on-site surveys, so that dissatisfied visitors may not be present to voice their dissatisfaction. Satisfaction also may remain high through coping, although visitors are encountering situations that are less than ideal. Coping on the individual level may be stressful; at the societal level, it may reduce recreational opportunities. As people adjust their definition of an area or are displaced from an area, satisfaction remains high, so managers continue a policy with negative effects to wilderness character. Nationwide recreational opportunities could move from low density of users and wilderness settings to higher density and a more developed setting. This is an interesting point and worth noting. As recreation use increases, recreational opportunities are likely to become increasingly developed to accommodate more visitors.

This trend of increasing use and development is occurring in Blackstone Bay and possibly the rest of Prince William Sound. Use data suggest more visitors are using the Sound, and development of recreational infrastructure is also increasing. The Forest Service began hardening campsites in 2003, and Alaska State Parks began in 1994. Both agencies have plans for more campsite hardening projects. In 1993, Blackstone Bay had no recreational infrastructure. It has moved from an undeveloped setting to a slightly more developed setting. With an increase of users, recreational opportunities have moved from lower visitor densities to higher visitor densities (Colt et al. 2002).

PERCEPTIONS OF HARDENED CAMPSITES

Many respondents expressed an understanding that campsite hardening is important to protect resources, and therefore they are supportive of campsite hardening.

This is consistent with other research. Bullock and Lawson (2007) found that visitors were generally supportive of resource protection interventions. Since visitors were supportive of campsite hardening as a mitigation measure, it makes sense to wait for these impacts to occur before initiating a hardening project. Hardening projects that are driven by campsite impacts, using reactive management, may be more likely to gain public support. Education concerning how hardened campsites protect the resource may also increase public support.

Five respondents of the 11 who were asked about campsite hardening were accepting of hardened campsites in Blackstone Bay but expressed an appreciation that there were other areas in Prince William Sound that were not developed. The cost to wilderness character in remote areas of the Sound was deemed too high. Interestingly, Blackstone Bay was perceived as something different from a wilderness setting. This is corroborated by the prevalence of expectations of crowding. Other research has shown that visitors perceptually zone wilderness (Lucas 1964; Stankey 1971). It seems most visitors have defined Blackstone Bay as a high-use, "periphery" zone, and they expect the things that come with high use, such as hardened campsites. Campsite hardening in more remote areas may be less acceptable to the public, as suspected by White et al. (2001).

Two interviewees at natural campsites suggested that their campsite might be improved by some sort of development. This sentiment may have been driven by the relatively crowded conditions at the time. Conversely, three visitors appreciated the undeveloped nature of their campsite, before campsite hardening was discussed.

Almost all of those (9 out of 11) camped at hardened campsites expressed an appreciation for the added functionality hardened campsites offer. Functionality of a campsite is an important determinant of perceptions of campsite quality. A minimal amount of unobtrusive development, which serves an important function, seems generally appreciated. In the abstract, lack of development seems to fit with individuals' idea of wilderness, but in reality, people enjoy amenities to make their experience easier in Blackstone Bay, which is consistent with the findings of Cole and Hall (2009).

Natural-looking sites are more consistent with what visitors feel is appropriate in wilderness. Bullock and Lawson (2007) found management structures that were constructed of natural materials and blended with the natural surroundings had little effect on visitors' experience. Marion (1987) recommends that treatment techniques used for mitigating impacts ought to be less visually and ecologically obtrusive than the original problem. As previously mentioned, three interviewees were not sure if Forest Service–hardened campsites were man made. The Forest Service may be able to reach goals of consolidating use without intruding into visitors' wilderness experience by making hardened campsites appear natural. Without the realization that these are man-made campsites, the conflict of infrastructure in wilderness would not be a consideration. This is an outstanding opportunity. Future hardened campsites, especially in the Wilderness Study Area, should strive to be as natural appearing as possible.

LESSONS LEARNED

Overall, this research suggests that campsite hardening is a publicly acceptable management tool, but it is not without cost. The perceived benefits of hardened campsites centered on the enhanced functionality (82 percent) but also included resource protection from recreational impacts (73 percent). Hardened campsites were considered acceptable if present in high-use areas but less acceptable in remote locations (45 percent). The perceived cost to wilderness character is likely higher in more remote areas of Prince William Sound. Many visitors camped at hardened campsites alluded to a cost to the wilderness character (64 percent). Hardened campsites were frequently perceived as inconsistent with the symbolic idea of wilderness. When planning for developed campsites, the cost to wilderness character of an area ought to be weighed carefully against benefits to the public and the resource.

The U.S. Forest Service is faced with a dual mandate when managing forest lands within the Nellie Juan-College Fjord Wilderness Study Area. The Wilderness Study Area is managed under the 1964 Wilderness Act, which requires agencies to preserve the natural integrity of the wilderness resource while providing for its continued recreation use. The act uses vague language when it defines wilderness as "untrammeled," "without permanent improvements," and "with the imprint of man's work substantially unnoticeable." The management agency is to "preserve the wilderness character." At times contradictory to this, wilderness is intended to "be administered for the use and enjoyment of the American people." Often these two mandates conflict. Campsite hardening is a specific example of this conflict. It is a tool used to manage use and enhance "enjoyment of the American people." It enhances social aspects of recreation by containing use in small areas and potentially separating user groups. But hardened campsites are an "imprint of man's work," and in a sense, they "trammel" a wilderness. Hardened campsites are man-made developments that may be perceived as contradictory to the concept of wilderness. It is not clear where to draw the line when developing the wilderness for "use and enjoyment" at the cost to "wilderness character." This research was intended to elucidate this issue though the voices of people out camping and enjoying Prince William Sound.

This work was intended to address common communication challenges, not only from the management agency to the public, but from the public to the management agency. Often, management agencies endeavor to build support for management actions through educational campaigns. This research offers insights into what is important to recreationists when discussing hardened campsites. This information could be used to focus educational efforts. Support for hardened campsites was driven by four factors: an understanding that hardened campsites are intended to protect the resource, they are limited to "high-use" areas, they offer functional amenities, and they are natural appearing. Opposition to campsite hardening centered on the perceived loss to wilderness character. Educational campaigns could focus around these themes.

It is important that the Wilderness Study Area is managed in accordance with the public's desires. There are many communication challenges associated with realizing users' desires. Community meetings are often poorly attended and are often composed of only a subset of the users of an area. By conducting on-site interviews, this research was able to reach out to those who are using a place but may be unlikely to express their opinion through the usual avenues. A good example of this was the British gentleman interviewed at Decision Point (the most developed campsite). When asked what could be better about this campsite, he commented that the outhouse was pretty stinky. It didn't occur to this gentleman that the place was "developed." This is not someone who would be present at a community meeting about the development of Prince William Sound, but European visitors are a significant set of Sound users. Through this research that user group was given a voice. Being able to reach people while they are at their campsite also puts management decisions in context and encourages relevant ideas that may not come up in the meeting hall.

Although this type of research is difficult and sample sizes in remote areas may be small, it is important in overcoming communication challenges among resource managers and recreation users of a place.

STAKEHOLDER ESSAY

TAKE ME FOR A HIKE IN THE WOODS (BUT NOT THOSE WOODS)

JEREMY ROBIDA

WHEN I WORKED AS a sea kayak guide in Prince William Sound, I had the opportunity to work with a variety of people on an almost daily basis. Clients ranged in age and ability, but most were in Alaska for the first time, most seemed to be in average or above-average fitness level, and most craved a "wilderness experience." The bulk of trips I guided were day trips to specific places and highlights such as Columbia or Valdez Glacier. We'd water taxi out, paddle for several hours, talk geology/plants/animals, drink some hot cocoa, and snap some pictures, and then I'd deliver them home safely and undress them from the boots and rain gear I had issued earlier that morning. There were also camping trips lasting between one to six nights out, which involved packing all food, gear, and equipment we'd need. Over the time I spent at this job (yes, "job," because believe me, it's not as easy as what you'd think), I noticed that people had significantly different attitudes as to what this "wilderness experience" meant.

This concept really struck home for me on a three-day outing. My particular clients were three middle-aged gentlemen who fell into what we in the guide world refer to as the "high-maintenance" category. This client category is generally not your favorite type, as they are demanding, require a lot of attention, and have a lot of expectations as to what the trip will, or should, be like. Categories such as the "nonstop picture taker" or "everything is spiritual and I feel blessed to be here" or "feed me little chocolate snacks and I'll be your friend" types are significantly easier to please. The flip side of the high-maintenance category is that if their needs are met, these clients tend to be good tippers at the end of the trip. So, even though you'd prefer to hang out and paddle with the "I'm broke but super cool and adventurous" category, it's generally more financially advantageous to deal with people who truly need your guide skills, trust your judgment, and pay you for it.

The night before our paddling adventure began, I had a chance to meet the clients. We used this time to discuss gear and trip logistics and look at maps. Clients and guides had a chance to get a feel for one another and cover last-minute details. I knew almost instantly these guys would be tough to please. No trouble though . . . the next three days would be perfect sunny weather and calm seas, and we'd be paddling some amazing coastline. They wanted to see animals and do some hiking in the remote places we'd visit; they wanted a true Alaska wilderness experience, they stressed.

With our water taxi laden with gear and kayaks, we departed the docks that following morning. Spirits were high, digital cameras snapped frame after frame, and the sun shone down. Perfect. Easy. We arrived at our drop-off point fairly quickly, unloaded, and watched as the water taxi roared off and left us alone on the beach. At that point, things changed.

I always cover the basics first: proper bathroom procedure. As I explained the finer points of doing your business in the intertidal zone, I received some puzzled looks. There was a long pause. "No outhouses anyplace?" they asked. "No," I replied with a puzzled look back.

Things kept getting stranger. We loaded boats and started paddling. Just a short distance later, the "are we there yet" question surfaced. I explained that our campsite was roughly one hour away. Okay, I'm thinking to myself, I'm screwed. We'd barely paddled for 15 minutes and they were already tired. This was a three-day trip, during which we'd actually have to paddle for hours and cover some bigger distances.

As promised, we soon set foot on the beach. I explained that we would head off for a hike after unloading boats and getting situated. They agreed with this plan and everyone got to work organizing gear. It was sunny, waves lapped the beach gently, and our camp looked like the cover of *Alaska* magazine. As they erected tents, I could overhear their excitement to go hiking.

"Where's the trail?" they asked as we ducked into the brush line off the beach. As I gingerly held back a large devil's club leaf, I explained they'd have to watch out for certain things, such as this particular plant. I further launched into my well-rehearsed "yell for bears" and "don't touch" speech as we prepared to climb a steep embankment and enter the old growth forest. Again, and with more urgency: "Where's the trail?"

Reality was setting in for me. These folks had said they wanted a "wilderness experience." But what they really wanted was a manicured, domesticated, and subdued version of "wilderness" containing park rangers, toilet seats, and concessions stands. In an attempt to salvage the hike, I explained that we were on a unique island, and that 50 people at most would set foot onto this particular beach the entire summer. Panic grew as they listened to their guide talk about the remote and desolate qualities of the place. "We'll get lost," they muttered, chastising me for even considering bringing them into the dark and scary woods. No luck. Our hike ended as soon as it began.

For the remaining two days I spent with these clients, I did everything I could to shield them from the "Alaskan wilderness experience" they had expressed such great interest in, initially. Instead, my challenge became keeping them pacified via copious amounts of food, keeping heartbeats at a resting tempo, and doing all the camp chores. I finished the trip, received a hearty cash tip, and wished them luck with the rest of their Alaskan adventure.

So just what is a "wilderness experience"? Authors—including Henry David Thoreau, Edward Abbey, Terry Tempest Williams, and John Muir—have wrestled with that question far more intimately than I ever desire to. And frankly, as I figured out guiding, we've all got our opinions and expectations about what we want to discover out in the woods. It was interesting for me to see the "wilderness experience" discussion manifest itself in a guide/client perspective—and then, of course, to try to meet those expectations and please those people.

CHAPTER 17

TRENDS AND CHARACTERISTICS OF CAMPSITE CONDITIONS IN PRINCE WILLIAM SOUND

PAUL TWARDOCK AND CHRISTOPHER MONZ

SOUND BITES

THE HARDENING OF sites by using beach gravel to make durable platforms is one tool that managers can use to direct use away from biological hot spots and possibly avoid campsite proliferation.

Leave No Trace education should continue to encourage visitors to use durable cobble sites that are devoid of vegetation and already impacted to avoid the spread of campsite impacts.

The use of condition class ratings should be augmented, or even replaced, with a three-tiered approach that reflects the most significant and easily measured impacts.

INTRODUCTION

The demand for wildland recreation and nature-based tourism opportunities continues to increase in many protected areas in North America (Cordell 2008) and worldwide (De Lacy and Whitmore 2006). Increased use can result in human disturbance and changes to the environmental conditions of protected areas and associated management efforts directed at minimizing these impacts. Being able to detect changes in patterns of resource impacts associated with campsite degradation is one tool that managers can use to indirectly gain insights about use levels in an area. Relationships between user experience and campsite character are complex (e.g., chapter 16). However, with an understanding of the types of impacts deemed unacceptable to users, managers can use condition trends at campsites as a possible early warning system for changes in user satisfaction.

Managers can also use evaluations of campsite conditions as one indicator of the effectiveness of management actions being taken to achieve resource protection goals at sites. For example, the effectiveness of Leave No Trace education efforts to mitigate resource damage in a particular bay or drainage might be assessed by an effort to monitor changes in resource conditions for sites in that region. Having this type of an understanding of effectiveness of such mitigation strategies is essential for many aspects of adaptive management in the Sound, and in other systems managed for sustainable human use. This chapter presents an analysis of a long-term data set for condition class of campsites in the Sound and a monitoring methodology appropriate for other coastal regions.

Given the increases in wildland use, a common concern among managers of protected areas is resource change due to backcountry camping. Primitive camping is a popular wildland recreation activity with the potential to affect resource condition intensively at the on-site scale and extensively due to site expansion and proliferation (Leung and Marion 1999b; Cole 2004a). Campsites are important from a managerial and visitor perspective, as they serve as focal points for visitor activities, thereby creating nodes of concentrated use. Moreover, this concentration of use frequently results in localized and intense resource impacts, which if severe are often judged as unacceptable by managers and recreationists. Studies of informal (visitor-created) campsites in parks and protected areas are common in the recreation ecology literature, often reporting the degree to which visitor use can affect change on site ecological conditions (e.g., Frissell 1978; Cole 1983a; Monz and Twardock 2004). Long-term studies, involving repeated examination of the trends of change over long periods, are more rare (Cole and Hall 1992; Marion and Cole 1996; Cole et al. 2008).

Reviewing the campsite monitoring literature reveals some commonalities and distinctions in approaches. Early approaches (Frissell 1978) typically employed condition class systems that evaluated campsites via an observer-based rating scale along a continuum of minimally impacted through highly impacted. Still applied frequently today, this approach can be performed very rapidly, provides some quantitative information, and is an easy way to classify individual sites. Alternatively, multiple-indicator methods (Cole 1989; Marion 1991; Newsome et al. 2001) have gained application as they address several limitations of condition class systems. First, they minimize overreliance on one measure that may suffer from observer bias and provide more information on specific indicators, which enhances sensitivity and precision of measurements. Some condition class systems assume that two or more variables (such as vegetation loss and soil loss) covary as degradation increases. If this does not consistently occur, proper classification of sites becomes more difficult for the observer, and even when performed reliably, the ability to draw conclusions on either variable is somewhat compromised. Various multiple-indicator approaches have been developed, with some approaches including multiple observation-based scales and others combining scale ratings with continuous-measurement indicators. Marion (1991, 1995) suggests combining condition class and multiple-indicator measurement approaches, thus providing robust data yielding a wide range of information to land managers. While more complex than condition class systems, multiple-indicator approaches can be performed fairly rapidly in the field and have also received widespread application.

Managers find campsite assessment studies useful as they often seek to minimize undesirable resource impacts and the associated aesthetic degradation of sites to maintain high quality wildland experiences for visitors. Although less common, other types of campsite studies have used experimental designs to examine functional relationships such as use impact (Cole 1995; Cole and Monz 2004a) and spatial patterns of impact (Cole and Monz 2004b). Overall, several generalizations about campsite condition trends can be drawn from this literature. First, over time on established sites, changes in the number and areal extent of impact tend to be more pronounced than changes in intensity of impact. For example, Cole and Hall (1992) studied campsites over an 11-year period in the Eagle Cap Wilderness in Oregon and found campsite size increased substantially, but mean vegetation cover was relatively stable. Similar results were found over a 20-year period in Grand Canyon National Park (Cole et al. 2008). Second, aggregate impact (increased number of sites and total area of disturbance) tends to increase over time and may be more of a management concern than the level of degradation at individual sites. For example, Cole's (1993) assessment of three wilderness areas in the western United States found that over a 12- to 14-year period, the total number of sites increased substantially in each area, but degraded in resource condition in only one area. Finally, assessment studies suggest and experimental studies confirm that on a given site, most impact occurs at low use levels, and subsequent increases in use do not result in proportional increases in impact (Leung and Marion 2000; Cole and Monz 2004a; Cole et al. 2008). Overall, these findings support the importance of campsite assessment studies in informing management actions to maintain the quality of resource conditions.

Research has also been conducted to evaluate the effectiveness of both at-large and confinement-camping management strategies in wildlands. Confinement seeks to limit the number and size of campsites. Typically, visitors are encouraged to camp on designated sites or in areas where campsites have previously been established and to concentrate traffic in the core areas of sites. Alternatively, at-large camping allows the visitor to select places to camp provided the locations are in accord with overall regulations, such as distance to water. Several studies have assessed the effectiveness of confinement strategies with somewhat contrasting results. One conclusion, based on research conducted in the eastern United States, has found confinement to be effective at limiting the formation of new sites and reducing existing site numbers to a few well maintained, desirable sites (Marion 1995; Marion and Farrell 2002; Reid and Marion 2004). This rationale is a well-accepted practice in some parks and wilderness areas, particularly in locations where visitor-use pressure is high (Leung and Marion 2000) and is central to current best practices for minimum-impact camping (Hampton and Cole 2003; Twardock 2004). In contrast, Cole et al. (2008), in a 20-year period in Grand Canyon National Park, found more new site formation in areas under a confinement strategy than in areas that encouraged at-large use. While they provide some speculation as to the cause of this phenomenon, overall, these results demonstrate the challenges of minimizing site proliferation, even while employing strategies shown to be effective elsewhere.

This project was initiated as a large-scale and long-term effort to assess the location, extent and condition of informal campsites in popular coastal camping areas of Prince William Sound, Alaska, and to track these characteristics over time. The project was conceptualized with four overall goals that will form the basis of the discussion in this chapter. First, since the literature on change in biophysical conditions of campsites over a prolonged period is currently limited to a few published studies (e.g., Cole and Hall 1992; Cole 1993; Cole et al. 2008), a primary objective was to characterize changes in campsite condition over a long period—13 years in this study. No studies of this kind have been conducted in northwest coastal environments, especially in coastal Alaska, with the exception of our previous work in Prince William Sound (Monz and Twardock 2004; Monz 1998). To fulfill this objective, we discuss changes in campsite size, vegetation cover, soil exposure, and other impacts commonly associated with campsites. A second objective was to evaluate the consequences of the current at-large camping management strategy from the perspective of new site creation and site abandonment. Current knowledge of the effectiveness of at-large management is also limited to the few studies and environments previously described. In these regards, we provide an analysis of 15 camping beaches where the total possible camping area is limited and where we have tracked site characteristics and proliferation over time. A third objective was to examine whether impacts vary by specific environments where campsites were found in the Sound. To accomplish this objective, we conducted an analysis of resource condition data by the primary campsite substrates. In an at-large camping strategy, these results can help visitors select sites where the potential for increases in impacts is lessened. A final objective was to examine established campsite protocols to determine whether improvements could be made for continued assessments in Prince William Sound and for the broader field of campsite assessment. For this, we conducted an in-depth statistical analysis of the multiparameter campsite condition data to examine whether a method of classifying sites could be developed that would be an advancement of established condition class approaches.

This study contrasts with many previously published works examining long-term trends in campsite conditions because the authors and associates conducted all assessments with the same protocol, likely making the findings more consistent over time. This chapter provides a summary of the overall findings of this project with a discussion of both the implications of our findings to management of Prince William Sound and suggestions for conducting future extensive campsite studies in the Sound and elsewhere. We also refer the reader to our previous papers from this project on trends in resource conditions in the Sound (Twardock et al. 2010) and on a site-classification analysis (Monz and Twardock 2010) for a more in-depth discussion of our findings.

METHODS

STUDY AREA

Prince William Sound, Alaska is located approximately at 61° north, 148° west and spans a large geographic area of over 10,000 km^2. Most areas in the Sound are remote and accessible by road only from the port towns of Valdez and Whittier. Over the last three decades, tourism and recreation have become principal economic activities, with commercial sightseeing tours, cruise lines, chartered hunting camps, personal watercraft tours, and sea kayak outfitters now operating extensively in the area (see chapter 3). The wild nature and wilderness character of Prince William Sound are a primary attraction for many visitors. The Chugach National Forest manages most of the uplands of Prince William Sound, including the 800,000 ha Nellie Juan Wilderness Study Area. In addition to the national forest, there are state marine parks, Alaska Native village and regional corporation lands, municipal lands, private lands, and state university lands adjacent to the Sound.

Visitor management is complicated by the large and geographically complex nature of Prince William Sound. There are few access points and over 4,800 km of shoreline consisting of rocky cliffs interspersed with beaches of a gray sandstone (graywacke) and slate (N. Lethcoe 1987; J. Lethcoe 1990). Upland visitation in the temperate spruce/hemlock rain forest is limited due to the boggy nature of most soils above the beaches. The majority of use in western Prince William Sound occurs in the summer months due to harsh conditions between September and April, although hunting may be a substantial use during this time.

Prince William Sound is surrounded by a closed Sitka spruce/western hemlock (*Picea sitchensis/Tsuga heterophylla*) forest (Viereck et al. 1992) typical of the northwest coastal region and Southeast and Southcentral coastal Alaska (Barbour and Billings 1991). The forest understory includes highbush blueberry (*Vaccinium ovalifolium*), ferns (*Athyrium filix* and *Dryopteris expansa*), horsetail (*Equisetium* spp.), dwarf dogwood (*Carnus canadensis*), devil's club (*Oplopanax horridus*), wintergreen (*Pyrola asarifolia*), skunk cabbage (*Lysichiten americanum*), and peat mosses (*Sphagnum* spp.). Thin, wet, acidic soils overlay bedrock and gravel substrates. Beaches consist of graywacke sand and gravel with plant communities of dune grass (*Elymus mollis*), and succulents such as seabeach sandwort (*Honkenya peploides*), beach pea (*Lathyrus maritimus*), and oysterleaf (*Mertensia maritima*).

VISITOR USE IN PRINCE WILLIAM SOUND

Although current use data are lacking, past-use trend analysis and anecdotal information suggest that the Sound has remained in high demand for backcountry camping experiences over the last two decades. For example, Twardock and Monz (2000) reported a near doubling of total kayak visitor use days during an 11-year period from 1987–98. Chandra Poe, a geographical information systems specialist with the U.S. Forest Service, noted that Forest Service outfitter/guide user day data for the area has shown a consistent increase since 2004 (C. Poe, personal communication). Moreover, relatively recent (2000) construction of road access to the port town of Whittier has increased tourism traffic, with available data indicating that between 2000 (the year access changed from rail only to road/rail) and 2007, vehicle use increased an average of 4 percent per year (ADOT 2009). These overall trends are suggestive of ongoing demand for backcountry experiences in Prince William Sound and corroborate findings such as those discussed in chapter 11.

CAMPSITE ASSESSMENT

We used campsite assessment protocols suggested by Cole (1989) and Marion (1991) with minor adaptations to coastal Alaskan environments. Over the duration of the study, we chose to consistently apply the same protocol to allow for an exact comparison of site attributes (table 17.1) over time. Assessments were performed during the summer growing season (June through August). Measurement of vegetation cover and soil exposure followed the ocular measurement approach suggested by Marion (1991), and for each campsite, an undisturbed adjacent area was selected as a control for vegetation loss calculations. We used the variable radial transect method for measurement of campsite areas (Marion 1991, 1995). Condition class measurements were obtained by ocular estimation on a standard condition class scale (e.g., one through five numerical ratings from minimal to severe impact) as suggested by Marion (1991).

TABLE 17.1. Site attributes, assessment methods, and measurement scale

SITE ATTRIBUTE	METHOD USED	MEASUREMENT SCALE
Area of observable impact	Radial transect measurement	Square area of campsite
Condition class	Ocular estimation	Five-level condition class scale
Fire sites	Counts	Total number of fire sites present
Informal trails	Counts	Total number of trails present
Mineral soil exposure on site	Ocular estimation	Six-level cover scale (%): 0–5, 6–25, 26–50, 51–75, 76–95, 96–100
Stumps/cut shrubs	Counts	Total number of cut stumps present
Vegetative ground cover on site and in control areas	Ocular estimation	Six-level cover scale (%): 0–5, 6–25, 26–50, 51–75, 76–95, 96–100
Campsite substrate type	Observation	Cobble, organic soil, sand
Human waste sites	Ocular estimation	Three-level human waste scale
Litter and trash	Ocular estimation	Four-level trash quantity scale
Root exposure on site	Ocular estimation	Three-level root exposure scale
Tree damage	Ocular estimation	Three-level tree damage scale

Slight modifications to the condition class definitions were made to be more applicable to Prince William Sound (see Kehoe 2002). Other site attributes (table 17.1) were assessed via the guidance suggested in Marion (1991). Additional procedural details for this study are available in Monz (1998) and Kehoe (2002).

Assessments were conducted via a census approach with beach locations informed by local guides, educators, charter boat operators, and Forest Service kayak rangers (figure 17.1). The assessment conducted in 2008 was a full remeasurement to the extent possible of all known camping beaches identified during the previous 12 years of the study. We also identified 15 camping beaches with multiple sites and tracked resource changes on these beaches over time. These beaches were selected based on their popularity, their proximity to well-travelled routes of wilderness users, and their attractive geographical features such as proximity to tidewater glaciers and camping amenities. In addition to providing site-level information, this approach provided an analysis at a larger spatial scale, allowing measurement of site proliferation and total aggregate impacted area at these beaches.

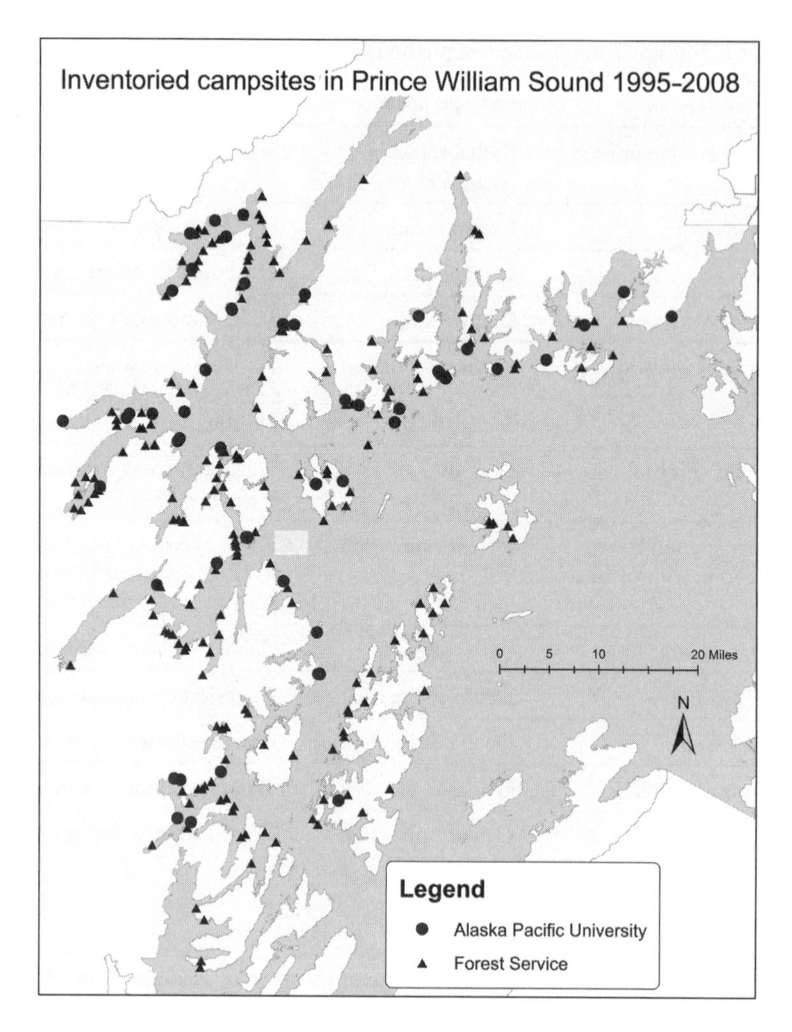

FIGURE 17.1. Inventoried campsites in Prince William Sound, 1995 to 2008.

DATA PROCESSING AND ANALYSIS

Campsite areas were determined geometrically from the variable radial transect data by utilizing a custom computer program (Dr. J. Marion, Virginia, 2008). Relative vegetation cover loss was calculated by the following formula:

$$\text{Vegetation Cover Loss} = 1 - \frac{\%\ \text{cover in campsite}}{\%\ \text{cover in control plots}} \times 100$$

All data were summarized and synthetic variables were calculated using Microsoft Excel (Microsoft 2007) and SPSS statistical software. We followed measurement scale definitions suggested by Vaske (2008) and used parametric approaches where variables could be defined as continuous. Where variables did not satisfy the criteria for continuous measures, we report non-parametric statistics. Paired T-tests, Wilcoxon Signed Rank tests, ANOVAs, and Kruskal-Wallis Tests were conducted using standard procedures.

For the multivariate analysis of the multiparameter assessment methodology, we followed suggestions by Jolliffe (2002) and Tabachnick and Fidell (2001) in terms of appropriateness of the variables and general approaches, and we used an analysis strategy similar to that of Leung and Marion (1999b). Initially, we conducted an exploratory-factor analysis (principal components extraction) on resource-assessment variables to reduce the 10 impact indicators into a meaningful set of interpretable factors. This was followed by a standard K-means cluster analysis on the factor scores from the three-factor solution in order to classify the sites in accord with the impact characteristics. The overall approach allowed for an examination of the most relevant impact indicators and for the development of an empirical classification of sites according to the multiparameter data.

RESULTS

CAMPSITE CONDITION TRENDS

For sites monitored for a 10- to 13-year period, paired T-tests and Wilcoxon tests comparing the initial measurement with the 2008 assessment indicate significant decreases in vegetation cover loss ($T = 2.77$, $p = .007$), condition class ($T = 3.35$, $p = .001$), trash rating ($Z = -4.31$, $p = .000$), and number of access trails ($T = 2.46$, $p = .017$), and a significant increase in area of observable impact ($T = -2.59$, $p = .019$). Trend data on changes in campsite condition are presented over a 10- to 13-year time frame due to variations in the time span from when sites were initially assessed to the comprehensive 2008 assessment (table 17.2). Differences in time span are more an artifact of when the initial assessments were conducted and do not necessarily imply that sites with less time between measurements are "newer" sites.

SITE PROLIFERATION AND ABANDONMENT

Conditions on the subset of 15 beaches were consistently monitored during the course of the study, and analysis of the trends in number of sites present and the total impact at this larger spatial scale provides a perspective on the proliferation of resource changes (table 17.3). The average number of sites per beach increased significantly during the study period, from 2.13 to 3.07 sites per beach ($T = -3.11$, $p = .008$). Total number of observable sites increased from 36 to 46, with 16 new sites being formed

TABLE 17.2. Change in conditions on campsites for which 10 to 13 years of monitoring data are available

IMPACT PARAMETER	ASSESSMENT		PAIRED T-TEST RESULTS	
	PRIMARY	2008	*T*	*p*
Continuous measures[a]				
Area of observable impact (m^2)	41.1	66.1	-2.59	0.019
Area of vegetation loss (m^2)	35.5	54.6	-1.38	0.174
Area of soil exposure (m^2)	25.1	45.1	-1.35	0.180
Condition class	3.5	2.8	3.35	0.001
Number of fire sites	0.32	0.21	1.43	0.146
Number of informal trails	2.89	2.27	2.46	0.017
Mineral soil exposure (%)	52.1	49.3	0.521	0.604
Number of stumps/cut shrubs	2.13	2.18	-0.113	0.911
Vegetation cover loss (%)	77.6	61.7	2.77	0.007
			WILCOX SIGNED RANK RESULTS	
			Z	*p*
Ordinal measures[b]				
Human waste	1	1	-0.905	0.366
Litter/trash	3	2	-4.31	0.000
Tree damage	2	1	0.171	0.864
Root exposure	1	1	-0.218	0.827

[a]Values are means. Primary assessment is the initial assessment; secondary assessment is measurements conducted 10 to 13 years afterward, $n = 65$.

[b]Values are medians, $n = 65$.

TABLE 17.3. Site proliferation and impact change on beaches consistently remeasured for duration of the study

IMPACT PARAMETER	ASSESSMENT[a]		PAIRED T-TEST RESULTS	
	PRIMARY	SECONDARY	*T*	*p*
Vegetation cover loss (%)	70.17	56.70	-1.41	0.180
Condition class	3.06	2.97	0.220	0.829
Total area of impact (m^2)	43.27	73.45	-2.12	0.043
Total area of vegetation loss (m^2)	34.2	45.2	-0.725	0.480
Number of sites per beach	2.13	3.07	-3.11	0.008
Total number of sites	36	46		
Number of new sites formed		16		
Number of sites revegetated		6		

[a]Values are means. Primary assessment is the initial assessment; secondary assessment is 2008 measurements conducted 7 to 13 years afterward, $n = 15$.

TABLE 17.4. Comparison of selected resource condition parameters on Prince William Sound campsites in 2008 on differing substrate types

IMPACT PARAMETER	SUBSTRATE TYPE			ANOVA RESULTS	
	ORGANIC SOIL	COBBLE	SAND	F	*p*
Vegetation cover loss (%)	38.2	71.8	57.0	17.1	0.000
Condition class	2.3	3.3	2.6	15.6	0.002
Mineral-soil exposure (%)	5.5	65.4	65.7	88.1	0.000
Area of observable impact (m^2)	43.4^{ns}	82.7^{ns}	52.27^{ns}	2.8	0.064
Area of vegetation loss	22.1	66.0	34.6	4.86	0.009
Area of soil exposure	3.0	59.9	36.7	8.81	0.000
N	66	69	11		

^{ns}No significant difference

and 6 sites abandoned and revegetated. Total impacted area per beach increased (T = -2.12, p = .043), with vegetation cover loss, condition class, and area of vegetation loss exhibiting no significant differences or strong trends.

SUBSTRATE TYPES

Campsites were found on three primary substrate types: (1) areas of beach cobble (stones 2 to 5 cm in diameter, also called "graywacke") dominated by beach grasses and succulents; (2) a highly organic soil in the forest understory dominated by mosses and forbs; and (3) sand with diverse plant cover. Sites on sand were the least frequent in the study area. The greatest vegetation loss and highest condition class ratings were observed on cobble substrates. Cobble was also significantly higher than organic soil in mineral-soil exposure, area of vegetation loss, and area of soil exposure (table 17.4). No significant differences were observed in the overall size of the impacted area among the three substrate types, although a trend was observed suggesting that cobble sites tended to be substantially larger than the other types.

MULTIVARIATE ANALYSIS AND CAMPSITE CLASSIFICATION

Exploratory-factor analysis of the 10 measured resource indicators (table 17.5) revealed an interpretable three-factor solution that accounted for 54.9 percent of the variation in the data. Factor loadings for individual items of less than 0.4 were considered less important and were eliminated from the results to aid in interpretation. Based on the factor loadings, we interpreted Factor 1 to be "tree and vegetation damage," since the tree damage, root exposure, and number of trails indicators loaded most substantially

TABLE 17.5. Factor analysis of 10 indicators of site impact in Prince William Sound

SITE ATTRIBUTE	ROTATED FACTOR LOADINGS[a]		
	FACTOR 1	FACTOR 2	FACTOR 3
Vegetation cover loss		0.786	
Mineral-soil exposure		0.830	
Tree damage	0.759		
Root exposure	0.653		
Number of tree stumps			0.457
Number of trails	0.633		
Fire impacts			0.483
Trash rating			0.492
Presence of human waste			0.818
Area of observable impact		0.585	
Eigen value	2.89	1.51	1.01
Percent of variation explained, cumulative	28.9	44.0	54.9

[a]Principal components extraction with varimax rotation. Loadings less than 0.4 were eliminated for ease of interpretation, $n = 146$. Factors are interpreted as 1 = tree and vegetation damage; 2 = areal disturbance; 3 = behavior-related disturbance.

on this factor. Trails leading from the site were most commonly indicated by vegetation loss and damage. We interpret Factor 2 as "areal disturbance" due to the loadings of indicators of area of impact, mineral soil exposure, and cover loss and Factor 3 as "behavior-related disturbance," since all indicators with high loadings are resultant from generally undesirable visitor behavior. The factor scores for each of these three factors were saved for each campsite (the cases in the analysis), and a K-means cluster analysis procedure was performed on these scores. After several trials, this process produced an interpretable three-cluster solution (table 17.6). Examining the mean factor scores of the final cluster centers reveals three distinct groupings: (1) *minimally impacted sites* with low mean factor scores on all factors; (2) *intensively impacted sites* with a high score on the areal disturbance factor; and (3) *comprehensively impacted sites* with a high mean factor score on the tree and vegetation factor and positive scores on the other factors. A total of 62 sites were identified as minimally impacted, with a total of 46 and 38 as intensively and comprehensively impacted, respectively.

TABLE 17.6. Final cluster centers[a] from analysis of factor scores of campsite impact

FACTOR NAME	CLUSTER, CAMPSITE TYPE[b]		
	1	2	3
Tree and vegetation damage	-0.310	-0.690	1.34
Areal disturbance	-0.850	1.02	0.147
Behavior-related disturbance	-0.051	-0.126	0.236
N	62	46	38

[a]Mean factor scores.

[b]Cluster names: 1 = minimally impacted campsites; 2 = intensively impacted campsites; 3 = comprehensively impacted campsites, *n* = 146.

DISCUSSION

The assessment and monitoring of campsite conditions is an important information-gathering step in the overall management of resource conditions. Similar monitoring programs have been applied in many natural areas (e.g., Cole 1983b; Frissell 1978; Marion and Leung 1997) and are fundamental components in the application of long-term planning frameworks such as limits of acceptable change (Stankey et al. 1985) or Visitor Experience and Resource Protection (National Park Service 1997). Determining trends in resource conditions often highlights the need for management actions, and monitoring can help ascertain their effectiveness. Nonetheless, difficult decisions must be made from a management standpoint, and monitoring can inform but not accomplish the decision process.

The trend analysis suggests that substantial increases in site size and the number of sites per beach are occurring in Prince William Sound (tables 17.2 and 17.3). This increase in areal extent of resource impact is associated with a decrease in intensity of impact as relative cover loss, condition class ratings, trash, and the number of trails per site are declining while other impact parameters are unchanged. These findings are counter to similar campsite studies (e.g., Cole et al. 2008), which have typically reported stability of resource conditions on long-established sites. This suggests that camping activities are not spatially concentrated and visitors are using multiple locations with less intensity and that use on established sites is expanding at site margins into previously undisturbed areas. In addition to increases in the areal extent of impact on long-term sites, an increased number of sites per beach and increased aggregate area of impact is present (table 17.3). This increase reflects that the pioneering of new sites is more prevalent than the abandonment of old sites. On the 15 beaches monitored intensively during the study, the number of total sites increased by 27 percent, which reflects an addition of 16 new sites, along with an abandonment of 6 previously established sites.

These two findings—the expansion of the areal extent of individual sites and a trend toward increased number of sites—suggest that the current at-large camping strategy has been ineffective at limiting the proliferation of campsite impacts. Experimental studies have consistently shown that a substantial amount of impact occurs with initial site use, but confining use to the same area can result in less overall aggregate area of impact (Hammitt and Cole 1998). Moreover, Cole (1981a) and Marion and Farrell (2002) suggest that the key to successful backcountry campsite management is to confine camping-related change to as small an area as possible. Managers should consider these trends in adopting future camping policies that seek to spatially limit camping activities to existing impacted sites or unvegetated cobble substrates, as the expansion of the areal extent of impact is arguably of high ecological and managerial importance.

We are limited in this study to infer relationships between resource condition trends and potential influential factors, such as soil/vegetation tolerance, use-impact relationships, and site accessibility. This limitation is not uncommon in monitoring studies. The findings are, however, at least suggestive of some important influences that have management significance and should be examined further in future work. First, we observed significant differences in vegetation cover loss, condition class, area of vegetation loss, and soil exposure between sites located on the three primary substrates found in Prince William Sound (table 17.4). In general, sites located on cobble substrates were higher in some impact parameters than sites located on other substrates, particularly organic soil. There was also a trend for cobble sites to be considerably larger than the other types. While we cannot be certain until an experimental study is performed, this finding suggests that these areas of very thin organic soils that harbor the beach grass/succulent plant community may be initially less tolerant of camping use, but once impact has occurred, the exposed cobble is likely to be highly resistant to camping impacts. These sites also tend to have fewer naturally occurring barriers to expansion, as they often occur at the interface of the beach berm and the forest community. Despite the susceptibility to initial impact, the use-impact theory and our experience indicate that camping on previously impacted cobble areas devoid of vegetation where possible and limiting activities to areas of existing impact is strongly suggested. This strategy will also potentially limit some other trends we report, such as the areal spread and proliferation of sites. Management policy advocating this practice will likely be successful in Prince William Sound, as these areas are abundant and cobble sites can provide relatively dry, comfortable campsites. Future research using experimental designs to examine the resistance and resilience of these areas is also needed in order to develop the most accurate management strategies for long-term sustainable use. These results have significant implications to camping management, as similar beach cobble/closed temperate rain forest environment is common throughout the Northwest coastal region.

Our multivariate analysis of the data suggests two overall conclusions. First, our work verifies the utility of Leung and Marion's (1999b) approach in identifying an interpretable structure in the campsite monitoring data and in empirically reducing the multiple-indicator approach to a few (in our case three) interpretable impact factors. Leung and Marion (1999b) also suggest that the implications of data reduction would be a commensurate reduction in the number and type of indicators measured in the field, thus increasing assessment efficiency. Applying this suggestion to our monitoring program, for example, one of the indicators on the areal disturbance factor, such as relative cover loss or mineral soil exposure, could be eliminated since they load similarly on the same factor.

Our second conclusion is that based on the findings, we question the utility of condition class ratings in campsite monitoring, especially when multiple-indicator systems are also used. This combined approach was forwarded by Marion (1995), perhaps due to the previous widespread application of conditions class systems developed in the 1960s and 1970s (e.g., Frissell 1978). In our experience, condition class estimates are perhaps the most subjective and difficult of all impact indicators, because they frequently ask the evaluator to examine multiple impact factors simultaneously (e.g., vegetation cover loss and soil loss). Often, these impacts do not covary in the field, further complicating the decision process. Moreover, condition class scales frequently just describe the intensity of impact along a continuum but do not necessarily address the types of impact that may be present in different classes of sites. In this study, we collected condition class ratings along with multiple-indicator data, and so this presents an opportunity to examine this issue, as Leung and Marion's (1999b) study did not directly report the condition classes of the site classification types. What condition classes lack is the ability to characterize the types of impact present, which may lead to very different management strategies for campsite types. Distinguishing between site types may be extremely important for managers. For instance, intensively impacted sites, being low on behavior-related impacts but of high areal impact, might respond to certain management actions that seek to contain impacts spatially. However, comprehensively impacted sites showing a high degree of behavior-related impacts may require additional actions to also limit undesirable visitor behaviors.

CHALLENGES, FUTURE STUDIES, AND CONCLUSIONS

Many challenges occurred during the 13 years of the study. For example, the large, remote nature of Prince William Sound made the one-time census of sites time consuming and expensive. Another example includes how to account for sites that had some slight evidence of previous use but no discernible difference in ground vegetation between on site and off site over the duration of the study.

Prince William Sound has a long history of human use, at least 3,000 years (Lethcoe and Lethcoe 1994). Many historical cultural sites have characteristics such as fresh water, dry camping above mean high tide, open views, good fishing, and intertidal zones that attract today's recreational user. Starting during World War II with the opening of the Whittier rail link, recreational use started to increase. Outfitters created camps and "improved" sites by hauling greywacke gravel from beach into the uplands. Many of these sites have become popular campsites. Other historical sites had little to no use during our study and have revegetated to a considerable degree. During the study, we found many of these old sites, either through local knowledge or by finding signs such as old fire rings, tree damage, or trash. For these sites, a condition class 0 was assigned according to Marion (1991) and Kehoe (2002) and referred to as potential sites. These sites may be occasionally used, and as use grows and displacement occurs, they may see significantly more use and related impacts. There are 34 potential, or condition class 0, sites (Twardock et al. 2010). Because over 13 years only a few of these sites showed signs of additional impact, we did not include them in our analysis. By not including them, however, means of condition class and other parameters may be higher. However, if they do show signs of impact in future studies, they will be included, and results will affect means accordingly. For now, the sites are worth monitoring, as they are the logical sites for displaced use.

The rate of site revegetation is also unknown, confounding attempts to determine the analytical value of potential sites. Cole (2004a) suggests that site revegetation occurs slowly, though he states that most studies have occurred at higher elevations. Furthermore, Monz et al. (2010) indicate many recreation ecology studies are located in montane environments. However, anecdotal evidence indicates sites can revegetate quickly in Prince William Sound's temperate rain forest ecosystem. For instance, fuel and equipment caches left over from the 1989 *Exxon Valdez* oil spill cleanup were completely covered with moss by 2008. Experimental studies on different soil types to determine which sites recover faster could help shape management decisions.

Another challenge is the proliferation of hardened sites. For millennia, individuals have used beach gravel to make dry, level camping spots. Since the 1980s, Alaska State Park and Chugach National Forest staff have developed popular sites with gravel pads, public recreation cabins, outhouses, tent platforms, and bear-proof food containers (see chapter 16). Some of this development has occurred since the study started, in 1995. Some sites could not be remeasured as tent platforms covered the site (e.g., Surprise Cove). If a site was developed after the initial site visit and remeasurement was possible, we remeasured without accounting for impact of the hardening.

The site work might have affected our results, with sites becoming larger with a few days of work or conversely shrinking as greywacke was added or subtracted to sites. The addition of tent platforms, cabins, and outhouses might alter use patterns and allow sites to revegetate, or adversely attract more use and cause site proliferation when users arrive, find the platforms occupied, and camp nearby. Though tent platforms and other development theoretically reduce impacts, additional study of how tent platforms and gravel pads affect site size and related impacts is needed.

Over the course of the study, methodology stayed relatively consistent. The research team was consistent, with at least one individual being present for almost all site visits. However, as the study progressed, technology improved, as did our efforts to make measurements more precise and repeatable. For instance, GPS coordinates went from an accuracy of 100+/-m to less than 2 m. Additionally, initial work done in 1995 used locator pins that rusted, and location maps were vague or nonexistent, making center-point relocation difficult or impossible. Over time we became much more precise in our mapping, using more accurate pictures and maps that were more useful in relocating sites. Also, the addition of standard protocol (Kehoe 2002) increased the consistency of data, making it less reliant on individual judgment. We also note that the field of campsite assessment has developed since the time of our initial study, in part due to some of the protocol refinements contributed by the authors. Studies commencing at present should incorporate these refinements.

Many opportunities exist to better understand site conditions in Prince William Sound. Further work includes experimental studies on site resilience and resistance; studying the effects of site development; looking for spatial impacts on a larger scale, including the introduction of invasive plant species; and a repeat of the 2008 study to inform managers of site changes for management purposes.

During the study's 13 years, we learned much about changes in conditions in Prince William Sound. Site size has increased, and we detected a trend of site proliferation. Other impact parameters such as trash, fire rings, root exposure, and tree damage have stayed the same or declined. Cobble soils had greater vegetation loss than organic soils, though cobble-beach sites still remain the preferred camping choice. Data analysis suggests that a multiple-indicator approach of campsite evaluation will be more useful than the current single, condition class approach. This data base will provide researchers and managers a valuable opportunity to study change to campsite conditions in Prince William Sound for years to come.

LESSONS LEARNED

The challenge to managers is to maintain and improve the recreationist's experience by protecting natural resources and respecting the visitor's wilderness experience. This study indicates that once a site in Prince William Sound has been impacted, little change occurs over time. Site proliferation was occurring at some sites, however. The hardening of sites by using beach gravel to make durable platforms is one possible tool that managers can use to address this trend. A hardened site should attract use and theoretically contain the use to the gravel platform. By selecting sites with low biological value, the manager can direct use to avoid damage to biological hot spots, and possibly minimize campsite expansion. However, this is challenging to do, since many biologically important attributes, such as salmon streams, are also important to visitors. Also, no study has been done on site proliferation at hardened sites. It is possible that visitors will expand hardened sites as use increases.

The use of Leave No Trace educational materials offers an opportunity to influence visitors' behaviors. The study indicated some changes in parameters that could be related to Leave No Trace practices. For instance, litter and trash, fire sites, and informal trailing have become less prevalent (table 17.2). However, some anecdotal evidence indicates an increase in the number of sites with signs of human waste. Educational efforts aimed at concentrating use on well-used gravel beaches would be most effective in promoting sustainable recreation in the Sound. Though condition class ratings offer an easy way to communicate a site's status, they do a poor job in explaining why the site is a certain class. We recommend a three-tiered approach to rating campsites using selected impact parameters that reflect the most important attributes to a site.

STAKEHOLDER ESSAY

EVALUATING IMPACT—A RANGER'S VIEW

JENNIFER GESSERT

MY FIRST EXPERIENCES in the Sound were in 2002, on the eastern side, as a wilderness sea kayaking guide. I immediately fell in love with that new corner of Alaska, so different from the Interior I had already grown to love so much in my five years in Alaska. Much more similar to my native Oregon, the Sound's prolific wildlife, untrammeled character, and vast temperate rain forest embraced me. During the years I guided there, that side of the Sound appeared to be minimally impacted by backcountry travelers. Speaking nothing of the tragedy of the *Exxon Valdez* oil spill, the tundra and beaches were nearly always trace free of human traffic. My fellow guides and I regularly brought groups to a number of camping sites that never revealed our presence. We often camped in the soft tundra above the rocky beaches, and none of our sites in the Chugach National Forest had tent pads or any improvements of any kind, nor did they need them. That was my impression of where Prince William Sound stood, as far as impact level due to backcountry travel.

After a few wonderful years of guiding out of Valdez, I was fortunate enough to have the opportunity to work as a backcountry kayak ranger for the Tongass National Forest, out of Juneau. The areas we patrolled on our eight-day paddling trips were all in designated Wilderness Areas: Tracy Arm-Ford's Terror Wilderness Area, Chuck River Wilderness Area, and Endicott River Wilderness Area. This experience opened my eyes not only to the impressive and dramatic beauty of Southeast Alaska, yet another corner of Alaska to fall in love with, but to the concept of designated Wilderness and the profound impact of the Wilderness Act of 1964. The way management was approached in the Wilderness Areas we patrolled was significantly defined by the Wilderness Act, and strictly so, I would say. We did not improve on any campsites, aside from eliminating trash and human traces, of which overall there was very little, as it was a relatively low-traffic area due to its remoteness. We spent a lot of our energy educating visitors about the Wilderness Act, its significance in preserving such pristine places, and the need to hold sacred the use limitations brought upon those areas both as visitors and managers for their preservation in perpetuity.

I was there only one year before I took the same position on the western side of Prince William Sound, in the Chugach National Forest. The portion of the Sound we helped to manage included the Nellie Juan-College Fjord Wilderness Study Area, which presented a lot of gray in comparison with the black-and-white-style Wilderness management I had encountered in Southeast Alaska. The "Wilderness Study Area" label meant that the area was not yet designated as Wilderness, but was basically on hold to be as such, and should thus be managed as Wilderness until designation or lack thereof was determined. To date it has been over 30 years that the area remains a Wilderness Study Area.

This was difficult, because it is no small thing to manage an area as Wilderness, especially not without the hard and cold official designation or mandate to do so. The most significant issue is that Whittier, the only road-access point of entry to Prince William Sound's western side, is only an hour away from Anchorage, the state's largest population base of more than 280,000 people. Furthermore, the tunnel that cuts through the Chugach Mountains from the Anchorage side to the Prince William Sound side was originally a railroad tunnel. It opened to vehicle traffic in 2000. Whittier is now the nearest road-accessible and navigable saltwater to Anchorage for recreational boaters, since most of the waters surrounding Anchorage have extreme tidal activity and dangerous mudflats that make it difficult at best for recreation.

The impacts I observed on the western side of Prince William Sound over my three years as a backcountry kayak ranger for the Chugach National Forest were profound compared to anything else I had seen in Alaska. Blackstone Bay is by far one of the most popular recreational destinations in the state, and I once camped at the Willard Island Spit with 21 other people on one Fourth of July. It can be positively crowded, although that is rare. Aside from Blackstone Bay, however, I found backcountry travelers' impacts to be minimal.

It is truly a conundrum for the Forest Service to know how to properly manage a place like Blackstone Bay. It would be irresponsible to allow the heavy traffic to make mud pits out of perfectly useful campsites when those campsites can be maintained with a few buckets of gravel from the beach. Using natural retaining logs here and there, even if a chainsaw is used in the process, hardly seems to be a major impact when the alternative is serious erosion and significant degradation of the topsoil. There are more difficult issues as well, such as tundra trails above the beach areas at the Thirteen-Mile site, especially. Whether it is appropriate to put gravel down, or build boardwalks, or even build an outhouse is a very difficult question when there is heavy use taking place in a designated area that is "supposed to be managed as Wilderness" but isn't actually Wilderness . . . yet.

When I worked in the Sound, managers had other ideas, like developing other sites around the Sound in efforts to attract visitors to those sites, thereby concentrating use. There was also talk about a kayak trail or yurt system through the Sound for the same reason, but I never felt very good about these ideas. I didn't see significantly harmful user impacts very frequently around the rest of the Sound, and I felt those developments would be putting the cart before the horse, as they were not in response to the use already taking place, but rather a proactive and perhaps unnecessary level of development in the Sound.

Opinions on these and many other matters ran hot among my fellow rangers. We all took our responsibility very seriously, and these are multifaceted issues with no clear answers.

Regardless of my own opinions, one thing is sure: the many visitors I spoke with who had been using Prince William Sound as a part of their subsistence and recreation over many decades had definitely noticed an increase in traffic in the Sound. This was always especially associated with the opening of the Whittier Tunnel.

Prince William Sound is one of my very favorite places on Earth. I can fly over the Sound in a jet and know what fjord I am over and what the tide level is at the time. It has been one of the gifts of my life not only to play, work, and become intimate with Prince William Sound but also to work on her behalf and protect her. That we are asking the difficult questions and discussing the complicated topics that this book addresses is profoundly important to the future wellness of this truly majestic area.

CHAPTER 18

SOUND STORIES

Understanding Prince William Sound Through the Stories of Those Who Live There

SARA BOARIO, AARON J. POE, CHRIS BECK,
TANYA IDEN, AND LISA OAKLEY

SOUND BITES

PUBLIC OUTREACH AND EDUCATION are important tools for reaching users in the Sound in order to minimize impacts on resources recovering from the *Exxon Valdez* oil spill.

Local voices can influence visitor expectations and foster stewardship as well as further understanding and appreciation of wildland gateway communities.

Public information and public participation methods can be reinvigorated through the use of new media and cocreation of products that build trust between agencies and stakeholders.

INTRODUCTION

As described in chapter 4, the *Exxon Valdez* oil spill (EVOS) affected a wide range of critical biological resources and habitats as well as four dependent human services—passive use, recreation/tourism, subsistence, and commercial fishing—which are still listed as "not fully recovered" (or "recovering") by the EVOS Trustee Council (chapter 4). This project began as an effort to address gaps in communication about the work that has been done by the Trustee Council and to engage current residents and visitors to the Sound in promoting stewardship messages and sharing their stories about the importance of the region to its communities. Based on the EVOS Trustee Council's Restoration Plan, we identified three areas where a community-based story project could address potential threats to recovery in the Prince William Sound region.

Area one: A comprehensive education effort is targeted at independent users of the region relative to minimizing their impacts on recovering resources.

As we learned in chapter 7, more than 60 percent of recreational boaters using the Sound during the summer were sportfishers traveling independently through the region. Based on 2006 Whittier Tunnel records, vehicles towing trailers were the fasting growing type of traffic accessing the Sound. The huge spatial extent of the Sound and relative lack of agency or enforcement presence through most of the region likely results in a majority of users never coming in contact with area managers. This fact was recognized by Murphy and others in chapter 11 when they recommended the best strategy for minimizing impacts to EVOS recovering resources would be increased efforts in environmental education targeted at Sound users. By convening Sound communities and stakeholders around the issue of stewardship of injured resources and services, we aimed to learn from their collective insights on the most effective education messages that make sense for the region and its users, but also on strategies for how we get those messages disseminated to those who most need them. This type of grassroots development of education messages has proven very effective in numerous wildland settings and often leads to greater stakeholder/community participation in subsequent collaborative resource stewardship efforts.

Area two: With the exception of making scientific information available to the public in individual report form, there have been few efforts to restore or enhance passive uses (including aesthetic and intrinsic values) damaged by this spill in the Sound.

There have been few comprehensive attempts to engage, exchange information with, and educate residents, visitors, and the public at large on the status of recovery in a widely accessible and broad-based manner. This proactive effort is necessary to communicate the results and status of restoration and recovery to date and to ensure the long-term restoration and protection of passive use values, which, like other human services, may be threatened by increasing human use and its resource impacts in the Sound. By presenting a contemporary view of the region's communities, sharing the findings of restoration efforts, and simultaneously sharing the effects of the spill in the voices of individuals who lived through it, the site shows the complex story of recovery for the region. Furthermore, by providing ways that people can learn more, including how to best visit the Sound and share a story of their own experience, the site furthers societal understanding of the region.

Area three: Public information and public participation methods can be reinvigorated.

There does not appear to be a record of engaging human users from *all* injured service categories together to communicate the story of recovery and stories of place that can help guide and shape future human use (behavior) to ensure recovery of human services and natural resources, protect against further impacts, and mitigate user conflicts. The 1994 plan states that "priority shall be given to strategies that involve multi-disciplinary, interagency, or collaborative partnerships." This effort aimed to meet all three components and ultimately hopes to produce a platform for better public participation in the messaging they want to share about their relationship and experiences in the Sound.

It is our hope that the final product will provide the broader public with an interactive visual representation of the Sound and its recovery since 1989, including the intersection of injured resources and services and the economic, cultural, and subsistence value of these resources and services to the communities. Through its story content and interactive functionality, it will also serve as an educational tool for guiding users in the Sound in such a manner that their apparently increasing numbers don't result in confounding the recovery of resources and services injured by the spill.

METHODS

The power of maps to educate and guide human use is well supported in the field of geography, which recognizes that direct experiences underlie one's understanding of geography and that place/map-making gives form and structure to intention and experience and can, therefore, facilitate and guide human behavior (Relph 1976). Ecoregional mapping, and the closely related bioregional mapping, incorporates biophysical (resource) and social-cultural (human) information across an ecosystem and involves stakeholders and communities in the map creation process to ensure traditional ecological and community knowledge informs its creation and as an opportunity to inform/educate participants. There are various models to draw from, including National Geographic's social geomapping initiative.

Additionally, traditional ecological and community knowledge—an essential piece of an ecoregional map—is a widely recognized tool in the fields of planning and geography to involve citizens and users in generating responsibility for place, creating community, and enlisting readers/broader public in that community in order to create, maintain and protect "places worth caring about" (Eckstein and Throgmorton 2003). Traditional ecological and community knowledge is particularly useful when paired with research results and visual and spatial analysis. Creating a visual and narrative record of the resources, human services, and stories was explicitly added to this project to communicate how they should be protected from potential impacts of human use. By informing visitors how to visit a place without damaging sensitive natural, cultural, and social resources by providing explicit education messages (e.g., wildlife viewing practices, anchoring practices that minimize damage to aquatic resources, camping practices which prevent disturbance to archeological resources, etc.), the map also facilitates stewardship of the mapped region.

In order to refine our community engagement approach, we tapped into an existing stewardship council which had been created to guide the exploration of a marine trail concept that had been proposed for Prince William Sound. This council was composed of a dozen regional representatives, including members representing major landowners in the region, regional stakeholders, communities, and user groups, and was engaged along with the general public in the story selection process. In addition to their roles advising on the marine trail, they agreed to advise our team, which was led by the Chugach National Forest and the statewide, nonprofit interpretative association Alaska Geographic. Council members volunteered their time to participate in meetings, helped host us in communities, and spread the word about the project. Through this arrangement, we developed a complementary engagement process where meetings were held and promoted jointly in hopes of attracting broad public interest in the two concepts.

In addition to the stewardship council, the Chugach National Forest contracted with Alaskan planning firm Agnew::Beck Consulting, specialists in community engagement and planning who had extensive experience working with our five target communities, to help design, promote, and facilitate public meetings. Together, the Chugach National Forest, Agnew::Beck, and Alaska Geographic implemented the strategy with assistance from local community liaisons who worked to identify meeting venues, promote, and host meetings, as well as be the face of the project within their communities.

We also relied on the liaisons to identify community-specific outreach techniques that would encourage the participation of as wide a range of citizens as possible. Meeting promotion took place through a variety of communication outlets including fliers in communities, advertisements and public service announcements on local media outlets, and in the two smallest communities of Tatitlek and Chenega Bay, by intentional word-of-mouth contact. A project website and e-newsletter was established to keep communities apprised of progress (figure 18.1).

Our basic approach was to implement two rounds of public meetings in each of the six target communities. The first meeting was to present the project concept and gather any feedback on how it might best serve the region and which regional organizations, groups, or individuals might be interested in follow-up contact from our team. We also wanted to learn how people might be encouraged to use the site and contribute to it as well as to hear about existing story archives and sources that might help generate content.

FIGURE 18.1. Screenshot from the project website used to promote participation in public meetings for the Sound Stories.

A significant part of each meeting was invested in brainstorming a list of stories that residents would like to see shared about their communities, and in so doing, we asked them to think of story types across several broad categories, including best visitor/stewardship practices (ranging from Leave No Trace to safety practices), history, cultural heritage, and wildlife and the environment, as well as a special category of individuals perceived as Sound Stewards who, through their actions, have promoted conservation of the region and its communities. We encouraged meeting participants to spread the word and provided them with materials, including story cards that they could disseminate to friends and neighbors to generate awareness about the project and ideally net additional story ideas.

When traveling to initial meetings, we brought media specialists capable of capturing stories people were willing to share during or after meetings and to capture some basic stock media for additional stories. Given that travel to four of the region's six communities had to be done by plane or boat, we took full advantage of these visits to generate potential site content.

Following feedback from our initial round of meetings, we began investing in development of story ideas with the goal of producing a few dozen iconic stories from the region that we could share back with communities as concrete examples of the site's content. Story collection was done through a series of small contracts with locals in communities, including both professional and amateur photographers/videographers and freelance journalists. We also made deliberate efforts to engage with high school students in their classrooms to develop stories about their community.

These initial efforts generated a variety of stories, which we broadly classified as a photo with brief text and links to more or related information sources (type one); a photo essay—a series of photos set to text and/or audio clips (type two); and video and audio production (type three).

Our intent was to show the range of story formats in such a manner that individuals with limited media production experience could contribute to the site (type one) but also to highlight the possibilities for posting more polished, high-end products from the type two and type three categories. In order to build capacity, we also invested in two multiday media training workshops in the larger Sound communities of Cordova and Valdez.

All stories on the site were assigned latitude and longitude, were tagged with place names (of the community, bay, island, etc.), were assigned an icon representing the type of story, and were published along with a short text description and links to related materials both internal and external to the site. Stories were then embedded in a Google Map platform that could be accessed by standard PC and Mac browsers, allowing users to click through the icon and pop-up title tag to the rich media content within. Text from individual stories was housed on the site in such a way that it could be searchable by an Internet search box. It was also indexed by a series of tags allowing for content to be queried three different ways: interactive map, site-based search, and site index.

Following results from these meetings, we also worked to refine the overall concept of the site as a virtual visitor center for the Sound that shared information on the resources and communities of the Sound in the voices of individuals who "live, work and play there." The mantra for the effort became to *Educate, Conserve, and Enjoy*—with the intent that the site educate visitors about the uniqueness of the Sound in order to elevate understanding of conservation practices and increase enjoyment as they travel though the region.

Feedback from community meetings helped us understand that significant effort would have to be invested in the site if we imagined it becoming a prominent information source in the region. It was suggested that we produce a tangible companion product that could be distributed in all Prince William Sound communities that would drive visitors and locals to explore the proposed website. By actively distributing a free, desirable informational map, which mirrored the site both in design and in intent (to convey stewardship information and best management practices for the Sound and for visiting its communities), it was suggested we could reach a broader visitor audience. In order to appeal to the types of tourism organizations likely to help with distribution of the map, a significant amount of this content was focused on town profiles and visitor information for the main access communities for the Sound and the two Alaska Native villages. The final information was vetted first by community leaders, then reviewed by communication and education specialists.

Finally, through the community engagement process we realized that in order for this site to be sustainable through time we would need to partner with local entities who had similar interests in getting best visiting practices and quality information about the region out to visitors and locals. Follow-up meetings and contacts were made with a variety of organizations in the Sound in an attempt to inventory the community of like minds around our objectives. These meetings were often used as an opportunity to capture stories about those organizations and promote awareness of their efforts in the Sound.

After a year of story collection and follow-up discussions with interested organizations and individuals, as well as development of a beta version of the website, we initiated a second round of community meetings. Example stories collected in communities were shared with participants. In addition to presenting the entire list of stories heard from all communities, we deliberately shared example favorites from each community to get a sense of what other regional stakeholders were interested in sharing.

We also shared the working model of the website, online in those communities where Internet was available and in example printed pages in those communities without a reliable connection. The working URL was offered, and meeting participants were asked to share with friends and neighbors in order to explore the type of content and the site interface. With a mind toward site functionality, we solicited feedback on the level of interactivity that communities would feel comfortable seeing from such a site—ranging from essentially a static page where only the publishers could produce content to something more interactive where stories could be uploaded by community members and visitors, commented upon, and shared with others. We also invited their insights on how communities might be engaged in helping with editorial review of site content.

Following insights from this second round of community meetings, we introduced new levels of interactivity, then solicited professional reviews of the site from other web designers and communication experts. We specifically asked for consultation on the content management system that powered the site, the quality of the content, design of the site, and overall functionality. Following these reviews, we made substantial changes with increased interactive function and simplified content management procedures.

RESULTS

During the spring of 2009, we were able to conduct initial community meetings in Anchorage, Whittier, Cordova, and Valdez, as well as the two Alaska Native communities of Tatitlek and Chenega. Participation was highly variable in communities, with meeting attendance ranging from a few individuals to 20 in Cordova. The second round of meetings in spring and summer of 2010 took place in all communities except Tatitlek, where weather and logistics prevented a follow-up community visit. Through the course of these meetings, we received 366 ideas for stories that community members thought best communicated the important natural, economic, and social values of Prince William Sound. To date, approximately 120 stories from this list have been produced by either the project organizers or on our behalf through a variety of incentive-based projects like small contracts to schools and freelance reporters or the products of media training programs. We received only one unsolicited story from a member of the public through the project website or the working beta version of Sound Stories. Story content ran across nine major topic areas and was presented on the beta site as seen in figure 18.2.

Communities seemed to enjoy the group process of reflecting on which stories best shared the experience of what it's like to live and work in Sound communities, as well as which things they wished visitors to be aware of when traveling through the region. In the latter category, stories focusing on awareness of safety hazards to travelers relative to weather and water conditions as well as an awareness of private property, local social norms, and environmental sensitivity were regular suggestions. Similarly, it was important for residents to impress upon visitors that the Sound was their home and a place deeply valued both for its apparent pristine, wildland nature *and* its ability to sustain their communities through local industry. It's a place they want to protect and use at the same time: a "working wilderness."

Our mantra of *Educate, Conserve, Enjoy* resonated with people, but residents seemed to have a difficult time understanding how a communication platform of the type proposed in the first meetings could really facilitate protection of the region. There was also confusion as to the intent of the site as being something that promoted use in the region or if we intended to promote specific businesses or tourism opportunities. On a similar note, some meeting participants had concerns about who would maintain and control site content—with some worried about it being

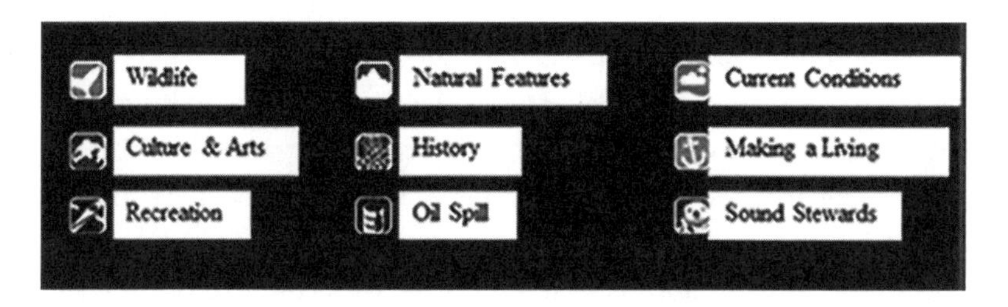

FIGURE 18.2. Icons used to categorize stories published on the beta version of the Sound Stories website.

a mechanism that could ultimately drive unwanted visitation to the region, which could be detrimental to residents' quality of life.

Somewhat conversely, representatives from chambers of commerce saw the potential for the site to facilitate more sustainable tourism, as it increased the potential for travelers to be better informed about the best ways to visit the region and experience more of the diversity of opportunities it has to offer. Museums saw the site as a beneficial investment in media development about the history of the region and as a way to get some information from their archives online. Similarly, nonprofit organizations with stewardship missions saw the site as a potential way to highlight their own activities and messaging to a broader audience.

Communities most interested in widely promoting tourism (Valdez and Whittier) were openly enthusiastic about the project, with Cordova being more cautious. Tatitlek and Chenega Bay, communities without much invested in visitor services at the time, also expressed concern and reflected on potential negative impacts of their stories promoting potential use around their remote communities. An unfortunate consequence of pairing this project with the more controversial proposed marine trail during our first round of community meetings (though a great way to share cost and audiences and benefit from that group's citizen advisory group) was that it somewhat complicated our message and likely resulted in less enthusiasm for the project, particularly in Tatitlek.

During the first round of community meetings it was clear that we were not particularly successful at conveying what the site interface might look like and how users would access content or submit their own stories. Similarly, it wasn't clear to many community members how stories and the site would ultimately look and feel, relative to other types of ways that their communities have been portrayed in other media. The second round of meetings was vital in overcoming both of these issues, as folks could see and hear their friends and neighbors featured in videos with backgrounds of their local landscapes and communities. Similarly, when they saw examples of peers and prominent citizens contributing to the site (e.g., Nancy Lethcoe, featured in chapter 1 of this volume), their comfort level with the project increased.

When presented with the beta version of the site, people could see how the story content was displayed in a map interface (figure 18.3) and how content could be accessed using an Internet browser (figure 18.4). The draft site allowed individuals to better reflect and comment on what had originally been an intangible concept during the first round of meetings. Being able to see the site also afforded insights to us on overall site functionality, ideal audiences for the site, and organizations that might be interested in supporting it, as well as appropriate levels of interactivity and the public's ability to share content. On this latter point, communities liked the idea of the public being able to share stories but wanted a site with custodial controls such that stories could not be immediately uploaded without review.

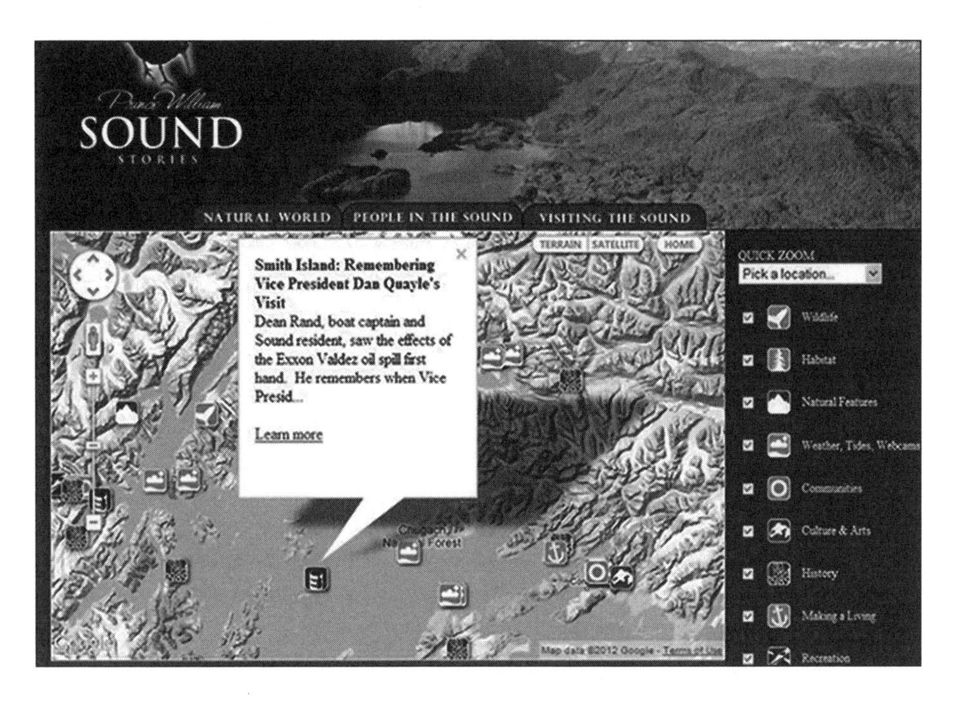

FIGURE 18.3. Screenshot from beta version of the website showing map-based story interface from the beta version of the website.

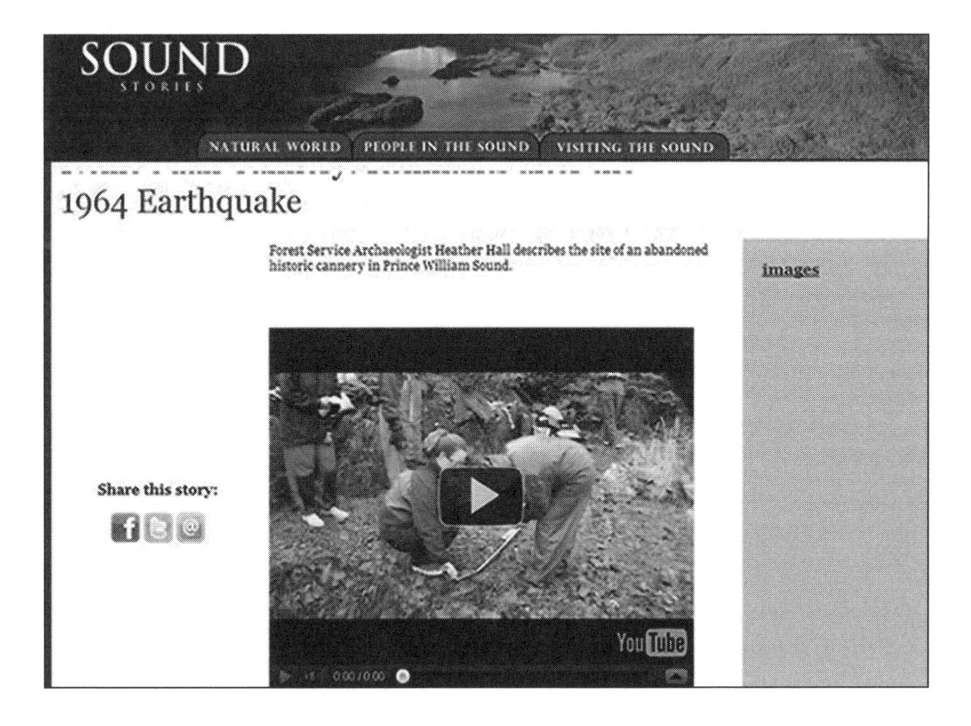

FIGURE 18.4. An example screenshot following a click through to story content housed under a history icon on the beta version of the Sound Stories website.

Following the second round of meetings and consultations with potential regional partners to explore potential financial support for sustaining the site, we recognized the need to sharpen our interface to make it more user friendly, with greater potential for users to submit and share their stories. This feedback, along with some external reviews by web designers and communications specialists, resulted in a complete redesign of the site, which was completed in 2012. In order to reach our largest target audience (visitors to the region accessing the area via Whittier, Valdez, and Cordova), we were also advised to launch a companion paper map that would advertise the site and its unique content, which could be distributed free of charge.

DISCUSSION

As Prince William Sound and the types of unique outdoor opportunities it provides continue to increase in popularity, communities will likely be serving more, and increasingly diverse, types of visitors. Recreation and tourism researchers have long understood that the use/impact relationship is highly complex and nonlinear. Use levels alone do not explain the level of impact on natural resources and other users; behavior of users, travel methods, group size, season of use, length of stay, and rules and regulations also affect impact (Borrie et al. 1999; McCool et al. 2007). These findings suggest, according to Borrie and his colleagues, that, "education and information programs and rules and regulations aimed at changing visitor behavior might be more effective" than simply limiting use. By soliciting Sound communities to identify the types of messaging they want to share with visitors, as well as how they feel the importance of the region and its people can best be conveyed, we hoped to provide compelling education and stewardship messages to visitors.

The behavior of independent users can be especially difficult to manage through traditional means of regulation in dispersed and remote settings like the Sound. A key recommendation for Prince William Sound, and a central theme for this volume, suggests a focus on educating users as a course of action for those federal agencies in the region aiming to minimize the effects of increasing human use on resources/species injured by the spill. Increased information sharing and education is especially important given the findings of more recent research, noted earlier, which showed that more than 60 percent of recreational boaters using the Sound during the summer were sportfishers who were traveling independently through the region (Wolfe et al. 2006). Our effort aims to provide resources like real-time information about weather and travel conditions as a way to attract interest from these independent travelers. Furthermore, by providing a platform upon which they can share their stories about their experiences and learn from the stories of others, we hope to create a community of users who will return and contribute to the site over time.

Our efforts underscore the critical link between passive use and recovery of injured resources and resource-dependent human services, including recreation/tourism, subsistence, and (to a lesser extent) commercial fishing, as introduced in chapter 4. Managing human use to avoid further damage to oil spill–injured natural resources, human services, and habitat is a critical goal of the "general restoration" category of the 1994 Restoration Plan. Passive use recovery methods, like this ecoregion-based story mapping project, can also serve to guide human use by engaging and educating human users.

The Sound Stories process and platform provides an opportunity to address some of the earliest suggestions for the restoration of the recreation/tourism human service. In 1995, Hennig and Menefee advised that efforts focus on "public education" and "visitor information" in order to guide and educate human users and minimize impacts on recovering resources and services. Their team returned suggestions from the public through a survey and series of focus groups in the mid-1990s, which included guiding use to known facilities, promoting opportunities near communities, protecting and *encouraging appreciation* for the aesthetics of key visual corridors and the region's natural, historic, and scenic features. The study concluded that while "recovery of injured natural resources will restore recreation resources to some extent . . . because the injury to recreation as a service was due in large part to changes in perception, restoration actions must also address the perception of lost or diminished recreation opportunities."

The platform provides an opportunity to highlight these facilities and features and improve perception of the Sound's recovery and recreation opportunities. Similarly, our efforts aim to foster appreciation for the uniqueness of the region using the voices of the people who live, work, and play in Prince William Sound. As the Hennig and Menefee report suggests, research alone is not enough to ensure restoration—communities, landowners, and agencies must educate and communicate more effectively, as well.

THE UPS . . .

The most valuable immediate products from the process of engaging these communities was agency personnel and our partners simply learning more about the communities through deliberate listening sessions. Telling the first-person stories of individuals who live, work, and play in the region helps to humanize the need for managers to take thoughtful steps as they contemplate sustainable human use management. Furthermore, this type of storytelling effort may serve to create a better communication conduit between individual managers and stakeholders, narrowing gaps in understanding between these two groups. This connection was best fostered when efforts were made to cocreate stories, deliberately intended to define a sense of place, from the perspective of individuals and organizations. The project management team and folks brought in to support story collection and production were essentially treated to a living, interactive museum experience of Prince William Sound. This learning exchange also afforded the Chugach Forest Service an opportunity to talk about how we see our role in the region and to help demonstrate our commitment to supporting its communities.

This project also allowed the U.S. Forest Service to conduct an informal inventory of conservation, industry, and governmental organizations from communities in the Sound. Many of these groups were able to get their messages out through contributed stories. But of greater importance, the process facilitated an inventory of potential partners in the region whose interests overlap our own. Similarly, by creating capacity/resources in outreach and education for organizations like museums, civic groups, and chambers of commerce, Forest Service managers leverage their ability to spread the word about issues and initiatives being sponsored in the region.

By working alongside local residents and organizations, we broadened our understanding of their values/missions and learned how they characterize the important aspects of their communities—something with clear indirect benefit as we look at addressing management challenges into the future. Particularly from those instances where we worked with community members and organizations to develop their stories came lasting relationships which have been tapped in subsequent years as other management issues and opportunities arose. For example, several of the individuals we met through the course of this project committed to volunteering their time on resource advisory committees established to help the agency prioritize the use of funds from the Secure Rural Schools Act in the region. This unanticipated benefit of the project created direct personnel support that benefitted other agency management objectives beyond the effort to build capacity in story sharing.

Certainly, individual stories resulting from the project serve to educate visitors, but perhaps of equal importance, these stories also have the potential to foster better understanding between the communities of this region. There appears to be a tendency for Prince William Sound communities, which are relatively isolated from one another, to have preconceptions about the values and tendencies of their regional neighbors. In very broad-brush terms, Valdez is seen to be about industry/oil money based on an extractive economy, where the majority of people are actually less interested in the Sound (e.g., the majority of Valdez Boat Owners Association members live several hours away in Fairbanks, Alaska); Cordova guards the Sound as its "working wilderness," where residents' livelihood is intricately connected to natural systems; and Whittier is seen as the "strangest town in Alaska" (e.g., Taylor 2000), where 90 percent of residents share the same roof.

Through the course of sharing multimedia stories between communities in our second round of meetings, long-term regional residents furthered their knowledge about one another. For example, the extremely limited availability of private land in Whittier essentially makes single-family housing impossible, which was something that was poorly understood by Cordova residents who only pass through that town when boarding or disembarking from state ferries. By sharing the diverse and yet often overlapping perspectives of the region's residents, we hope to foster a common understanding between communities that are all inextricably linked to the health of a naturally functioning and unique ecosystem.

Though individual communities have stewardship plans and a number of nonprofit organizations and agencies serve the region, there is currently little in terms of regional efforts aiming to foster a collaborative visioning process for the future of the Sound. The resulting communication conduit may help in convening larger conversations and addressing larger regional issues.

THE DOWNS . . .

A number of questions remain about the overall impact this effort will have on the sustainable development of the Sound, but questions also remain about the specific sustainability of the website itself. The process for engaging communities through public meetings, collecting and producing dozens of stories, was certainly costly. Our initial expectation of individuals from these communities making their own stories and publishing them through the site without assistance from the project team was not realized. Rather, we needed to invest funding in trainings and story collection using a variety of surrogates within the communities.

Another challenge for the project was capturing the attention of stakeholders in the target communities. Objectives advertised in advance of the meetings were somewhat vague and esoteric compared to the usual type of public meetings hosted by land management agencies. In this instance, the Chugach National Forest was not proposing any specific management action for which we were soliciting formal comments, nor were we inviting them to participate in a clear planning process. Though people became interested in the process once they participated in the meeting, and more importantly after helping with story production, getting the initial word out to attract interest was challenging. For example, gains were made in participation and interest when we invested in sending media trainers and collection crews to communities in advance of our second round of public meetings.

Similarly, we struggled with engaging Alaska Native entities in participating in story collection. Certainly given the living tradition of storytelling within their cultures and communities, recognizing the value of stories to promote civic behavior and sustainable use of land and resources is central to their way of life. However, bridging a gap of trust between tribal leadership and a federal land manager is a barrier, further complicated by perceptions about outsiders asking them to share their stories. Alaska

Native institutions already invested in outreach, like the Ilanka Cultural Center in Cordova, were interested participants and saw the site as an opportunity to spread awareness about their efforts. We also had some success in approaches where trusted local individuals collected interviews, but unfortunately, the site does not yet fully represent the perspectives of Alaska Natives in the Sound overall.

At this point in the site's development, it is difficult to know how successful we will be at capturing and keeping the attention of Prince William Sound visitors and residents with this website. Ultimately, the long-term success rests on attracting users who are willing and able to take the time to post new content, allowing the site to be a recognized source of new, relevant information. Since the project began in 2009, there has been an explosion in the diversity of social media tools, with current standards like Facebook pages, YouTube channels and Instagram/Flickr photo streams establishing models that the general public is comfortable using. Though we have renewed our investment in interactivity with the second release of the website, it's not clear whether our interface will remain relevant as an easily accessible platform or if we need to find ways to integrate our content into these commercially available platforms.

A project attempting to engage six communities in an ongoing conversation across wide topic areas has capacity issues at a variety of levels. We were only able to produce approximately one-third of the stories suggested by residents, and even this required significant staff work and funding resources. At a less quantifiable level was the snowball effect of correspondence with residents and organizations participating in the effort. For example, our first public meeting netted around 50 participants, but by the end of the project, almost 250 individuals were on our active mailing list. A number of these people/organizations helped with story collection, contributed footage or photos, supported our media crews with logistics, or took our media trainings and were very valuable assets to the project. However, given the limits of staff time, it was difficult to engage with them at a level that would likely maintain their ongoing interest.

At the time of writing, we are still struggling with the exact process or model for engaging the communities in helping us review potential future story submissions and seek out new material. The original vision had been for a Sound-wide editorial board made up of local citizens who would commit to serving a term helping us to find and approve new stories as they were submitted. We have been unable to establish this process, wanting to be cautious about creating a gatekeeper effect on site content, while at the same time being true to the overarching theme of the site being community driven.

The challenges associated with repeating the Sound Stories project would be great, given the extent of effort required: deliberate community involvement to identify stories, fostering of media production, and carving out a niche as a communication tool for communities in an evolving social networking environment. Similar efforts that aim to promote regional tourism through a series of paper maps, including those undertaken by National Geographic referred to as geotourism (http://www.nationalgeographic.com/travel/geotourism/about/), have struggled to be as broadly inclusive of the public as Sound Stories and ultimately have returned products that promote individual landmark businesses. This latter piece was something that would have proven difficult for a federal agency to implement and ultimately limits some of the financial incentives that Prince William Sound businesses might have had when it comes to contributing content to the site.

Other commercially driven tourism sites like chambers of commerce have the backing from this portion of the private sector and thus typically have greater financial sustainability. The Chugach National Forest does have more than 60 outfitters and guides under permit to operate on the forest within the Sound. These partnership arrangements might make it possible for these companies to contribute content without compromising ethical constraints around public/private sector limitations, but that approach was not explored. Further complexity arises with the fact that the site is hosted by Alaska Geographic, a nonprofit that has a mission to further appreciation of public land in Alaska but has to be cautious about promoting commercial business. Finding an approach that draws financial support from business interest that is also in line with both organizations' foundational principles *and* is true to the desires of multiple stakeholders in the Sound is challenging.

LESSONS LEARNED

Our engagement efforts with communities allowed us to begin to better understand Sound communities as they see themselves. Typically, when public land managers approach communities, we are asking them to specifically reflect on *our* management strategies or proposed actions. The experience and relationships gained from the public engagement and collaborative media production aspects of this project are perhaps its most unique contribution to furthering sustainable management in the Sound. When managers learn from stakeholders by working alongside them on a project that is about defining and sharing sense-of-place stories, understanding is furthered between the two parties. Beyond the relationships created between managers and individuals or organizations, these efforts result in tangible management outcomes for agencies.

Outreach and education efforts that are communicated in the words of the local public have tremendous ability to further an agency's stewardship goals. Individual stories contributed by residents have greater potential to resonate with their peers than what is often perceived (or ignored) as institutional and impersonal. For purely logistical reasons, managers have a difficult time in projecting their messaging and presence across an area as large as the Sound. In addition to this challenge, managers can have a significant credibility deficit when it comes to suggested approaches about personal behaviors and choices in a wildland setting. This effort allowed us to sense the values held by community members and produce a platform for them to share best use practices information that is specific to the Sound with those visiting the region as well as with each other.

Beyond leveraging the public to support agency stewardship efforts, this forum allows a direct view of the values held by local constituents. The study of landscape values has become an increasingly important aspect of land management planning. As agencies ramp up efforts to understand the perspectives of their communities, efforts that share personal stories and ideally result in work toward their production are invaluable. The language that communities use to describe the importance of their local resources and landscapes can be learned and incorporated, allowing agencies to better communicate with constituents.

This work represents an uncommon and complex endeavor that pushes traditional boundaries in the ways agencies relate to their constituents. In this example, the Chugach National Forest aimed to promote story sharing by individuals and organizations to engage and educate a broader audience and foster connections between local communities. The traditional land management agency target of furthering land and resource stewardship in the region was the explicit product of the project, but equal importance was placed on the outcomes of the process itself. Showing an interest in opening a dialogue about the future of the region (not just the forest) with the goal of working alongside citizens to communicate the importance of the Sound, from a community perspective, is a new approach for land management agencies. Even without technological complexities, this type of community engagement is difficult territory for managers who are more comfortable working within traditionally defined jurisdictions and norms. As wildland managers attempt to better integrate public perspectives into their planning and communication efforts, the benefits gained and challenges faced within the Sound Stories project are worth close examination. Despite the enormous obstacles, the project inspired passionate participation and ultimately good will among participants, as well as cooperation between stakeholders and managers—a notable achievement for a public land planning effort.

RESEARCH NOTE

PREDICTING THE FUTURE—SOUND STYLE

ROSA H. MEEHAN

PRINCE WILLIAM SOUND is one of the most studied regions along Alaska's coast, due in part to the 1989 *Exxon Valdez* oil spill and the extraordinary efforts taken to understand the effects of the spill. The Sound is also home to many communities that depend upon the bounty of the Sound as well as a popular tourist and recreation destination. The Alaska Ocean Observing System (AOOS) is a consortium of federal and state agencies and research institutions, whose mission is to increase observations about Alaska's coasts and oceans—and turn that observational data into information products for users of the marine environment. We chose the Sound to develop an interface that provides real-time environmental information as a way to engage this region's users. Using a network of ocean buoys, weather stations, and forecast models, we hoped to demonstrate how a comprehensive ocean observing system might work under Alaska conditions and to learn how it might be used by stakeholders.

Key to accomplishing the AOOS mission is linking diverse information gathered by scientists, and monitored by weather stations and other instruments, to those who live, work, and recreate in this vast and complex region. User needs are varied, although a common thread is the desire to predict the future: Will the winds help me sail to Montague Island? Will I be able to safely kayak up College Fjord? How big are the waves and will I be able to fish and not get seasick—and where and how many fish will there be? The engine quit; where will we drift and can they find us?

Questions such as these—asking for environmental information to make real-time decisions—inform AOOS website development at http://www.aoos.org. The website, or data portal, provides primary access for users to real-time information, as well as the ability to query predictive models on a time- and site-specific basis. The collection of this type of information also became the basis for a large-scale experiment in Prince William Sound that was developed over a period of years. The challenge: how well do existing monitoring systems and associated forecast models really work?

Key AOOS scientists and partners initially looked at existing predictive models for weather, circulation, waves, and trajectories to determine if the models could be refined and if so, how. Given the small size of the region (approximately 100 square kilometers), it seems like this should be easy, but in reality, the Sound is a complex area with significant freshwater input from melting glaciers, an intricate coastline with hundreds of islands and small embayments, and oceanic currents entering and exiting through Montague Strait and Hinchinbrook Entrance. Recognizing that all of these factors affect predictions and must be accounted for in any model underscored the need for further study and resulted in the design and implementation of two major field experiments—one in 2004 and one in 2009.

During a two-week period in each year, scientists focused on gaining a better understanding and quantification of variability throughout the Sound. Utilizing instruments and monitoring stations, several projects collaborated to develop a better understanding of environmental processes. The 2009 experiment saw the addition of new technologies—underwater vehicles and shore-based radar stations—to the more established drifter and buoy instruments. Together, they obtained a real-time measurement (snapshot) of the Sound. The suite of drifters (passive measuring instruments that travel with the currents at different depths) needed to be retrieved and redeployed at various locations to capture all the nuances of the Sound's currents. Researchers needed patience, a good GPS signal, and a keen eye to track and find all the drifters.

Communication during the experiment was vital. Collaborators included researchers from a variety of universities, agencies, and other science organizations during the experiment—some in the field and some working from computers in Los Angeles or Fairbanks. To make things interesting, some of the worst storm series of the summer hit during the 2009 experiment. Fortunately, a mix of the original plan with some adaptive sampling resulted in the acquisition of enough data to assess the accuracy of the predictive models.

What did we learn? Prince William Sound is a complicated place, and the experiment demonstrated how much a variety of physical processes affects currents, waves, and local weather conditions. Not surprisingly, forecast models designed for other parts of the world did not catch the fine scale variation in the Sound. Effective predictive models depend on numerous data points—like the data gathered from weather stations and buoys. Resources and partnerships keep the network running, but more data platforms are needed, and the existing network must be maintained. Bottom line: additional resources and partnerships are needed to maintain and improve the data network that supports forecasts.

Refinements to the ocean and atmospheric models made possible by this study have led to improved trajectory projections, useful for predicting where an oil spill might go or for helping rescuers find people swept overboard. Improved weather predictions are now available—and much appreciated by boaters, fishermen, and folks who live and depend upon resources in the Sound.

Want to check conditions before you go? Pull up the Alaska Ocean Observing System website at http://www.aoos.org and check the real-time sensors for current conditions, including temperature, wind, precipitation from weather stations, and real-time images from a broad network of webcams. Check the model explorer to obtain a site-specific prediction of your favorite fishing spot or kayaking destination.

What is the big takeaway message from this experiment? Models are only as good as the data that feeds into them. Prince William Sound is complex, and many factors affect local conditions. Residents and travelers to the Sound are fortunate that such an extensive network of weather stations and other data stations feed real-time data into models and also that they have easy access to the information through AOOS. The remainder of Alaska does not have nearly the coverage as the Sound, but, compared to the amount of information collected off the coast of the Lower 48 states, we have a long way to go.

CHAPTER 19

TOWARD SUSTAINABLE HUMAN USE MANAGEMENT IN PRINCE WILLIAM SOUND

A Synthesis of Data and Management Recommendations

AARON J. POE, COURTNEY BROWN, DALE J. BLAHNA,
CLARE M. RYAN, AND RANDY GIMBLETT

SOUND BITES

THOUGH RECREATION USE has increased in the Sound, it is not at a crisis point. This presents an opportunity to apply a proactive, systematic planning process focused on specific issues to promote sustainable human use strategies.

Management techniques implemented in partnership with stakeholders, including local residents, commercial operators, and other agencies, will be the most effective at addressing issues in a sprawling system like the Sound that is expensive and challenging to access.

Claims of adaptive management are impossible without deliberate investment in monitoring efforts designed to track progress on *specific* human use and resource management issues.

INTRODUCTION

Since the *Exxon Valdez* oil spill, the Sound has experienced increased human use activity, especially activities occurring on salt water. The growth of the recreation and tourism sector statewide has been accompanied by improved access to the region. In the western Sound in particular, the opening of the Anton Anderson Memorial Tunnel (Whittier Tunnel) access road in 2000 has led to both increased personal and commercial recreation/tourism use (Colt et al. 2002). The previous chapters illustrate that the human uses and benefits of the Sound are many and make clear the need for planning for sustainable human use.

Managers must address concerns about the implications of increasing human use (chapter 11) while at the same time finding ways to provide and protect the diversity of recreation opportunities described in chapter 7. Efforts have been made to identify potential effects of increased human use on wildlife species by exploring overlap at a regional scale (chapter 12) as well as evaluating potential impacts based on analyses of site characteristics for beaches favored by both humans and sensitive wildlife species (chapters 13 through 15). Such efforts are challenging and can yield more questions than answers when it comes to offering guidance on best approaches for species conservation. These investigations have been simultaneously considered alongside potential impacts to subsistence harvesters (chapter 9) and local community members (chapter 8) from new users coming from outside of the Sound.

Forest Service managers needed a way to get ahead of potentially controversial issues like hardening campsites to concentrate use in remote areas managed for "wilderness character" (chapter 16) and evaluating the likelihood of conflicts between bear hunters and other shoreline users (chapter 10). The Forest Service recognized the need for a planning process that could allow for, and even promote, opportunities for outfitters and guides to support the economic stability of the Sound's communities without impacting an ecosystem that had been so severely injured by the oil spill.

This chapter represents the culmination of the Prince William Sound Framework process that was initiated by Forest Service leadership in 2006 (chapter 4). The chapter presents how the data and context described in previous chapters were used to develop strategies to address human use management in the Sound through an issue-driven planning process. The results provide guidance for providing "keystone" recreation experiences in the Sound while simultaneously protecting social and ecological values of the region. The chapter also provides monitoring recommendations that can be used to inform an adaptive management strategy for human use. It is our hope that the process we describe can serve as a model for other managers struggling to balance simultaneous mandates for conservation and providing for human use of wildlands.

EXISTING CHUGACH NATIONAL FOREST RECREATION MANAGEMENT DIRECTION

The Chugach National Forest manages about 80 percent of upland Prince William Sound, including the 2.1-million-acre Nellie Juan Wilderness Study Area. The state of Alaska and Alaska Native Corporations representing the Chenega, Tatitlek, and Eyak tribes manage about 20 percent of land in the region, with less than 1 percent owned by private individuals. The Sound is predominantly remote and wild. Road access is limited to the communities of Whittier and Valdez, and the transportation corridors consist solely of marine waterways.

Most human use in the western Sound takes place from May through September. The majority of use takes place within one-half mile of the saltwater shoreline, which is primarily accessed by kayak or motorboat. Upland use is limited due to the boggy nature of most soils above the beaches and the steep, rugged terrain. Activities are mostly low impact, such as small group overnight use, day hiking, sightseeing, hunting, and fishing. Human use is not evenly distributed in the Sound, and certain locations are more desirable because of their distance from communities, presence of glaciers and postglacial landscapes, availability of landing areas, protected anchorages, sport fish streams, cabins and other recreation infrastructure, and wild game concentrations (chapter 6). Human use is often concentrated in locations that are physiographic bottlenecks, which restrict access to desirable intertidal areas and their associated accessible uplands.

Human use management on U.S. Forest Service lands within the Sound is broadly guided by the 2002 Land and Resource Management Plan, or Forest Plan, and two supporting carrying capacity analyses framed within the Recreational Opportunity Spectrum approach (chapter 4). These documents laid the foundation for human and resource management in the region but lacked the specificity necessary to address changing issues in the region and were inadequate to address what the Chugach National Forest Service saw as its evolving stewardship role within the Sound.

Forest-wide direction applicable to Prince William Sound is guided by the 2002 Forest Plan. Specifically, desired conditions are described as "*managed primarily to maintain the wild character of this area and its unique wildlife. Human access will remain almost exclusively by boat or aircraft with the exception of the road accessed portals of Whittier and Valdez.*" Of particular significance is the fact that about 54 percent of the land in the region is currently designated as Recommended Wilderness (USDA Forest Service 2002). The majority of the western Sound area falls within the 2,116,000-acre Nellie Juan-College Fjord Wilderness Study Area. Desired future conditions for this part of the Sound are identified as "providing outstanding opportunities for solitude, quiet and isolation when traveling cross-country. Scenery will be natural in appearance. . . . People should expect to use primitive skills in an environment that offers a moderate to high level of challenge and risk."

The Forest Plan established the fundamental values for which uplands were to be managed but in many cases failed to take into consideration activity occurring in adjacent areas, including the salt water that supports the majority of human use activity in the region. Furthermore, the management prescriptions defined by the Forest Plan are generalizations that pertain to large spatial extents and do not specifically attempt to address the variation in uses between specific areas. The supporting carrying capacity documents had similar limitations and were developed in the absence of specific information about existing patterns of human use within the region. The existing carrying capacity analyses were developed to spread out human use throughout the Sound as the number of visitors increased by limiting use in higher-use areas (chapter 4). This can increase the ecological impacts of human uses as visitors are displaced to pristine, sensitive areas and can produce social impacts, as visitors with relatively high contact thresholds move into low-use areas that put them in conflict with visitors who went there specifically for solitude. In this way, carrying capacity could eventually homogenize the human use experiences offered in the Sound, rather than provide for a spectrum of recreational use and solitude experiences.

EVOLVING RECREATION MANAGEMENT DIRECTION

In national discussions regarding visitor management, the Forest Service recognizes that protecting resources and visitor experiences requires going beyond specifying how much visitor use is too much. It requires understanding visitor experiences and user activity types; predicting and monitoring effects on resources, social settings, and economic conditions; and identifying and employing a variety of management strategies and tools. In order to meet ecosystem management goals, recreation resource literature emphasizes the need to incorporate relevant social and biophysical data and to understand the importance of strong connections between humans and their environment rather than focusing solely on humans as disturbance factors to the environment (Meffe et al. 2002; Keough and Blahna 2006; Blahna 2007).

The Forest Service has a national and regional focus on promoting sustainable recreation, and reconnecting people to nature is an important theme (Kimbell et al. 2009; Collins and Brown 2007). Several studies show declining contacts with nature in American society, and specifically declining rates of visitation to National Park Service, Forest Service, and state parks lands since the early 1990s (Louv 2005; Pergams and Zaradic 2008; Walls and Siikamaki 2010). Pergams and Zaradic (2008) also found declining sales of fishing licenses and several other indicators of decreasing rates of outdoor participation. Some long-term ramifications of these trends may include decreasing health and well-being of citizens, reductions in proenvironmental behaviors, and declining support for conservation and land management activities (Collins and Brown 2007). While there is some debate about these trends (Cordell 2008), what is clear is that encouraging outdoor recreation activities and contact with nature has become a goal of Forest Service leadership and even the Obama administration (Walls and Siikamaki 2010).

From a land management perspective, the challenge is to provide exceptional recreational opportunities and high-quality experiences while simultaneously protecting resources (Keough and Blahna 2006; Blahna 2007). This is the essence and challenge of sustainable human use, which is the goal of the Forest Service Sustainable Recreation Framework (USDA Forest Service 2010). The Sustainable Recreation Framework also identifies the need for regional-scale, integrative analyses; public engagement and partnerships; and restoration of impacted sites. But it does not provide a process or method for meeting these critical needs. Nor do existing recreation planning and management tools, most of which were developed decades ago or focus on specific site or recreation experience needs (McCool et al. 2007; Cerveny et al. 2011). Existing forest planning tools tend to be highly detailed, environmentally focused, data-driven methods that rarely provide specific guidance or the monitoring necessary for implementing adaptive management. As a result, the Prince William Sound Framework, or "the Framework," was born out of the Chugach National Forest Service's need for refined management to address what it saw as emerging human use issues in the Sound at multiple scales and a national push to move to more sophisticated models of recreation planning.

APPROACH

A regional-scale, issue-driven planning approach was adopted to develop a framework for sustainably managing commercial and noncommercial visitors to the Sound. The process for developing the Prince William Sound Framework focused on identifying issues and framing each problem to determine its causal factors in order to build an understanding of the problem (Clark and Stankey 2006). Inadequate problem framing can result in managers addressing the wrong problem, collecting the wrong data, or stating the problem so it cannot be solved (Bardwell 1991). Issue-driven planning can more effectively consider all management tools before selecting the most appropriate tool for the problem. This approach allows managers to frame the scope and focus the analysis for identifying data needs and monitoring efforts. It also avoids the sometimes complex technical-rational planning process that can be too broad, diffuse, and nontargeted to achieve meaningful results that can be used by managers.

The Prince William Sound Framework approach is based on four planning principles: integrating data from social and ecological systems; using multiple analytic scales of analysis (ranging from regional- to landscape- to site-specific levels); applying an issue-based planning approach; and using stakeholder collaboration. We begin by identifying the relevant spatial and temporal scales of analysis. Then we identify key recreation experiences and critical human use issues based on the social and ecological information covered in previous chapters. The resulting analysis of issues leads to recommended management strategies and practical monitoring measures and practices that can be used to evaluate effectiveness and implement adaptive management.

ANALYSIS SCALES

The Prince William Sound Framework considers three scales of analysis: regional (the Sound as a whole); landscape (Analysis Areas); and site level (General Areas). The use of physiographic boundaries relevant to use patterns, which also allowed summary of existing human use data sets maintained by various land and resource managers in the region, was key to our multiple-scale approach (chapter 6). The results of that effort divided the study area into 31 Analysis Areas and 552 General Areas. Human use of Prince William Sound is strongly seasonal due to variable weather and extreme winter conditions. Recreational use of uplands (excluding hunting), recreational boating, and sportfishing occurs mainly from May through September, with the bulk of use occurring June 15 to August 31 (chapter 6 and chapter 7); therefore, summer became the focal season for analysis.

IDENTIFYING KEYSTONE RECREATION EXPERIENCES

We identified four key recreation activities, or "keystone" experiences, based on existing literature and the recreation use data collected with questionnaires and focus groups (chapter 5; chapter 7; chapter 8; Colt et al. 2002). Managing for "special" or "unique" experiences is a common theme in the literature, but there is no widely accepted definition. For the purposes of this project, we defined *keystone experience* as a relatively specific and popular visitor activity that is unique to the area and a primary visitor attraction rather than something people do as they travel to some other destination. In short, it is what makes the area a special place to visit and provides experiences that are rare or not readily found elsewhere. Borrowing from the wildlife literature on keystone species, managing for keystone recreation experiences allows managers to protect critical and potentially scarce experiences first and secondarily to manage for other activities that are less critical. This approach recognizes that it is impossible to manage for all activities and experiences simultaneously and that judgment must be used to identify an area's critical contribution (or niche) in the provision of recreation and tourism opportunities in the broader region. It also recognizes that focusing on a relatively few specific recreation experiences is necessary for establishing management direction and setting standards, as well as implementing practical recreation monitoring efforts.

Four keystone experiences were identified for the Sound: tidewater glacier day trip; overnight tidewater glacier and solitude; big game hunting; and wildlife watching. These experiences have a seasonal component, and many areas likely offer varying levels of quality of experience and may simultaneously offer more than one key experience.

TIDEWATER GLACIER DAY TRIP

Visiting tidewater glaciers is a primary goal for many day trip visitors out of Whittier or Valdez. During the summer months, three to four tour boats provide these opportunities daily, and passengers can view one or more tidewater glaciers with a relatively short time investment. Similarly, this is an experience sought by some special use permit holder kayak and charter boat clients in Blackstone and Columbia Bays. Comparable opportunities are available at Shoup Bay inside the port of Valdez (chapter 7). Additionally, significant numbers of private recreationists are accessing some of these areas, especially Blackstone Bay, by motorboat (chapter 8).

OVERNIGHT TIDEWATER GLACIER AND SOLITUDE

The opportunity to camp or anchor overnight in the vicinity of a tidewater glacier is a key experience sought by kayakers and recreational boaters (chapters 7 and 8). An integral component of this experience, particularly for kayakers, is a complementary feeling of relative solitude. The noise and wakes emitted by motorized vessels have negative impacts on feelings of solitude, especially in the perception of kayakers. Of primary concern are the presence and availability of adequate campsites and overnight anchorages. This was evaluated using a relative numeric range of campsite and anchorages from GIS layers synthesized in chapter 6.

An additional concern relative to solitude as a component of these overnight experiences comes from the overall vessel activity within an area. In each Analysis Area with overnight tidewater camping or anchoring opportunities, the potential for solitude was characterized as (1) *unlikely*, (2) *potential*, and (3) *likely*. This was evaluated using a relative ranking system (i.e., *high*, *medium*, *low*) applied to each Analysis Area based on overall vessel density predictions from chapter 7. Areas with high vessel activity were considered to have *unlikely opportunities for solitude* whereas medium offered *potential solitude* and low offered areas of *likely solitude*.

Further consideration was given to the presence and relative number of known daily tour or cruise boats and commercial fishing fleets given the noise footprint and large size of these vessels. In cases where more than two tour vessels were likely per day in an Analysis Area (chapter 6) with otherwise low or medium vessel density (College Fjord and Columbia Bay, respectively), the potential for solitude was considered to be diminished to the next categorical level. This was also considered to be true if commercial fishing activity category was defined as high or highest within the Analysis Area (chapter 6).

BIG GAME HUNTING

Analyses of recreation use patterns from chapter 6, as well as data collected by the household harvest study in chapter 9, clearly demonstrate that hunting is a key experience for many users of the Sound. During spring and fall months this group represents the majority of recreation use occurring in the region. Focus group discussions conducted with hunters (chapter 8) show that this group seeks areas with plentiful game, has a preference for reduced overall competition from other hunters, and also desires areas with reduced overall use (i.e., solitude).

WILDLIFE WATCHING

While wildlife viewing is not a rare activity, the Sound provides opportunities to view rare and distinctive wildlife, especially for visitors from outside of Alaska (e.g., grizzly bears, walruses, sea otters, humpback whales, and orcas). All recreation user groups identified wildlife watching as a *very important* aspect of their recreation experience in Prince William Sound (chapter 7). Certainly, wildlife abounds throughout the region, though specific General Areas were identified in the human use hot spots project (chapter 6) as having known optimal wildlife watching opportunities.

ISSUE FRAMING

We framed the issues in two steps—issue identification and issue analysis—by considering the Framework management goals and objectives (chapter 4), available data and insights from analyses presented in previous chapters, and manager judgment. A number of issues were considered based on (1) the perceived level of threat; (2) the ability to rigorously define the issue, identify objectives, and monitor indicators of progress; and (3) Forest Service ability to implement management to address the issue given Forest Service jurisdiction and capacity. The issue-framing process was designed to help managers deal with three basic challenges inherent to recreation management planning: (1) identifying a manageable number of priority problems or issues (identification); (2) determining the underlying causal factors of each problem (description); and (3) selecting management tools that address the specific problems and meet regional and site specific management objectives (management and monitoring). The description of the issues incorporated management judgment, stakeholder input, and results of data presented in previous chapters of this volume. Throughout this process, special attention was paid to consideration of the regional context and the inclusion of managing for both resource protection and providing quality recreation experiences.

The five specific management issues selected are listed in table 19.1. One topic that was identified as a potential issue but was not fully developed for inclusion in this process relates to potential impacts of human use on wildlife species. The research efforts in chapters 13 and 15 demonstrate the difficulty of equating use overlap with disturbance effect and, in fact, in multiple instances indicate this may not be a problem under current human use regimes. Though the efforts described in chapter 14

TABLE 19.1. Five visitor management issues to be addressed by the Chugach National Forest in Prince William Sound

PWS VISITOR MANAGEMENT KEY ISSUES
1. Reduce degradation at culturally sensitive sites
2. Protect solitude opportunities in tidewater glacier areas
3. Maintain large-scale, regional solitude opportunities
4. Assess and collaboratively manage big game hunting
5. Develop comprehensive commercial outfitter-guide use allocation

indicate there are some species that may be at risk from recreation, Forest Service managers did not feel they had enough information to develop the issue and follow it with specific management or monitoring action. Though the results of the effort described in chapter 12 could inform the development of disturbance studies, Forest Service managers felt that specific information about disturbance effects could not be clearly linked to recreation. Furthermore, many of the species *potentially* being impacted are marine species, managed by other agencies. Similarly, the regulation of saltwater boat traffic is not a focus of the Forest Service. As a result, managers felt that it would be inappropriate for a single agency to develop the issue, propose management actions, and conduct follow-up monitoring for success.

ISSUE ANALYSIS: DESCRIPTION, MANAGEMENT, AND MONITORING

REDUCE DEGRADATION AT CULTURALLY SENSITIVE SITES

Analysis of GIS human use layers produced in chapter 6 and completed by Fidel (chapter 16) showed nine primitive campsites on the Chugach National Forest that overlap with historic or otherwise culturally sensitive sites. An additional 49 inventoried primitive campsites are adjacent to private lands owned by Alaska Native entities—the majority of which are also of cultural significance to the tribes of the Sound. The management objective is to decrease overall visitor use at these shoreline sites beginning with those at greatest risk for degradation. This will be accomplished by working in collaboration with Alaska Native entities in the region and using data, including human use density layers (chapter 7) and the campsite hardening analysis completed by Fidel (chapter 16), as well as local knowledge to set priorities for management actions.

One management option includes restoring campsites overlapping with heritage sites to natural conditions, so that each becomes a less desirable site for camping.

Nearby alternative campsites could be hardened in order to concentrate use in areas that protect cultural sites and are resilient to recreational use (chapter 16). If necessary, new primitive campsites could be established to concentrate use in areas away from sensitive sites. Finally, Forest Service staff and partners can work to publicize these more appropriate campsites to shoreline recreationists, further ensuring that use is directed away from sensitive sites.

Monitoring activities linked to this management issue could include (1) soliciting annual reports from key informants (water taxi operators, outfitters and guides, Forest Service law enforcement staff, backcountry rangers, etc.) regarding observations of site use; (2) conducting annual visual site inspections by kayak rangers and partners (ideally including Alaska Native entities) using the protocol developed for campsite monitoring like those described in chapter 17 (e.g., evidence of fires, garbage, trampled vegetation, bare soil, etc.); (3) measuring cultural site degradation at the Forest Service–managed cultural sites; and (4) conducting annual surveys of key stakeholder perspectives (e.g., Alaska Native groups) to assess whether use levels are perceived to have changed following implementation of management actions.

PROTECT SOLITUDE OPPORTUNITIES IN TIDEWATER GLACIER AREAS

In tidewater glacier areas, solitude is a keystone experience for the kayak user group and is also important for recreational boaters. Literature suggests that, like resource management in general, key experiences can be lost through incremental decision-making over time (Kahn 1966; Odum 1982). Kayakers who camp overnight are especially sensitive to and dependent on solitude (i.e., quiet, low boat wake, low number of motorized users) for their overall satisfaction levels. To retain solitude experiences for kayakers, managers should manage certain areas *explicitly* to prevent the loss of solitude and to preserve low density visitor opportunities. Based on summer vessel density predictions in chapter 7 and known areas of commercial fishing and tourism traffic identified in chapter 6, key areas where solitude is likely include Icy Bay and Port Nellie Juan. Harriman Fjord/Barry Arm, College Fjord, and Unakwik Inlet represent areas where solitude is potential. Three areas where solitude is unlikely are Blackstone Bay, Columbia Bay, and Shoup Bay in Valdez Arm; each of these is within relatively close proximity to port access from Whittier or Valdez.

Many individuals visiting the Sound want to be able to visit and experience the natural beauty and ecological setting of a tidewater glacier as part of a day trip. This experience is only available within one day's travel distance from Whittier or Valdez, and as a result, these areas (Blackstone Bay, Columbia Bay, and Shoup Bay in Valdez Arm) typically receive a high amount of vessel traffic. Individuals experiencing these areas are not necessarily dependent on the availability of infrastructure (i.e., overnight anchorages or campsites) and many spend no time or very limited time on shore. Promoting day use at tidewater glacier areas where solitude is unlikely could concentrate use and limit dispersion of visitor use into tidewater glacier areas with greater potential for solitude. This strategy could also include permitting additional outfitter and guide use into these same higher-use areas for day use and for overnight use if adequate campsites are available. An additional action would be to allow non-motorized use (primarily kayaks) to increase up to thresholds consistent with Forest Service guidance for the Wilderness Study Area, at least in areas where adequate numbers of campsites exist in order to provide for this keystone experience. Similarly, ensuring that adequate numbers of campsites and overnight anchorages exist in high-use areas is another way to keep use from spilling into other areas that currently have better potential for solitude.

In areas identified as having likely or potential solitude, measures can be taken to avoid the development of facilities such as permanent camps, cabins, group sites, and trails unless they are specifically employed to concentrate use and promote solitude within that area. Managers should also scrutinize additional proposed motorized uses of likely and potential solitude areas. For example, when making decisions about newly proposed charter boat use, they might consider the levels of use in these areas according to work in chapters 6 and 7. Those contemporary numbers could be considered as maximum thresholds for allocation of new use to avoid increasing the overall volume of motorized traffic in these areas with higher potential for solitude.

Monitoring activities linked to this management issue could include (1) conducting an annual summary of visitor use data reported by outfitters and guides for seven tidewater glacier areas, as part of Chugach National Forest's annual commercial use monitoring, as well as yearly data from water taxi operators and transporters who drop clients in these areas; (2) conducting a user experience survey every five years targeting individuals using the seven tidewater glacier areas to evaluate visitor expectations relative to the ability to experience solitude (and other factors considered important relative to the availability of solitude); (3) conducting summer vessel density surveys every five years using randomly established transects stratified along a gradient of use (high, medium, low) to detect changes in vessel activity in tidewater glacier areas; and (4) conducting seasonal ambient noise monitoring in a subset of these seven tidewater locations stratified along a gradient of solitude potential (likely, potential, unlikely).

MAINTAIN LARGE-SCALE REGIONAL SOLITUDE OPPORTUNITIES

Beyond the site-specific scale of tidewater glaciers, managers need to explicitly plan for protecting large regions of solitude so that these regions do not disappear incrementally through site-by-site decision-making processes. The analysis of recreation vessel density discussed in chapter 7 indicates that the majority of the eastern and southern parts of the Sound have overall low levels of human activity. Access to this area is difficult, resulting in a "natural" overall lower level of use. Such large regions where use is not encouraged allow for the maintenance of diversity in recreation user experiences and offer opportunities for individuals (such as hunters and solitude-seeking kayakers) in search of highly solitude-oriented experiences with very low contact levels. They also provide an area where a greater focus can be given to ecological values.

To maintain a large contiguous region of relatively low levels of human use while simultaneously directing use to areas of high and some medium use, managers should concentrate use in high- and some medium-use areas to avoid spreading use into low-use areas. This includes allowing an increase in commercial outfitter and guide use in those areas, but also employing indirect management techniques such as outreach/education and promotion of recreation opportunities in high- and some medium-use areas. In areas with low overall use, managers should avoid advertising these areas and developing facilities that may impact the potential for solitude, such as permanent camps, cabins, group sites, and trails, unless they are specifically employed to concentrate use and promote solitude within that area.

Monitoring activities linked with this management issue should include (1) conducting annual summary of visitor use data reported by permitted commercial outfitters and guides for low-use areas as part of Chugach's annual commercial use monitoring; (2) collecting annual data from water taxi operators and transporters who drop clients to these areas; and (3) periodically conducting vessel density surveys using randomly established transects stratified along a gradient of use (high, medium, low) to detect changes in vessel activity in areas currently defined as low.

ASSESS AND COLLABORATIVELY MANAGE BIG GAME HUNTING

Big game hunting is a contentious issue in the Sound, in part due to competition over what some perceive as limited resources. Jurisdictional considerations between the Forest Service and the Alaska Department of Fish and Game (ADF&G), and differing norms of sport and subsistence hunters, further complicates management. There are also different social norms associated with local hunters from communities in the Sound compared to others from outside the region and concerns about commercial hunting permit allocations and locations of use. Furthermore, species availability is restricted to particular locations in the Sound, which can cause bottlenecking and crowding issues for hunters. For example, goat hunting opportunities are limited by accessible goat habitat (Columbia Bay and Port Bainbridge) as well as administratively by a limited number of harvest tags that hunters are competing to fill. This leads to intense pressure to be the first to harvest in a particular game management unit and increases the potential for conflict.

Competition between user groups occurs in the spring between kayak use and black bear hunting in the western Sound (chapter 10). New hunting pressure has occurred here as a result of increased access through Whittier. This has resulted in new hunters to the area with different behavioral norms and expectations relative to Sound-resident hunters which may increase intragroup conflict (chapter 8). This influx of new users may also increase potential for conflicts between kayakers and hunters (chapter 10). A Forest Service moratorium on issuance of new permits for commercial hunting in the eastern Sound has also resulted in demand displacement, increasing commercial hunting permit requests in the western Sound (chapter 4). Most of the large game species harvest is by private individuals, but Chugach National Forest decisions on commercial permit allocations may increase competition for game, reduce hunters' feelings of solitude, and compromise user experience. This is probably most important in areas where there is limited overall use, as well as those areas used most heavily by local community residents, as identified in chapter 9.

Land and resource managers need to develop a more comprehensive understanding of hunting practices in the Sound for at least three specialized groups: subsistence/local sport hunters, commercial sport hunters, and sport hunters from outside the Sound communities. They all have different motivations, expectations, and norms. Managers must work to understand the differences between these groups and assist in developing a code of practice *with* the hunting community that complements local traditional hunting and fishing practices. The Chugach National Forest has limited contact with this important group of stakeholders; therefore, efforts are needed (ideally in partnership with ADF&G) to engage them in a dialogue in order to better understand use patterns and hunter expectations.

In order to provide for this keystone recreation experience, a management partnership should be initiated between ADF&G and the Chugach National Forest with the aim of promoting both sustainable harvest and quality recreation hunter experiences. Some joint activities could include (1) identifying areas of high hunting use and concentrate other uses in different locations during hunting seasons to avoid potential conflicts between these user groups; (2) elevating awareness of the timing and locations of hunting use within other recreation user groups; (3) identifying areas with high likely hunter competition for key species and attempting to balance use between guided and private hunters in those areas; (4) conducting an assessment of hunting behavioral norms to identify potential causes of conflict; and (5) providing training for commercial operators in local hunting norms to help reduce conflict between them and private users in the Sound.

Monitoring activities should also be a cooperative effort between ADF&G and Forest Service, and could include (1) annually refreshing spatial information about household harvest of big game species using the spatial and seasonal framework and procedures identified in chapter 6; (2) engaging with key informants, including commercial outfitters and guides and experienced private hunters, about the quality and management of hunter experiences; and (3) conducting a targeted, representative survey of all hunter groups within the Sound to establish a baseline user experience for this key recreation group.

DEVELOP COMPREHENSIVE OUTFITTER-GUIDE COMMERCIAL USE ALLOCATION PROCESS

The methods used by the Forest Service to allocate commercial use across the Sound are inconsistent. Due to the different approaches in use allocation, a higher number of permits are being issued in the western Sound, which is managed as a Wilderness Study Area (chapter 4). Due to the more restrictive management style of the eastern Sound, the indirect effect has been to displace many permit holders' use to the western Sound. Site- and area-specific decisions have regional implications and consequences and therefore need to be coordinated.

While there is controversy about the amount of permitted guiding in the Sound, new social and biophysical data, and a better understanding of visitor use density, provide important information for managers. Special use permit holders represent a small part (less than 10 percent) of overall use during summer months, and visitors do not feel crowded in the Sound during the summer (chapter 7). These results suggest that commercial use permitted by the Chugach is unlikely to have wide-ranging negative effects on user experiences in the region. In fact, guides may play an integral and positive role in providing the public access to recreation opportunities in the challenging conditions of the Sound. Additionally, outfitters and guides can serve to demonstrate proper behavior to their clients with Leave No Trace principles and responsible wildlife viewing practices. Commercial use also contributes to the economic sustainability of local communities, which is a Forest Service priority for the region.

Therefore, we recommend the Chugach develop a plan for sustainable *growth* of outfitter and guide opportunities during the summer months, with a consistent and transparent commercial permit allocation process for charter and kayak outfitters and guides for the peak summer months. The process should be developed collaboratively and based on current human use data described in this volume (principally chapters 6 and 7). Drawing from those efforts, special use permit allocation could be based on a combination of existing total-use thresholds (high, medium, low), key recreation experiences, and new information about the availability of anchorages and campsites, as well as the presence of other commercial activity. Finally, managers should strive towards standardized use reporting, data entry, and summary to feed into monitoring efforts, allowing permit allocations that can be revisited as needed.

Monitoring activities linked to this management issue could include (1) annually summarizing use reported by permitted outfitter and guides to Analysis Areas to ensure commercial use levels do not exceed Wilderness Study Area standards; (2) conducting social survey and boat counts every five years in order to validate classification of high, medium, and low for total use comparison to commercial use; and (3) analyzing public use relative to commercial use to measure effectiveness of management and use allocation strategies.

DISCUSSION

Throughout the issue analysis process, four broader management themes emerged as necessary to address the issues: (1) managing for user expectations about their experience; (2) concentrating recreation use where appropriate; (3) forming management partnerships with other agencies and stakeholders; and (4) recognizing that progress on the issues can't be demonstrated without investments in practical and targeted monitoring. These same themes likely underscore a number of the issues faced by land and resource managers in other wildland settings and warrant further deliberation.

MANAGING FOR USER EXPECTATIONS

Much of wilderness recreation research and management focuses on evaluating or limiting numbers of individuals using the landscape in order to reduce social and resource impacts. The Prince William Sound Framework highlights some of the complexities that can be missed by taking such a singular focus. In the Sound, the perception of crowding is not a widespread issue, nor does it negatively impact user experiences, even during peak use summer months. Visitor perceptions of satisfaction and solitude have more do with their expectations, behaviors of others, and the spectrum of opportunities available to them. These dynamics go beyond simple measures for user experience that often focus on numbers of encounters with other recreationists. Managers should (1) help visitors to the region understand seasonal and geographic patterns of use so that they can develop realistic expectations about their recreation experience relative to the other users they may encounter and (2) promote awareness of the types of diverse experience opportunities sought by different groups using the Sound and how they contribute to the region's unique working wilderness character and add to the ranks of those invested in its conservation future.

CONCENTRATING RECREATION USE

Too often wildland managers let a couple of normative experiences, based largely on keeping different groups of people from coming into contact with one another, drive planning efforts (Blahna 2007, McCool and Cole 2001; Borrie et al. 1999). This results in the temptation to distribute use relatively equally throughout a region, an approach that had been used by the Forest Service in Prince William Sound in the past (chapter 4). With a better understanding of existing use dynamics (chapters 6 and 7) the Chugach National Forest can find ways to concentrate recreation in areas already experiencing relatively high levels of human use. Specific actions could include the following:

1. Increases should be allowed in permitted commercial operator use in areas that have already high relative overall vessel density up to 30 percent of overall use (per existing Forest Plan guidelines), if adequate anchorages and campsites exist.
2. The existence of campsites in high- and medium-use Analysis Areas should be published and publicized to identify those sites capable of supporting more use based on the analysis by Fidel (chapter 16) and using a platform like that described in chapter 18 so it can be readily available to those planning trips to the region.
3. Analyses described in chapters 7 and 16 should be used to guide a program of targeted campsite hardening in high- and medium-use Analysis Areas. This might include partnering with other land managers and stakeholders to develop or enhance a few targeted multigroup camp or day use sites in high-use Analysis Areas on nonnational forest lands. Such developments might deflect large groups from the Chugach National Forest's more primitive and low-use shoreline sites.
4. Collaboration with regional partners and other managers should be sought to encourage the use of vessel traffic routes within Analysis Areas with existing high and medium levels of use.

ESTABLISH MANAGEMENT PARTNERSHIPS

A key to managing for sustainable human uses in such a huge region with challenging access and multiple jurisdictions is better collaboration with other land managers and key stakeholders. Some specific partnerships could include the following:

1. Engagement should be sought with commercial operators in regular moderated sessions to detect potential changes in use dynamics and identify possible areas or user behaviors of concern within the region. The forums should be two-way information exchange, encouraging the sharing of management concerns and perspectives. Incentives offered for participation could include sponsored guide training or additional access to the latest mapped information about human use patterns, resource distribution, or other value-added information.
2. The perspectives and advice of local citizens in best-use practices should be shared to help guide newer users of the Sound. Such an effort could be combined with collecting systematic input on emerging issues in the region from local experts or key informants

3. Special use permit operators should be trained to engage their clients in the use of citizen science techniques to collect simple data to help meet issue monitoring identified above.
4. Citizen science partnerships should be incentivized and the value of data contributions from these efforts should be demonstrated by formally recognizing participants.

COMMITMENTS TO TARGETED MONITORING

The Forest Service is one of three *Exxon Valdez* oil spill federal trustees and is responsible for oversight of the recovery of resources and human services injured by the spill. In this role, the U.S. Forest Service is able to initiate monitoring and recovery efforts that look beyond the forest boundary for a more regional perspective—an approach now mandated by the agency under its new planning regulations. In the years since the spill, substantial resources have been expended monitoring species and habitat condition, but little effort has been spent to monitor patterns of human use. The Chugach National Forest has a significant opportunity and obligation to play a role in human use monitoring of the region in support of their management objectives and continuing spill recovery efforts.

Some components of the efforts proposed under this section could be conducted in association with the monitoring being developed in accordance with the Chugach Forest Plan revision launched in 2012. The issue monitoring efforts in this chapter, identified through issue analysis, would help the Forest Service implement a model of sustainable recreation management. These efforts range in rigor from key informant and visual inspections conducted in a consistent and repeatable way to statistically representative sampling of the region and its recreation users. Suggested data collection approaches common to addressing a number of the issues in this chapter include (1) gaining assistance in data collection from the many commercial operators active in the region and working in partnership with agency and university scientists with experience and a demonstrated interest in the region to complete analyses; (2) linking monitoring efforts to the comprehensive databases created and used in chapters 6 through 10 and 12 through 14 to assess overall recreation use and sensitive resource areas in the Sound; and (3) making specific efforts to share the results of monitoring efforts with regional managers to better inform the decisions that *all* managers make in the Sound relative to human use.

LESSONS LEARNED

In developing the Prince William Sound Framework, we realized there were several other broad lessons learned that apply to human use and recreation sustainability planning in general. These lessons also apply to other large landscape conservation planning situations, regardless of purpose, ranging from timber sales to state and federal wildlife habitat plans to national park area and management plans. The process we recommended in this chapter illustrates the many and varied social, economic, and environmental factors that must be considered in planning for the human uses of natural landscapes.

No quantitative process can weigh or "balance" the wide range of complex factors nor include the diversity of human values and value judgments necessary to address the multitude of social and environmental needs. Instead, we developed a broader, process-oriented framework that we believe can be tailored to other large landscape conservation situations and contexts. In our framework, identifying keystone human activities played a central role in allowing us to work through an issue-driven, rather than data-driven, process. This allowed us to identify the most relevant data to consider that could help address specific objectives and management strategies—then link those to related monitoring needs. We did not try to cover *all* needs, quantify *all* values, and simulate *all* outcomes. Such approaches too often lead down data-driven rabbit holes of "analysis paralysis," the likes of which have plagued natural resource agency planning for decades (Blahna 2007).

We believe this framework meets many of the ideals of planning processes that are rarely incorporated in large-scale conservation planning. First, it is both strategic and tactical. There are broad and general objectives for human use throughout the Sound and very specific recommendations for how to meet those goals. Second, it is (or can be) a living document, not a static plan that is difficult to implement and update. It is possible to track the specific issues, and management assumptions can be tested. Third, it is truly interdisciplinary, with a multitude of empirical data across many disciplines being used together in the framing and analysis of specific issues. Fourth, it provides practical monitoring guidance so that adaptive management is really possible. Finally, it combines, rather than conflates, social science *and* public engagement. Both were used in the development of the Framework, and each is identified as having complementary but different roles in supporting future management efforts.

Some of the issues and strategies identified in the Framework are straightforward and likely can be adopted immediately. Others can be partially adopted, while more controversial aspects of the issues will likely enter the formal planning process. Still other issues will require additional analysis, broad community engagement, and thoughtful collaboration with key stakeholders. Assuming some of the issues identified in this volume are carried forward and addressed (ideally in partnership with other managers and stakeholders), the effectiveness of the Prince William Sound Framework can be evaluated. The evaluation will show that some issues were successfully addressed, some partially addressed, and some not so well addressed—but the key is that this information *enters the next round of planning*, along with new issues that emerge. This is the *ideal* of adaptive management, and it is the last and most critical planning step that too often eludes natural resources managers.

We believe it's vital for the Sound communities, federal and state managers, industry, and visitors to the Sound to continue engaging in a comprehensive, proactive planning process that seeks to protect the diverse ways in which this region is valued by all. Without a serious dialogue and visioning about the sustainable growth of human use, incremental changes made by individual actors may overwhelm the very attributes of the Sound that make it so special to residents and visitors. The Chugach National Forest's Prince William Sound Framework, built upon the knowledge contained in this volume, provides a beginning point for an effort to ensure this unique and special place continues to be a "working wilderness."

STAKEHOLDER ESSAY

THINKING AND MANAGING ON A WATERSHED SCALE

KRISTIN CARPENTER

MY HUSBAND WEARS A SUIT TO WORK. His commute ranges from 60 to 120 miles. Sometimes he's away for days, or a month, at a time. Most of the time, he runs his small business by himself, though at times he has help from his sister and their nephew.

My daily orbit is a bit smaller, more on the order of within a mile or so. I walk down the hill to my office, around town at lunchtime to go to the post office and run other errands, and back to the house after work to get the dog and continue farther up to the ski hill for some exercise.

Pacific salmon travel hundreds and possibly thousands of miles during their migrations. Their lives begin in small streams that are freshwater tributaries to larger rivers. In this part of Southcentral Alaska, salmon fry migrate several hundred miles downstream to the saltwater Gulf of Alaska and Prince William Sound. From there, we don't know how far they travel in the ocean during their two- to five-year adult lives. Their annual migrations, which occurred throughout the North Pacific and southern Arctic oceans for centuries, follow a vast sweep of topography, and season and tide cycles.

At my office at 511 First Street in Cordova, I am asked frequently how much oil from the *Exxon Valdez* washed up on Cordova's shores. Cordova sits on the eastern edge of Prince William Sound, 80 miles as the crow flies southeast of Bligh Reef, where the *Exxon Valdez* struck a charted rock. Oil flowed south and west from the ship's cargo hold.

The answer to spilled oil on Cordova beaches is "none"—that should be good, right? No, it was devastating. Cordova lost three of its six processing plants after the spill. Families and businesses fled from town because they had lost their livelihoods.

Getting back to that suit my husband wears: it's safety-orange rain gear, because he is a commercial fisherman. He goes where the fish are, and that could be miles afield. Parts of western Prince William Sound, where he often fishes for sockeye salmon, were oiled by the *Exxon Valdez*, and the Cordova fishing fleet lost its buyer markets that summer of 1989. In 1991 and 1992, herring and pink salmon returns, which typically had made up 30 to 100 percent of many Cordova families' income, failed completely.

While our human scale of daily routine is generally limited to a handful of miles, the natural resource world—on which we depend for food, the water cycle, and building materials—functions on a watershed scale. Watershed drainage patterns are what drive nutrient cycling and habitat productivity, and for these natural processes to sustain us, we need to manage them on a scale that takes human activity into account for the entire system.

At my own job, as executive director of the Copper River Watershed Project (CRWP), another question I hear often about the Copper River is, "What are the main threats to the Copper River?"

If I have to sum it up in one phrase, my answer is "fragmented resource management."

The Copper River is described by those who raft it as world-class wilderness. And it is! The high bluffs through the Copper River Basin are the rim of ancient Lake Atna, and you can see bison grazing along their shoulders. Farther down, the river cuts a canyon through the Chugach Mountains and all the water from the Copper River's dozens of tributaries and the Chitina River is forced through Woods and Baird Canyons. You almost always see bear tracks at campsites along the river. They're after the same fish that subsistence fish wheel users, dip netters, and commercial fishermen are trying to catch.

But is this watershed pristine, public, and protected, as so many seem to think? It seems more like Swiss cheese to me.

Throughout the 26,500 square miles of the drainage, there are two regional native corporation land owners, six tribal councils, five federal land and resource managers, and three state agencies involved in resource management.

As an example of how the layers interact, when the CRWP works on marshaling resources to replace a culvert to restore fish passage, we talk to the Alaska Department of Transportation & Public Facilities and the landowner first. Depending on the site conditions, we may also need to consult with neighbors—a tribal council and/or private landowners, and possibly a utility cooperative. Then we consult with the permit agencies: Alaska Department of Fish and Game, U.S. Fish and Wildlife Service, and the Army Corps of Engineers.

And that's the crux of the issue. Fish and wildlife are managed by state and federal agencies, land and some water bodies by other state agencies, roads and the culverts and bridges that cross those water bodies by another entity. How do you integrate all those decision-making factors? By thinking, and planning, on a watershed scale.

My job is to push decision makers throughout the region to take fish habitat into account, to consider the cumulative impacts of development on the fish productivity of the entire watershed. We know that diversity is critical in biological systems, and that nicks in the region's habitat integrity here and there add up to losses system wide over time.

Managing on a watershed scale helps to take into account events that are as cataclysmic as the *Exxon Valdez* oil spill as well as small-scale habitat degradation, like adding fill for a building pad in the mouth of a salmon spawning stream. By ignoring the bigger picture—the watershed scale—in resource management, I fear that we risk becoming disconnected from how rivers and streams feed us, buffer flooding, provide recreation, and supply clean water. Salmon are so connected to their home watershed that they return to the stream where they were born. What's happening in your home watershed?

CONCLUSION

WHAT WE'VE LEARNED FROM THE SOUND

A Broader Perspective on Sustainable Human Use of Wildlands

AARON J. POE, RANDY GIMBLETT, AND DALE J. BLAHNA

THE GOAL OF THIS BOOK is to provide insights for planning and managing large, natural landscapes for sustainable human uses. The challenge of applying science to manage for sustainable human uses is that it also depends upon the integration of perspectives of stakeholders who may operate from different bases of knowledge. In particular, we wanted to present a case that managing human uses in a sustainable way is possible even while human uses increase. But this won't happen if conservation planners only focus on the negative effects of human activities. This approach leads managers to focus on limiting the number of visitors in order to "protect" wildland settings.

This volume should make it clear this traditional management focus is far too simplistic and can actually be counterproductive. In the short term, it may disperse people farther across the landscape, reducing the diversity of recreational opportunities and actually *increasing* the cumulative social and physical impacts as people are redistributed from higher to lower use areas. In the long term, it can have even worse consequences by restricting the numbers of people who can come to experience and appreciate wildlands and thus broaden the disconnect between society and the value of such places.

In this volume, we suggest a very different approach to planning and managing sustainable human use of a wildland. We recognized the environmental effects of recreation but also considered the broader array of human uses and values of the landscape, including social, community, and stakeholder impacts, both positive and negative. Recreation activities, experiences, attitudes, and sense of place considerations were also integrated with biophysical considerations at different scales of analysis.

The previous chapters in this book build on over three decades of work studying human-landscape interactions and associated conflicts, as well as policy related to protection of special places, wildland values, and human experiences. We believe that the approach recommended to the Forest Service in chapter 19 has merits beyond the region of Prince William Sound. Finally, through the course of putting this volume together, three interconnected themes emerged that are vital when managing for the sustainable use of wildlands.

SUSTAINABLE USE MANAGEMENT DEPENDS ON RIGOROUS SOCIAL SCIENCE

The Forest Service and many other land managers have attempted to use ecosystem management principles to guide management planning since the mid-1990s. However, the science driving these plans has been heavily focused on understanding ecological and biological components of landscapes. When considered at all, human components of ecosystems tend to be evaluated as separate from ecological components and as *impacts* on naturally functioning systems (Alberti et al. 2003). There is a large body of literature that identifies the need for integrated approaches to planning, but the use of social science still lags far behind the use of physical and ecological data (Endter-Wada et al. 1998).

Land management agencies often confuse public engagement with social science in an attempt to meet this need, but it is important to recognize that public engagement is not a proxy for the substantive, empirical science that brings data on user or stakeholder perspectives into the conservation planning processes (Endter-Wada et al. 1998). Human uses are part of all ecosystems, and landscape planning efforts, even for natural reserves, are pursued primarily to meet human goals. For this reason, human use sustainability is not just a "recreation" planning question. It must be part of all landscape conservation efforts.

When the Chugach National Forest initiated their Prince William Sound Framework effort in 2006 (chapter 19), the overarching objective was to create a recreation management strategy that protects resources and enhances user experiences in the Sound. This could only succeed if the strategy was based on a rigorous understanding of the distribution and dynamics of human use, and this would only come from specific investments in social science.

Subsequent analyses of those efforts, captured in this volume, illustrate that there are keystone experiences sought by recreationists in the Sound and that the system is not yet in "crisis" mode in terms of conflict, crowding, or impacts to wildlife and other resources. This reality and the depth of information provided by their investments in social science research and analysis provided the Forest Service with an opportunity to pursue a stepwise approach to adapting their management. The issue-based planning approach brought together social science data on use patterns and recreation experience opportunities with potential resource concerns to recommend management strategies and follow up monitoring.

The resulting specificity in these recommendations was only possible with the investments made in social science data that is both spatially and temporally explicit. In this case, GIS layers mapping the seasonality of use, by user type (chapters 6, 7, and 9), were integrated with research on the motivations and expectations of stakeholders (chapters 7 and 8). This allows for recommendations tailored specifically to address well-defined issues at the appropriate scale—be it Sound wide, for a bay or island complex, or for an individual sensitive site. The identified issues and related management strategies that came from this process provide guidance for managers to systematically assess, manage, and engage stakeholders to provide a diversity of sustainable recreation opportunities across the Sound.

SUSTAINABLE USE REQUIRES BROADENING THE CONVERSATION BETWEEN MANAGERS AND STAKEHOLDERS

Sustainable recreation planning calls for attempts to address increasing human use by investing in greater efforts to involve stakeholders in management issues and concerns. Rapid changes in human uses in the Sound have not been accompanied by proportionate efforts to connect with users and the public at large about the Sound, its communities and the role of managers in providing for many uses in the region. Some of the earliest human use assessments in the Sound (chapter 12) recognized that management efforts needed support from stakeholders to more effectively promote best practices and direct or deflect use (i.e., campsite hardening or outreach efforts aiming to concentrate use). Since private users comprise the majority of all users (chapter 7), effective communication efforts with this group of recreation users will contribute the most when aiming to protect recreation opportunities and sensitive resources in the region.

For purely logistical reasons, managers have a difficult time in projecting their messaging across an area as large as the Sound. Given the region's vast extent, opportunities for managers to engage users must be focused on ports of entry, where there are many potential partners to help with messaging. Outreach and education efforts communicated in the words of the local public have tremendous utility for furthering an agency's stewardship goals. When individual stories about local land ethics or sustainable use practices are contributed by residents (chapter 18 and many of the essays in this volume), they have greater potential to resonate with fellow users and visitors than management agency messages that are often perceived, or ignored, as institutional and impersonal. Furthermore, when there is a history of negative interactions between agencies and the public, messages from a stakeholder group may prove to be more effective.

Engagement efforts with communities during the planning process and implementation of the Framework, as well as other chapters in this volume (e.g., chapters 8 and 18), allowed managers to begin to better understand the Sound communities as they see themselves. The experiences and relationships gained from the public engagement and collaborative work that produces something with shared value are a vital contribution to furthering sustainable recreation management in the Sound—and likely other wildland settings.

Sustainable recreation management in the Sound hinges on sincere and effective community engagement. True system-wide approaches require land and resource managers to shed some constraints of their organization cultures and look beyond the traditional missions and jurisdictions of their agencies (Thomas and Thorne 2003). This may take generational change among management professionals, or legislative action, both of which will likely be pushed for by stakeholders demanding common sense approaches to management—approaches free from artificial administrative constraints that don't make sense for the communities, users, or ecosystems of wildlands.

SUSTAINABLE USE: A WORKING WILDERNESS PROMOTES SOCIETAL RELEVANCE

The Sound, while being a remote, rugged, and pristine wild place, is a "working wilderness," where wildness thrives not in the absence of human use or in spite of it, but as part of it. As stakeholders from chapter 18 describe it, the Sound is a thriving wildland setting sustained by the human communities that live, work, and play there. The Sound is recovering from a devastating oil spill, which had both social and ecological consequences. But the chapters in this book show that in contemporary terms, there is no lasting reduction in recreation use 25 years after the spill. Evidence of high user satisfaction (chapter 7), successful coping strategies from one of the most heavily used areas (chapter 16), and increased use since the spill all provide evidence that recreation user experience in the Sound is strong and viable.

In fact, some speculate that recreation may now be putting pressure on other system components—namely species and habitats expected to be vulnerable to human disturbance (e.g., chapter 15). This leads to an interesting point of further reflection. Could the sustainable future of species and habitats injured by the spill *actually depend upon growth* of a healthy recreation and tourism stakeholder community to add numbers, depth, and diversity to those who would advocate for the Sound? In the same way that it hosts crucial activities like subsistence harvest and commercial fishing, the Sound is a wildland setting that provides people with an invaluable recreation experience in the outdoors. In turn, this visitation builds a constituency of people who have experienced for themselves the true value of this place and who may be called upon to champion its future *as a wild place*. Just as commercial fishing continues as it has for generations, carried on by individuals who work in the Sound to carve out a living, subsistence harvesters from communities in the Sound likewise depend on the working wilderness for the resources it provides. Year after year, the numbers of these two groups may remain steady (or could even decrease, see chapter 10) but are likely to become a less significant proportion of not only Alaska's population but also of the nation as a whole (Walls and Siikamaki 2010).

Sustainable recreation that allows people to appreciate the Sound can provide a supporting voice when it comes to future policy decisions affecting the region. It has been recognized nationally by the U.S. Forest Service in the new objectives around sustainable recreation aimed at getting people out to use public lands (USDA Forest Service 2010). Indeed, encouraging political support was one reason U.S. Forest Service leadership implemented several new initiatives to increase the use of national forests and "reconnect" Americans with the outdoors (Kimbell et al. 2009). By facilitating increased and diverse recreation use, with a careful eye toward sustainability, managers can find ways to increase the numbers of those who can find common cause alongside commercial fishers and subsistence harvesters when it comes to protecting the Sound.

This sentiment runs counter to the tradition of much wildland recreation management—limiting numbers of individuals using the landscape in order to reduce social or ecological impacts. This model of only considering human use as a negative impact has limited the thinking of managers in Prince William Sound and likely has in other wildland settings too. Management approaches, commonly used for evaluating and limiting numbers of individuals using a landscape, have been shown by Cole (2001) and others to favor certain users and certain types of experiences, leading to decisions that may be inherently biased against visitors seeking an alternative experience from the perceived "norm." This approach typically translates into a fixation by managers on minimizing encounters between groups of users and keeping group sizes small regardless of the recreation activity, desired experience, or specific setting (e.g., chapter 4).

This volume highlights some of the complexities missed by approaches of such singular focus. In the Sound, perceptions about encounters seem to have more to do with user expectations, behaviors exhibited by others during encounters, and the specific opportunities sought by different groups of recreationists. This is likely the case in many other wildland settings and suggests that attempting to understand different experience objectives, behaviors, preferences, and tolerances to visitor densities should be central to every sustainable recreation planning effort.

More visitor-based research is required that focuses on employing techniques that move away from a single normative metric for defining recreation carrying capacity to one that explores the acceptability of a range of social and biophysical conditions related to visitor use. The Prince William Sound Framework described in chapter 19 is one such attempt to deal with a large wildland setting and address issues associated with human use. By looking for ways to provide for the diversity of types of user experiences, the Forest Service and its partners can help to protect this working wilderness for generations into the future.

CONCLUSIONS

The planning process launched by the Chugach National Forest in Prince William Sound is emblematic of the types of efforts pursued by managers attempting to address issues associated with human use management. It began with a narrow focus on understanding human use in terms of potential resource and social impacts that might be exacerbating the original effects of the oil spill. As researchers and managers began their work, they quickly realized they needed human use patterns to be more rigorously characterized in a spatially and seasonally explicit way. They also uncovered complexities that challenged previous assumptions about user conflict and resource impacts. In part by testing some of those assumptions, it seems the Sound, though changing in use patterns, was not necessarily in great jeopardy from human use. Furthermore, the community engagement efforts launched to find new ways to communicate about resource conservation helped the Forest Service gain new perspective on the various ways that people value the Sound and its resources. Research results combined with this newfound perspective and new personal connections with stakeholders, including several who contributed to this volume, allowed managers to develop specific recommendations for issues that will help the Sound be managed to support a diversity of uses for generations to come.

This connection to stakeholders also affords managers the opportunity to gain support in stewardship efforts from stakeholder groups who, when better understood, have been shown to have similar conservation values. This is certainly true in Prince William Sound, where a still somewhat recent tragic event like the *Exxon Valdez* oil spill wreaked havoc on species and habitats, which in turn devastated communities that depended on those resources. On the heels of such a terrible human-caused disaster, it should surprise no one that those who live and work in the Sound are concerned about its sustainable use. By working alongside stakeholders, managers can gain an understanding of the importance of this landscape to all people who live, work, and play in the Sound.

The objective of managing for sustainable use is a tricky endeavor. Investments in rigorous social science provide managers with the tools to understand how humans use, experience, and value wildlands. Social science is also an essential component to exploring questions of potential resource or social impacts from human use in these settings. Too often, such efforts involve only biological or ecological science, and evaluation of social impacts or conflicts regularly only consist of expert opinion. This was the case in the Sound, and sadly, it is likely the case in many wildland settings. Social science combined with thoughtful community engagement can be a bridge to help managers understand science-based knowledge in the context of stakeholder perspective and to see how changes in management might best be approached. When done well, community engagement can help managers learn how to develop scientifically defensible plans that are inclusive of a diversity of users. This allows wildlands to have a better chance of maintaining societal relevance in a world where threats to these landscapes extend far beyond even those of a catastrophic oil spill.

REFERENCES

Adams, P. A., G. J. Robertson, and I. L. Jones. 2000. "Time-Activity Budgets of Harlequin Ducks Molting in the Gannet Islands, Labrador." *Condor* 102:703–8.

ADCCED (Alaska Department of Commerce, Community, and Economic Development). 2010. *2010 Alaska Economic Performance Report.* Alaska Department of Commerce, Community, and Economic Development. https://www.commerce.alaska.gov/web/Portals/6/pub/2010_Alaska_Economic_Performance_Report.pdf.

———. 2011. "Chenega Bay." Alaska Department of Commerce, Community, and Economic Development.

ADF&G (Alaska Department of Fish and Game). 1982. *Black Bear Movements and Home Range Study.* ADF&G Federal Aid in Wildlife Restoration Report. Juneau, AK: Alaska Department of Fish and Game.

———. 2008. *Black Bear Management Report of Survey-Inventory Activities 1 July 2004–30 June 2007.* Juneau, AK: Alaska Department of Fish and Game.

ADNR (Alaska Department of Natural Resources). 1988. *Prince William Sound Area Plan.* Anchorage, AK: Alaska Department of Natural Resources.

———. 2009. *Alaska's Outdoor Legacy: Statewide Comprehensive Outdoor Recreation Plan (SCORP) 2009–2014.* Anchorage, AK: Alaska Department of Natural Resources. http://dnr.alaska.gov/Assets/uploads/DNRPublic/parks/plans/scorp/scorp0914.pdf.

ADOT (Alaska Department of Transportation). 1995. *Whittier Access Project: Final Environmental Impact Statement and Final Section 4(f) Evaluation.* FHWA-AK-EIS-94-02-DR. Anchorage, AK: Alaska Department of Transportation and Public Facilities and Federal Highway Administration.

———. 2009. "Whittier Tunnel Date Traffic." Alaska Department of Transportation and Public Facilities and Federal Highway Administration.

Afton, A. D., and S. L. Paulus. 1992. "Incubation and Brood Care." In *Ecology and Management of Breeding Waterfowl,* edited by B. D. J. Batt, A. D. Afton, M. G. Anderson,

C. D. Ankney, D. H. Johnson, J. A. Kadlec, and G. L. Krapu, 62–108. Minneapolis: University of Minnesota Press.

Agness, A. M. 2006. "Effects and Impacts of Vessel Activity on the Kittlitz's Murrelet in Glacier Bay, Alaska." MS thesis, University of Washington.

Agness, A. M., J. F. Piatt, J. C. Ha, and G. R. VanBlaricom. 2008. "Effects of Vessel Activity on the Near-Shore Ecology of Kittlitz's Murrelets (*Brachyramphus brevirostris*) in Glacier Bay, Alaska." *Auk* 125:346–53.

Alberti, M., J. M. Marzluff, E. Shulenberger, G. Bradley, C. Ryan, and C. Zumbrunnen. 2003. "Integrating Humans into Ecology: Opportunities and Challenges for Studying Urban Ecosystems." *Bioscience* 53:1169–79.

Anderson, D. R., K. P. Burnham, and W. L. Thompson. 2000. "Null Hypothesis Testing: Problems, Prevalence, and an Alternative." *Journal of Wildlife Management* 64:912–23.

Andres, B. A. 1997. "The *Exxon Valdez* Oil Spill Disrupted the Breeding of Black Oystercatchers." *Journal of Wildlife Management* 61:1332–28.

———. 1998. "Shoreline Habitat Use of Black Oystercatchers Breeding in Prince William Sound, Alaska." *Journal of Field Ornithology* 69:626–34.

———. 1999. "Effects of Persistent Shoreline Oil on Breeding Success and Chick Growth in Black Oystercatchers." *Auk* 116:640–50.

Andres, B. A., and G. A. Falxa. 1995. "Black Oystercatcher (*Haematopus bachmani*)." In *The Birds of North America*, No. 155. Philadelphia, PA: Academy of Natural Sciences; Washington, DC: American Ornithologists' Union.

Anthony, R. G., and F. B. Isaacs. 1989. "Characteristics of Bald Eagle Nest Sites in Oregon." *Journal of Wildlife Management* 53:148–59.

Arimitsu, M. L., J. F. Piatt, and M. D. Romano. 2007. "Distribution of Ground-Nesting Marine Birds Along Shorelines in Glacier Bay, Southeastern Alaska: An Assessment Related to Potential Disturbance by Back-Country Users." In *USDI Geological Survey Scientific Investigations Report 2007–5278*.

Arimitsu, M. L., M. D. Romano, J. F. Piatt, and D. F. Tessler. 2005. *Ground-Nesting Marine Bird Distribution and Potential for Human Disturbance in Glacier Bay National Park and Preserve, Alaska*. Anchorage, AK: USDI Geological Survey.

Atkinson, S., D. P. Demaster, and D. G. Calkins. 2008. "Anthropogenic Causes of the Western Steller Sea Lion *Eumetopias jubatus* Population Decline and Their Threat to Recovery." *Mammal Review* 38:1–18.

Barbosa, A. M., R. Real, A. L. Marquez, and M. A. Rendon. 2001. "Spatial, Environmental and Human Influences on the Distribution of Otter (*Lutra lutra*) in the Spanish Provinces." *Diversity and Distribution* 7:137–44.

Barbour, M. G., and W. D. Billings. 1991. *North American Terrestrial Vegetation*. Cambridge: Cambridge University Press.

Bardwell, L. 1991. "Problem Framing: A Perspective on Environmental Problem-Solving." *Environmental Management* 15 (5): 603–12.

Beale, C. M., and P. Monaghan. 2004. Human disturbance: People as predation-free predators? Journal of Applied Ecology 41:335–343.

Beaulieu, J. T., and R. Schreyer. 1985. "Choices of Wilderness Environments: Differences Between Real and Hypothetical Choice Situations." In *Proceedings: Symposium on Recreation Choice Behavior*, edited by G. H. Stankey and S. F. McCool, 38–45. Ogden, UT: USDA Forest Service.

Becker, J. M. 2002. "Response of Wintering Bald Eagles to Industrial Construction in Southeastern Washington." *Wildlife Society Bulletin* 30:875–78.

Belanger, L., and J. Bedard. 1989. "Responses of Staging Greater Snow Geese to Human Disturbance." *Journal of Wildlife Management* 53:713–19.

Bell, S. S., R. A. Brooks, B. D. Robbins, M. S. Fonseca, and M. O. Hall. 2001. "Faunal Response to Fragmentation in Seagrass Habitats: Implications for Seagrass Conservation." *Biological Conservation* 100:115–23.

Bellefleur, D., L. Philip, and R. A. Ronconi. 2009. "The Impact of Recreational Boat Traffic on Marbled Murrelets (*Brachyramphus marmoratus*)." *Journal of Environmental Management* 90:531–38.

Ben-David, M., G. M. Blundell, J. W. Kern, J. A. K. Maier, E. D. Brown, and S. C. Jewett. 2005. "Communication in River Otters: Creation of Variable Resource Sheds for Terrestrial Communities." *Ecology* 86:1331–45.

Ben-David, M., and H. Golden. 2007a. "River Otters (*Lontra canadensis*) in Southcentral Alaska: Distribution, Relative Abundance, and Minimum Population Size Based on Coastal Latrine Site Surveys." In *Southwest Alaska Inventory and Monitoring Network, Inventory and Monitoring Program Report.* Anchorage, AK: National Park Service.

———. 2007b. "River Otter Survey Protocols for Coastal Habitat." In *Southwest Alaska Inventory and Monitoring Network, Inventory and Monitoring Program Report.* Anchorage, AK: National Park Service.

Benowitz-Fredericks, M., A. S. Kitaysky, and A. Springer. 2007. "Seabirds." In *Long-Term Ecological Change in the Northern Gulf of Alaska*, edited by R. B. Spies, 94–113. Amsterdam: Elsevier.

Bishop, M. A., and S. P. Green. 2001. "Predation on Pacific Herring (*Clupea pallasi*) Spawn by Birds in Prince William Sound, Alaska." *Fisheries Oceanography* 10 (S1.1): 149–58.

Blaber, S. J., D. P. Cyrus, J. Albaret, C. Ching, J. W. Day, M. Elliott, M. S. Fonseca, D. E. Hoss, J. Orensanz, I. C. Potter, and W. Silvert. 2000. "Effects of Fishing on the Structure and Functioning of Estuarine and Nearshore Ecosystems." *ICES Journal of Marine Science* 57:590–602.

Blahna, D. J. 2007. "Introduction: Recreation Management." In *Proceedings: National Workshop on Recreation Research and Management*, edited by L. E. Kruger, R. Mazza, and K. Lawrence, 101–13. Portland, OR: U.S. Department of Agriculture, Forest Service, Pacific Northwest Research Station.

Blamires, S. J., M. B Thompson, and D. F. Hochuli. 2007. "Habitat Selection and Web Plasticity by the Orb Spider *Argiope keyserlingi* (Argiopidae): Do They Compromise Foraging Success for Predator Avoidance?" *Austral Ecology* 32:551–63.

Blumstein, D. T. 2003. "Flight-Initiation Distance in Birds Is Dependent on Intruder Starting Distance." *Journal of Wildlife Management* 67:852–57.

Blundell, G. M., M. Ben-David, and R. T. Bowyer. 2002a. "Sociality in River Otters: Cooperative Foraging or Reproductive Strategies?" *Behavioral Ecology* 13:134–41.

Blundell, G. M., M. Ben-David, P. Groves, R. T. Bowyer, and E. Geffen. 2002b. "Characteristics of Sex-Biased Dispersal and Gene Flow in Coastal River Otters: Implications for Natural Recolonization of Extirpated Populations." *Molecular Ecology* 11:289–303.

Borrie, W. T., S. F. McCool, and G. H. Stankey. 1999. "Protected Area Planning Principles and Strategies." In *Ecotourism: A Guide for Planners and Managers*, vol. II, edited by K. Lindberg, M. E. Wood, and D. Engeldrum, 133–54. North Bennington, VT: The Ecotourism Society.

Bosley, H. E. 2005. "Techniques for Estimating Boating Carrying Capacity: A Literature Review." Accessed March 21, 2011. http://www.trpa.org/wp-content/uploads/2005-Bosley-techniques-for-estimating-carrying-capacity.pdf.

Bowker, J. M. 2001. *Outdoor Recreation by Alaskans: Projections for 2000 Through 2020*. General Technical Report PNW-GTR-527. Portland, OR: USDA Forest Service.

Bowyer, R. T., G. M. Blundell, M. Ben-David, S. C. Jewett, T. A. Dean, and L. K. Duffy. 2003. "Effects of the *Exxon Valdez* Oil Spill on River Otters: Injury and Recovery of a Sentinel Species." *Wildlife Monographs* 153:1–53.

Bowyer, R. T., J. W. Testa, and J. B. Faro. 1995. "Habitat Selection and Home Ranges of River Otters in a Marine Environment: Effects of the *Exxon Valdez* Oil Spill." *Journal of Mammalogy* 76:1–11.

Bowyer, R. T., J. W. Testa, J. B. Faro, C. C. Schwartz, and J. B. Browning. 1994. "Changes in Diets of River Otters in Prince William Sound, Alaska: Effects of the Exxon Valdez Oil Spill." *Canadian Journal of Zoology* 72:970–76.

Boyle, S. A., and F. B. Samson. 1985. "Effects of Nonconsumptive Recreation on Wildlife: A Review." *Wildlife Society Bulletin* 13:110–16.

Brooks, D. J., and R. W. Haynes. 2001. *Recreation and Tourism in Southcentral Alaska: Synthesis of Recent Trends and Prospects*. General Technical Report PNW-GTR-511. Portland, OR: USDA Forest Service.

Brown, B. T., and L. E. Stevens. 1997. "Winter Bald Eagle Distribution Is Inversely Correlated with Human Activity Along the Colorado River, Arizona." *Journal of Raptor Research* 31:7–10.

Brown, G. 2005. "Mapping Spatial Attributes in Survey Research for Natural Resource Management: Methods and Applications." *Society & Natural Resources* 18 (1): 1–23.

Brown, G., P. Reed, and C. C. Harris. 2002. "Testing a Place-Based Theory for Environmental Evaluation: An Alaska Case Study." *Applied Geography* 22 (1): 49–77.

Brown, G., C. Smith, L. Alessa, and A. Kliskey. 2004. "A Comparison of Perceptions of Biological Value with Scientific Assessment of Biological Importance." *Applied Geography* 24 (2): 161–80.

Brunson, M., and B. Shelby. 1990. "A Hierarchy of Campsite Attributes in Dispersed Recreation Settings." *Leisure Sciences* 12:197–209.

Buehler, D. A. 2000. "Bald Eagle (*Haliaeetus leucocephalus*)." In *The Birds of North America*, edited by P. G. Rodewald. Ithaca, NY: Cornell Lab of Ornithology. https://birdsna.org/Species-Account/bna/species/baleag.

Buehler, D. A., T. J. Mersmann, J. D Fraser, and J. K. D. Seegar. 1991a. "Nonbreeding Bald Eagle Communal and Solitary Roosting Behavior and Roost Habitat on the Northern Chesapeake Bay." *Journal of Wildlife Management* 55:273–81.

———. 1991b. "Effects of Human Activity on Bald Eagle Distribution on the Northern Chesapeake Bay." *Journal of Wildlife Management* 55:282–90.

Bullock, S. D., and S. R. Lawson. 2007. "Examining the Potential Effects of Management Actions on Visitor Experience on the Summit of Cadillac Mountain, Acadia National Park." *Human Ecology Review* 14 (2): 140–56.

Burger, A. E. 1997. "Distribution and Abundance of Marbled Murrelets and Other Seabirds off the West Coast Trail and in the Broken Group Islands, Pacific Rim National Park in 1996." Unpublished report. Ucluelet, BC: Pacific Rim National Park.

———. 2002. *Conservation Assessment of Marbled Murrelets in British Columbia: A Review of the Biology, Populations, Habitat Associations and Conservation*. Delta, BC: Canadian Wildlife Service.

Bustnes, J. O., K. V. Galaktionov, and S. W. B. Irwin. 2000. "Potential Threat to Littoral Biodiversity: Is Increased Parasitism a Consequence of Human Activity?" *Oikos* 90:189–90.

Calkins, D. G. 1986. "Marine Mammals." In *The Gulf of Alaska: Physical Environment and Biological Resources*, edited by D. W. Hood and S. T. Zimmerman, 527–58. Anchorage, AK: National Oceanic and Atmospheric Administration.

Carney, K. M., and W. J. Sydeman. 1999. "A Review of Human Disturbance Effects on Nesting Colonial Waterbirds." *Waterbirds* 22:68–79.

Caron, J. A., and W. L. Robinson. 1994. "Responses of Breeding Loons to Human Activity in Upper Michigan." *Hydrobiologia* 279/280:431–38.

Carter, H. R., M. L. C. McAllister, and M. E. P. Isleib. 1995. "Mortality of Marbled Murrelets in Gill Nets in North America." In *Ecology and Conservation of the Marbled Murrelet*, edited by C. J. Ralph, G. L. Hunt Jr., M. G. Raphael, and J. F. Piatt, 271–84. Albany, CA: USDA Forest Service, Pacific Coast Research Station.

Carter, H. R., and S. G. Sealy. 1984. "Marbled Murrelet (*Brachyramphus marmoratus*) Mortality Due to Gill-Net Fishing in Barkley Sound, British Columbia." In *Marine Birds: Their Feeding Ecology and Commercial Fisheries Relationships*, edited by D. N. Nettleship, G. A. Sanger, and P. F. Springer, 212–20. Ottawa, ON: Canadian Wildlife Service.

Catton, T. 1997. *Inhabited Wilderness*. Albuquerque: University of New Mexico Press.

Cerveny, L., D. J. Blahna, M. Stern, and M. Mortimer. 2011. "The Use of Recreation Planning Tools in US Forest Service NEPA Assessments." *Environmental Management* 48:644–57.

Chabreck, R. H., T. L. Edwards, and G. Linscombe. 1985. "Factors Affecting the Distribution and Harvest of River Otters in Louisiana." In *Proceedings of the Annual Conference of Southeastern Fish and Wildlife Agencies* 39:520–27.

Chenega Corporation. 2008. "Resource Management Plan for the Village of Chenega Bay (Draft)." Anchorage, AK: Chenega Village Corporation.

Chilelli, M., B. Griffeth, and D. J. Harrison. 1996. "Interstate Comparisons of River Otter Harvest Data." *Wildlife Society Bulletin* 24:238–46.

Chugach Alaska Corporation. 2011. "History and Culture." Accessed October 10. http://www.chugach.com/who-we-are/history-culture.

Clark, R. N., and G. H. Stankey. 1979. *The Recreation Opportunity Spectrum: A Framework for Planning, Management, and Research.* General Technical Report PNW-GTR-098. Portland, OR: USDA Forest Service, Pacific Northwest Research Station.

———. 2006. *Integrated Research in Natural Resources: The Key Role of Problem Framing.* General Technical Report PNW-GTR-678. Portland, OR: USDA Forest Service, Pacific Northwest Research Station.

Cole, D. N. 1981a. "Managing Ecological Impacts at Wilderness Campsites and Evaluation of Techniques." *Journal of Forestry* 79:86–89.

———. 1983a. *Monitoring the Condition of Wilderness Campsites.* Research Paper INT-302. Ogden, UT: USDA Forest Service, Intermountain Forest and Range Experiment Station.

———. 1983b. *Campsite Conditions in the Bob Marshall Wilderness, Montana.* Research Paper INT-312. Ogden, UT: USDA Forest Service, Intermountain Forest and Range Experiment Station.

———. 1989. *Wilderness Campsite Monitoring Methods: A Sourcebook.* General Technical Report INT-259. Ogden, UT: USDA Forest Service, Intermountain Research Station.

———. 1993. *Campsites in Three Western Wildernesses: Proliferation and Changes in Condition over 12 to 16 Years.* Research Paper INT-463. Ogden, UT: USDA Forest Service, Intermountain Research Station.

———. 1995. "Disturbance of Natural Vegetation by Camping: Experimental Applications of Low Level Stress." *Environmental Management* 19:405–16.

———. 2001. "Visitor Use Density and Wilderness Experiences: A Historical Review of Research." In *Visitor Use Density and Wilderness Experience: Proceedings, Missoula, Montana, June 1–3, 2000*, compiled by W. A. Freimund and D. N. Cole, 11–20. RMRS-P-20. Ogden, UT: USDA Forest Service, Rocky Mountain Research Station.

———. 2004a. "Monitoring and Management of Recreation in Protected Areas: The Contributions and Limitations of Science." In *Proceedings of the Second International Conference on Monitoring and Management of Visitor Flows in Recreational and Protected Areas*, edited by T. Sievanen, J. Erkkonen, J. Jokimaki, J. Saarinen, S. Tuulentie, and E. Virtanen,

9–16. Working Papers of the Finnish Forest Research Institute. Metla, Vantaa, Finland. http://mmv.boku.ac.at/downloads/mmv2-proceedings.pdf.

Cole, D. N., and T. E. Hall. 1992. *Trends in Campsite Condition: Eagle Cap Wilderness, Bob Marshall Wilderness, and Grand Canyon National Park.* USDA Forest Service Research Paper INT-453. Ogden, UT: Intermountain Research Station.

———. 2009. "Perceived Effects of Setting Attributes on Visitor Experiences in Wilderness: Variation with Situational Context and Visitor Characteristics." *Environmental Management* 44:24–36.

Cole, D. N., P. Foti, and M. Brown. 2008. "Twenty Years of Change on Campsites in the Backcountry of Grand Canyon National Park." *Environmental Management* 41:959–70.

Cole, D. N., and C. A. Monz. 2004a. "Impacts of Camping on Vegetation: Response of Acute and Chronic Disturbance on Vegetation." *Environmental Management* 32:693–705.

———. 2004b. "Spatial Patterns of Recreation Impact on Experimental Campsites." *Journal of Environmental Management* 70:73–84.

Cole, D. N., and L. Yung, eds. 2010. *Beyond Naturalness: Rethinking Park and Wilderness Stewardship in an Era of Rapid Change.* Washington, DC: Island Press.

Collins, S., and H. Brown. 2007. "The Growing Challenge of Managing Outdoor Recreation." *Journal of Forestry* October/November: 371–75.

Colt, S., S. Martin, J. Mieren, and M. Tomeo. 2002. *Recreation and Tourism in Southcentral Alaska: Patterns and Prospects.* General Technical Report PNW-GTR-551. Portland, OR: USDA Forest Service, Pacific Northwest Research Station.

Cooke, S. J., and I. G. Cowx. 2006. "Contrasting Recreational and Commercial Fishing: Searching for Common Issues to Promote Unified Conservation of Fisheries Resources and Aquatic Environments." *Biological Conservation* 128:93–108.

Cooney, R. T. 2005. "Biological and Chemical Oceanography." In *The Gulf of Alaska: Biology and Oceanography*, edited by P. R. Mundy, 49–57. Fairbanks, AL: Alaska Sea Grant College Program.

Cooney, R. T., and K. O. Coyle. 1988. "Water Column Production." In *Environmental Studies in Port Valdez, Alaska: A Basis for Management*, edited by D. G. Shaw and M. J. Hameedi, 93–116. Lecture Notes on Coastal and Estuarine Studies 24. New York: Springer-Verlag.

Cooney, R. T., K. O. Coyle, E. Stockmar, and C. Stark. 2001. "Seasonality in Surface-Layer Net Zooplankton Communities in Prince William Sound, Alaska." *Fisheries Oceanography* 10 (S1): 97–109.

Cordell, H. K. 2008. "The Latest on Trends in Nature-Based Outdoor Recreation and Tourism." *Forest History Today* Spring: 4–10.

Cordell, H. K., C. Betz, and G. Green. 2008. "Nature-Based Outdoor Recreation Trends and Wilderness." *International Journal of Wilderness* 14 (2): 7–13.

Craig, R. J., E. S. Mitchell, and J. E. Mitchell. 1988. "Time and Energy Budgets of Bald Eagles Wintering Along the Connecticut River." *Journal of Field Ornithology* 59:22–32.

Crawford, H. M., D. A. Jensen, B. Peichel, P. M. Charlebois, B. A. Doll, S. H. Kay, V. A. Ramey, and M. B. O'Leary. 2001. "Sea Grant and Invasive Aquatic Plants: A National Outreach Initiative." *Journal of Aquatic Plant Management* 39:8–11.

Crawford, W. R., P. J. Brickley, and A. C. Thomas. 2007. "Mesoscale Eddies Dominate the Surface Plankton in Northern Gulf of Alaska." *Progress in Oceanography* 75:287–303.

Curland, J. M. 1997. "Effects of Disturbance on Sea Otters (*Enhydra lutris*) near Monterey, California." MS thesis, San José State University.

Czech, B., P. R. Krausman, and P. K. Devers. 2000. "Economic Associations Among Causes of Species Endangerment in the United States." *BioScience* 50:593–601.

Daniels, M. L. and J. L. Marion. 2006. "Visitor Evaluations of Management Actions at a Highly Impacted Appalachian Trail Camping Area." *Environmental Management* 38:1006–19.

De Lacy, T., and M. Whitmore. 2006. "Tourism and Recreation." In *Managing Protected Areas: A Global Guide*, edited by M. Lockwood, G. L. Worboys, and A. Kothari, 497–527. London: Earthscan.

Dean, T. A., L. Haldorson, D. R. Laur, S. C. Jewett, and A. Blanchard. 2000. "The Distribution of Nearshore Fishes in Kelp and Eelgrass Communities in Prince William Sound, Alaska: Associations with Vegetation and Physical Habitat Characteristics." *Environmental Biology of Fishes* 57:271–87.

del Hoyo, J., A. Elliot, and J. Sargatal. 1996. *Hoatzin to Auks*. Vol. 3 of *Handbook of the Birds of the World*. Barcelona, Spain: Lynx Edicions.

DesGranges, J., and A. Reed. 1981. "Disturbance and Control of Selected Colonies of Double-Crested Cormorants in Quebec." *Colonial Waterbirds* 4:12–19.

Dillman, D. A. 1978. *Mail and Telephone Surveys: The Total Design Method*. New York: Wiley.

Duffus, D. A., and P. Dearden. 1993. "Recreational Use, Valuation, and Management, of Killer Whales (*Orcinus orca*) on Canada's Pacific Coast." *Environmental Conservation* 20:149–56.

Eckstein, B., and J. A. Throgmorton. 2003. *Story and Sustainability: Planning, Practice, and Possibility for American Cities*. Cambridge, MA: MIT Press.

Ellison, L. N., and L. Cleary. 1978. "Effects of Human Disturbance on Breeding of Double-Crested Cormorants." *Auk* 95:510–17.

Endler, J. A. 1978. "A Predator's View of Animal Color Patterns." *Evolutionary Biology* 11:319–64.

Endter-Wada, J., D. Blahna, R. Krannich, and M. Brunson. 1998. "A Framework for Understanding Social Science Contributions to Ecosystem Management." *Ecological Applications* 8:891–904.

ERM (Environmental Resources Management) Inc. 2004. *Deep Creek Lake Boating and Commercial Use Carrying Capacity Study*. Proposal no. KOOR2200624. Annapolis, MD: ERM.

Erskine, J., ed. 1999. *Alaska Economic Trends* 19 (12). Accessed October 10, 2011. http://www.labor.state.ak.us/trends/trendspdf/dec99.pdf.

Erwin, M. R. 1989. "Responses to Human Intruders by Birds Nesting in Colonies: Experimental Results and Management Guidelines." *Colonial Waterbirds* 12:104–8.

Esler, D., J. A. Schmutz, R. L. Jarvis, and D. M. Mulcahy. 2000. "Winter Survival of Adult Female Harlequin Ducks in Relation to History of Contamination by the *Exxon Valdez* Oil Spill." *Journal of Wildlife Management* 64:839–47.

Eslinger, D. L., R. T. Cooney, C. P. McRoy, A. Ward, T. C. Kline Jr., E. P. Simpson, J. Wang, and J. Allen. 2001. "Plankton Dynamics: Observed and Modeled Responses to Physical Conditions in Prince William Sound, Alaska." *Fisheries Oceanography* 10 (S1): 81–96.

ESRI (Environmental Systems Research Institute). 2008. *ArcMap* (version 9.3). Redlands, CA: ESRI.

ESRI (Environmental Systems Research Institute). 2009. *ArcMap* (version 9.3.1). Redlands, CA: ESRI.

EVOS (*Exxon Valdez* Oil Spill) Trustee Council. 1994. *Summary of the Final Environmental Impact Statement for the* Exxon Valdez *Oil Spill Restoration Plan.* Anchorage, AK: *Exxon Valdez* Oil Spill Trustee Council. http://www.evostc.state.ak.us/Universal/Documents/Restoration/1994RestorationPlanEISSummary.pdf

EVOS (*Exxon Valdez* Oil Spill) Trustee Council. 2010. Exxon Valdez *Oil Spill Restoration Plan: Update of Injured Resources and Services.* Anchorage, AK: *Exxon Valdez* Oil Spill Trustee Council.

Ewins, P. J. 1993. "Pigeon Guillemot (*Cepphus columba*)." In *The Birds of North America,* No. 49, edited by A. Poole and F. Gill. Philadelphia, PA: Academy of Natural Sciences; Washington, DC: American Ornithologists' Union.

Fall, J. A., ed. 2006. *Update of the Status of Subsistence Uses in Exxon Valdez Oil Spill Area Communities. Exxon Valdez* Oil Spill Restoration Project Final Report (Restoration Project 040471). Anchorage, AK: Alaska Department of Fish and Game, Subsistence Division.

Fall, J. A., L. Stratton, P. Coiley, L. Brown, C. J. Utermohle, and G. Jennings. 1996. *Subsistence Harvests and Uses in Chenega Bay and Tatitlek in the Year Following the* Exxon Valdez *Oil Spill.* Technical Paper No. 199. Anchorage, AK: Alaska Department of Fish and Game, Division of Subsistence.

Fall, J. A., and C. Utermohle. 1999. *Subsistence Harvest and Uses in Eight Communities Ten Years After the* Exxon Valdez *Oil Spill.* Technical Paper No. 252. Juneau, AK: Alaska Department of Fish and Game.

Farrell, T., T. E. Hall, and D. D. White. 2001. "Wilderness Campers' Perception and Evaluation of Campsite Impacts." *Journal of Leisure Research* 33 (3): 229–50.

Fay, G. 2008. *Prince William Sound Tourism Economic Indicators.* Anchorage, AK: National Wildlife Federation.

Feder, H. M., and B. Bryson-Schwafel. 1988. "The Intertidal Zone." In *Environmental Studies in Port Valdez, Alaska: A Basis for Management,* edited by D. G. Shaw and M. J. Hameedi, 117–64. Lecture Notes on Coastal and Estuarine Studies 24. New York: Springer-Verlag.

Feder, H. M., and S. C. Jewett. 1988. "The Subtidal Benthos." In *Environmental Studies in Port Valdez, Alaska: A Basis for Management*, edited by D. G. Shaw and M. J. Hammedi, 165–202. Lecture Notes on Coastal and Estuarine Studies 24. New York: Springer-Verlag.

Festinger, L. 1957. *A Theory of Cognitive Dissonance.* Stanford, CA: Stanford University Press.

Fletcher, R. J., S. T. McKinney, and C. E. Bock. 1999. "Effects of Recreational Trails on Wintering Diurnal Raptors Along Riparian Corridors in a Colorado Grassland." *Journal of Raptor Research* 33:233–39.

Flint, P. L., D. L. LaCroix, J. A. Reed, and R. B. Lanctot. 2004. "Movements of Flightless Long-Tailed Ducks During Wing Molt." *Waterbirds* 27:35–40.

Francour, P., A. Ganteaume, and M. Poulain. 1999. "Effects of Boat Anchoring in *Posidonia oceanica* Seagrass Beds in the Port-Cros National Park (North-Western Mediterranean Sea)." *Aquatic Conservation: Marine and Freshwater Ecosystems* 9:391–400.

Fraser, J. D., L. D. Frenzel, and J. E. Mathisen. 1985. "The Impact of Human Activities on Breeding Bald Eagles in North-Central Minnesota." *Journal of Wildlife Management* 49:585–92.

Frid, A., and L. Dill. 2002. "Human-Caused Disturbance Stimuli as a Form of Predation Risk." *Conservation Ecology* 6:11–26.

Frissell, S. S. 1978. "Judging Recreational Impacts on Wilderness Campsites." *Journal of Forestry* 76:481–83.

Fry, D. M. 1995. "Pollution and Fishing Threats to Marbled Murrelets." In *Ecology and Conservation of the Marbled Murrelet*, edited by C. J. Ralph, G. L. Hunt Jr., M. G. Raphael, and J. F. Piatt, 257–60. General Technical Report PSW-GTR-152. Portland, OR: USDA Forest Service.

Garshelis, D. L., and J. A. Garshelis. 1984. "Movements and Management of Sea Otters in Alaska." *Journal of Wildlife Management* 48:665–78.

Giere, P., and D. S. Eastman. 2000. "American River Otters, *Lontra canadensis*, and Humans: Occurrence in a Coastal Urban Habitat and Reaction to Increased Levels of Disturbance." In *Mustelids in a Modern World: Management and Conservation Aspects of a Small Carnivore: Human Interactions*, edited by H. I. Griffiths, 107–25. Leiden, Netherlands: Backhuys.

Gilliland, S., G. J. Robertson, M. Robert, J.-P. L. Savard, D. Amirault, P. Laporte, and P. Lamothe. 2002. "Abundance and Distribution of Harlequin Ducks Molting in Eastern Canada." *Waterbirds* 25:333–39.

Gimblett, H. R., ed. 2002. *Integrating GIS and Agent Based Modeling Techniques for Understanding Social and Ecological Processes.* London: Oxford University Press.

Gimblett, H. R., S. Cable, D. Cole, and R. M. Itami. 2005a. "Recreation Visitation and Impacts in the Bighorn Crags Portion of the Frank Church—River of No Return Wilderness." In *Computer Simulation Modeling of Recreation Use: Current Status, Case Studies, and Future Directions*, edited by D. N. Cole, 18–21. General Technical Report RMRS-GTR-143. Ogden, UT: USDA Forest Service.

Gimblett, H. R., and R. M. Itami. 1997. "Modeling the Spatial Dynamics and Social Interaction of Human Recreationists' Using GIS and Intelligent Agents." Paper presented at the MODSIM 97 International Congress on Modeling and Simulation, Hobart, Tasmania, December 8–11.

———. 2006. *Modeling the Distribution of Human Use in Prince William Sound Using the Recreation Behavior Simulator*. Report prepared for the Chugach National Forest, Girdwood, Alaska.

Gimblett, H. R., and S. Lace. 2005. *Spring Black Bear Harvest Simulation Modeling in Prince William Sound, Alaska*. Final report prepared for the Glacier Ranger District. Girdwood, AK: Glacier Ranger District, Chugach National Forest. https://cals.arizona.edu/~gimblett/PWS_Report_2005_Final.pdf.

Gimblett, H. R., M. T. Richards, and R. M. Itami. 2001. "RBSim: Geographic Simulation of Wilderness Recreation Behavior." *Journal of Forestry* 99 (4): 36–42.

Gimblett, H. R., and H. Skov-Petersen, eds. 2008. *Monitoring, Simulation and Management of Visitor Landscapes*. Tucson: University of Arizona Press.

Gimblett, H. R., P. Wolfe, L. Kennedy, R. M. Itami, and B. Garber-Yonts. 2007. *Prince William Sound Human Use Study (PWSHUS)*. Report for the Glacier Ranger District. Girdwood, AK: Glacier Ranger District, Chugach National Forest.

Goering, J. J., C. J. Patton, and W. E. Shiels. 1973. "Nutrient Cycles." In *Environmental Studies of Port Valdez*, edited by D. W. Hood, W. E. Shiels, and E. J. Kelley, 225–48. Institute of Marine Science Occasional Publication No. 3. Fairbanks: University of Alaska.

Golden, H. N. 1996. *Furbearer Management Technique Development*. Federal Aid in Wildlife Restoration Research Progress Report, Grants W-24-3 and W-24-4. Juneau, AK: Alaska Department of Fish and Game.

Golumbia, T. E., D. R. Nysewander, R. W. Butler, R. L. Milner, T. A. Cyra, and J. R. Evenson. 2009. "Status of Breeding Black Oystercatchers *Haematopus bachmani* in the Salish Sea." *Marine Ornithology* 37:29–32.

Goudie, R. I. 2006. "Multivariate Behavioral Response of Harlequin Ducks to Aircraft Disturbance in Labrador." *Environmental Conservation* 33:28–35.

Goudie, R. I., and C. D. Ankney. 1986. "Body Size, Activity Budgets, and Diets of Sea Ducks Wintering in Newfoundland." *Ecology* 67:1475–82.

Goudie, R. I., and I. L. Jones. 2004. "Dose-Response Relationships of Harlequin Duck Behaviour to Noise from Low-Level Military Jet Over-Flights in Central Labrador." *Environmental Conservation* 31:289–98.

Grubb, T. G., W. W. Bowerman, J. P. Geisy, and G. A. Dawson. 1992. "Responses of Breeding Bald Eagles, *Haliaeetus leucephalus*, to Human Activities in Northcentral Michigan." *Canadian Field Naturalist* 106:443–53.

Grubb, T. G., and R. M. King. 1991. "Assessing Human Disturbance of Breeding Bald Eagles with Classification Tree Models." *Journal of Wildlife Management* 55:500–511.

Gunderson, L. H., and C. S. Holling. 2002. *Panarchy: Understanding Transformations in Human and Natural Systems*. Washington, DC: Island Press.

Haas, G., R. Aukerman, V. Lovejoy, and D. Welch. 2004. *Water Recreation Opportunity Spectrum (WROS) Users' Guidebook*. Lakewood, CO: U.S. Bureau of Reclamation.

Haines, A. M., M. Leu, L. K. Svancara, G. Wilson, and J. M. Scott. 2010. "Using a Distribution and Conservation Status Weighted Hotspot Approach to Identify Areas in Need of Conservation Action to Benefit Idaho Bird Species." *Northwest Science* 84:170–82.

Hall, T. E., and D. N. Cole. 2007. *Changes in the Motivations, Perceptions, and Behaviors of Recreation Users: Displacement and Coping in Wilderness*. Research Paper RMRS-RP-63. Fort Collins, CO: USDA Forest Service, Rocky Mountain Research Station.

Halstead, B. J., E. D. McCoy, T. A. Stilson, and H. R. Mushinsky. 2007. "Alternative Foraging Tactics of Juvenile Gopher Tortoises (*Gopherus polyphemus*) Examined Using Correlated Random Walk Models." *Herpetologica* 63:472–81.

Hamilton, D. J. 2000. "Direct and Indirect Effects of Predation by Common Eiders and Abiotic Disturbance in an Intertidal Community." *Ecological Monographs* 70:21–43.

Hammitt, W. E., and D. N. Cole. 1998. *Wildland Recreation: Ecology and Management*. 2nd ed. New York: Wiley.

Hampton, B., and D. N. Cole. 2003. *Soft Paths: How to Enjoy the Wilderness Without Harming It*. Mechanicsburg, PA: Stackpole Books.

Hampton, M. A., P. R. Carlson, H. J. Lee, and R. A. Feely. 1986. "Geomorphology, Sediment and Sedimentary Processes." In *The Gulf of Alaska: Physical Environment and Biological Resources*, edited by D. W. Hood and S. T. Zimmerman, 93–143. MMS Publication No. 86–0095. Anchorage, AK: National Oceanic and Atmospheric Administration.

Hansen, H., M. Ben-David, and D. B. McDonald. 2007. "Effects of Genotyping Protocols on Success and Errors in Identifying Individual River Otters (*Lontra canadensis*) from Their Faeces." *Molecular Ecology Notes* 8:282–89.

Harney, J. N., M. Morris, and J. R. Harper. 2008. *ShoreZone Coastal Habitat Mapping Protocol for the Gulf of Alaska*. Project 08–01. Sidney, BC: Coastal and Ocean Resources.

Hastings, K., P. Hesp, and G. A. Kendrick. 1995. "Seagrass Loss Associated with Boat Moorings at Rottnest Island, Western Australia." *Ocean and Coastal Management* 26:225–46.

Heberlein, T. A., and P. Dunwiddie. 1979. "Systematic Observation of Use Levels, Campsite Selection and Visitor Characteristics at a High Mountain Lake." *Journal of Leisure Research* 11 (4): 307–16.

Hébert, P. N., and R. T. Golightly. 2006. *Movements, Nesting, and Response to Anthropogenic Disturbance of Marbled Murrelets (*Brachyramphus marmoratus*) in Redwood National and State Parks, California*. Species Conservation and Recovery Program Report 2006-02. Arcata, CA: Department of Wildlife, Humboldt State University.

Heimberger, M., D. Euler, and J. Barr. 1983. "The Impact of Cottage Development on Common Loon Reproductive Success in Central Ontario." *Wilson Bulletin* 95:431–39.

Hendee, J. C., G. H. Stankey, and R. C. Lucas. 1990. *Wilderness Management*. Golden, CO: North American Press.

Hennig, S., and W. Menefee. 1995. *Prince William Sound Recreation Project*. *Exxon Valdez* Oil Spill Restoration Project Final Reports 93065 and 94217. Anchorage: Alaska Department of Natural Resources and USDA.

Henny, C. J., L. J. Blus, S. P. Thompson, and U. W. Wilson. 1989. "Environmental Contaminants, Human Disturbance and Nesting of Double-Crested Cormorants in Northwestern Washington." *Colonial Waterbirds* 12:198–206.

Henry, E., and M. O. Hammill. 2001. "Impact of Small Boats on the Haulout Activity of Harbour Seals (*Phoca vitulina*) in Métis Bay, Saint Lawrence Estuary, Québec, Canada." *Aquatic Mammalogy* 27:140–48.

Hockey, P. A. R. 1987. "The Influence of Coastal Utilization by Man on the Presumed Extinction of the Canadian Black Oystercatcher *Haematopus meadewaldoi* Bannerman." *Biological Conservation* 39:49–62.

Hood, D. W. 1986. "Physical Setting and Scientific History." In *The Gulf of Alaska: Physical Environment and Biological Resources*, edited by D. W. Hood and S. T. Zimmerman, 5–27. MMS Publication No. 86–0095. Anchorage, AK: National Oceanic and Atmospheric Administration.

Hood, D. W., and S. T. Zimmerman, eds. 1986. *The Gulf of Alaska: Physical Environment and Biological Resources*. MMS Publication No. 86–0095. Anchorage, AK: National Oceanic and Atmospheric Administration.

Irons, D. B., S. J. Kendall, W. P. Erickson, L. L. McDonald, and B. K. Lance. 2000. "Chronic Effects of the *Exxon Valdez* Oil Spill on Summer Marine Birds in Prince William Sound, Alaska." *Condor* 102:723–37.

Itami, R. M. 2002. "Mobile Agents with Spatial Intelligence." In *Integrating GIS and Agent Based Modeling Techniques for Understanding Social and Ecological Processes*, edited by H. R. Gimblett, 191–210. London: Oxford University Press.

———. 2003. *Estimating Capacities for Pedestrian Walkways and Viewing Platforms*. Report to Parks Victoria. Brunswick, Australia: GeoDimensions Pty Ltd.

———. 2008. "Level of Sustainable Activity: Moving Visitor Simulation from Description to Management for an Urban Waterway in Australia." In *Monitoring, Simulation and Management of Visitor Landscapes*, edited by H. R. Gimblett and H. Skov-Petersen, 331–48. Tucson: University of Arizona Press.

Itami, R. M., and H. R. Gimblett. 2001. "Intelligent Recreation Agents in a Virtual GIS World." *Complexity International* 8. https://arizona.pure.elsevier.com/en/publications/intelligent-recreation-agents-in-a-virtual-gis-world.

Itami, R. M., R. Raulings, G. MacLaren, K. Hirst, R. Gimblett, D. Zanon, and P. Chladek. 2003. "RBSim 2: Simulating the Complex Interactions Between Human Movement and the Outdoor Recreation Environment." *Journal for Nature Conservation* 11:278–86.

Iverson, S. J., A. M. Springer, and J. Bodkin. 2007. "Marine Mammals." In *Long-Term Ecological Change in the Northern Gulf of Alaska*, edited by R. B. Spies, 114–34. Amsterdam, Netherlands: Elsevier.

Jacobs, J. 1974. "Quantitative Measurement of Food Selection: A Modification of the Forage Ratio and Ivlev's Electivity Index." *Oecologia* 14:413–17.

Johannsen, N., and E. Johannsen. 1975. *Exploring Prince William Sound: Its Fiords, Islands, Glaciers, and Wildlife*. Anchorage: Alaska Travel Publications.

Johnson, A., and A. Acevedo-Gutierrez. 2007. "Regulation Compliance by Vessels and Disturbance of Harbor Seals (*Phoca vitulina*)." *Canadian Journal of Zoology* 85:290–94.

Johnson, A. K., and C. P. Dawson. 2004. "An Exploratory Study of the Complexities of Coping Behavior in Adirondack Wilderness." *Leisure Sciences* 26:281–93.

Johnson, D. H. 1980. "The Comparison of Usage and Availability Measurements for Evaluating Resource Preference." *Ecology* 61:65–71.

Johnson, S. W., M. L. Murphy, D. J. Csepp, P. M. Harris, and J. F. Thedinga. 2003. *A Survey of Fish Assemblages in Eelgrass and Kelp Habitats of Southeastern Alaska*. NOAA Technical Memorandum NMFS-AFSC-139. Springfield, VA: U.S. Department of Commerce.

Jolliffe, I. T. 2002. *Principal Component Analysis*. 2nd ed. New York: Springer-Verlag.

Jørgensen, R., K. K. Olsen, I. B. Falk-Petersen, and P. Kanapthippilai. 2005. *Investigations of Potential Effects of Low Frequency Sonar Signals on Survival, Development and Behaviour of Fish Larvae and Juveniles*. Tromsø, Norway: Norwegian College of Fishery Science, University of Tromsø.

Kahn, A. E. 1966. "The Tyranny of Small Decisions: Market Failures, Imperfections, and the Limits of Economics." *Kyklos* 19:23–47.

Kaplan, J. D. 2003. "Human Recreation and Loon Productivity in a Protected Area, Isle Royale National Park." MS thesis, Michigan Technological University.

Kehoe, S. 2002. *Campsite Inventory and Monitoring Program for Prince William Sound, Alaska: A Procedural Manual*. Lander, WY: NOLS Research. Accessed May 15, 2009. http://rendezvous.nols.edu/files/Curriculum/research_projects/Recreational%20Ecology%20Reports/PWS_Camp_procedures_.pdf.

Keller, V. E. 1991. "Effects of Human Disturbance on Eider Ducklings *Somateria mollissima* in an Estuarine Habitat in Scotland." *Biological Conservation* 58:213–28.

Keough, H., and D. J. Blahna. 2006. "Achieving Integrative, Collaborative Ecosystem Management." *Conservation Biology* 20 (5): 1373–82.

Kimbell, A. R., A. Shuhman, and H. Brown. 2009. "More Kids in the Woods: Reconnecting Americans with Nature." *Journal of Forestry* 107 (7): 373–77.

Kissner, J., ed. 1973. *Fabulous Folbot Holidays*. Charleston, SC: Creative Holiday Guides.

Kline, T. C., Jr. 1999. "Temporal and Spatial Variability of $^{13}C/^{12}C$ and $^{15}N/^{14}N$ of Pelagic Biota of Prince William Sound, Alaska." *Canadian Journal of Fisheries and Aquatic Sciences* 56 (S1): 94–117.

Knight, R. L., and D. N. Cole. 1995. "Wildlife Responses to Recreationists." In *Wildlife and Recreationists: Coexistence Through Management and Research*, edited by R. L. Knight and K. J. Gutzwiller, 51–69. Washington, DC: Island Press.

Knight, R. L., and K. J. Gutzwiller, eds. 1995. *Wildlife and Recreationists: Coexistence Through Management and Research*. Washington, DC: Island Press.

Korschgen, C. E., and R. B. Dahlgren. 1992. *Human Disturbances of Waterfowl: Causes, Effects and Management*. Fish and Wildlife Leaflet 13.2.15. Washington, DC: U.S. Fish and Wildlife Service.

Korschgen, C. E., L. S. George, and W. L Green. 1985. "Disturbance of Diving Ducks by Boaters on a Migrational Staging Area." *Wildlife Society Bulletin* 13:290–96.

Kruse, G. H. 2007. "Crabs and Shrimps." In *Long-Term Ecological Change in the Northern Gulf of Alaska*, edited by R. B. Spies, 135–45. Amsterdam, Netherlands: Elsevier.

Kruuk, H. 1992. "Scent Marking by Otters (*Lutra lutra*): Signaling the Use of Resources." *Behavioral Ecology* 3:133–40.

———. 2006. *Otters: Ecology, Behaviour and Conservation*. Oxford: Oxford University Press.

Kucey, L. 2005. "Human Disturbance and the Hauling Out Behavior of Steller Sea Lions (*Eumetopias jubatus*)." MS thesis, University of British Columbia.

Kucey, L., and A. W. Trites. 2006. "A Review of the Potential Effects of Disturbance on Sea Lions: Assessing Response and Recovery." In *Sea Lions of the World*, edited by A. W. Trites, S. Atkinson, D. P. DeMaster, L. W. Fritz, T. S. Gelatt, L. D. Rea, and K. Wynne, 581–89. Alaska Sea Grant Program AK-SG-06–01. Fairbanks, AK: Alaska Sea Grant.

Kury, C. R., and M. Gochfeld. 1975. "Human Interference and Gull Predation in Cormorant Colonies." *Biological Conservation* 8:23–34.

Kvadsheim, P. H., and E. M. Sevaldsen. 2005. *The Potential Impact of 1–8 kHz Active Sonar on Stocks of Juvenile Fish During Sonar Exercises*. Norwegian Defence Research Establishment FFI/Report-2005/01027.

Lace, S. G. 2005. "An Analysis of Spring Black Bear Harvest in Western Prince William Sound, Alaska, Using Geographic Information Systems (GIS) and Agent-Based Simulation Modeling." MS thesis, University of Arizona.

Lace, S. G., H. R. Gimblett, and A. Poe. 2008. "Applying an Agent-Based Modeling Approach to Simulating Spring Black Bear Hunting Activities in Prince William Sound, Alaska." In *Monitoring, Simulation and Management of Visitor Landscapes*, edited by H. R. Gimblett and H. Skov-Petersen, 295–313. Tucson: University of Arizona Press.

Lance, B. K., D. B. Irons, S. J. Kendall, and L. L. McDonald. 2001. "An Evaluation of Marine Bird Population Trends Following the *Exxon Valdez* Oil Spill, Prince William Sound, Alaska." *Marine Pollution Bulletin* 42:298–309.

Lance, M. M., S. Richardson, and H. Allen. 2004. *Washington State Recovery Plan for the Sea Otter*. Olympia, WA: Washington Department of Fish and Wildlife.

Lanszki, J., A. Hidas, K. Szentes, T. Revay, I. Lehoczky, and S. Weiss. 2008. "Relative Spraint Density and Genetic Structure of Otter (*Lutra lutra*) Along the Drava River in Hungary." *Mammalian Biology* 73:40–47.

Larsen, D. N. 1983. "Habitats, Movements, and Foods of River Otters in Coastal Southeastern Alaska." MS thesis, University of Alaska.

———. 1984. "Feeding Habits of River Otters in Coastal Southeastern Alaska." *Journal of Wildlife Management* 48:1446–52.

Lawson, S. R., and R. E. Manning. 2001. "Solitude Versus Access: A Study of Tradeoffs in Outdoor Recreation Using Indifference Curve Analysis." *Leisure Sciences* 23:179–91.

Lethcoe, J. 1990. *An Observer's Guide to the Geology of Prince William Sound, Alaska*. Valdez, AK: Prince William Sound Books.

Lethcoe, J., and N. Lethcoe. 1994. *A History of Prince William Sound, Alaska*. Valdez, AK: Prince William Sound Books.

———. 1998. *A Cruising Guide to Prince William Sound, Alaska*. 2 vols. Valdez, AK: Prince William Sound Books.

Lethcoe, N. 1987. *An Observer's Guide to the Glaciers of Prince William Sound, Alaska*. Valdez, AK: Prince William Sound Books.

Leung, Y.-F., and J. L. Marion. 1999a. "Spatial Strategies for Managing Visitor Impacts in National Parks." *Journal of Park and Recreation Administration* 17 (4): 20–38.

———. 1999b. "Characterizing Backcountry Camping Impacts in Great Smokey Mountains National Park, USA." *Journal of Environmental Management* 57:193–203.

———. 2000. "Recreation Impacts and Management in Wilderness: A State-of-Knowledge Review." In *Wilderness Science in a Time of Change Conference—Volume 5: Wilderness Ecosystems, Threats, and Management; 1999 May 23–27; Missoula, MT*, compiled by D. N. Cole, S. F. McCool, W. T. Borrie, and J. O'Loughlin, 23–48. Proceedings RMRS-P-15-VOL-5. Ogden, UT: USDA Forest Service, Rocky Mountain Research Station.

Lewin, W. C., R. Arlinghaus, and T. Mehner. 2006. "Documented and Potential Biological Impacts of Recreational Fishing: Insights for Management and Conservation." *Reviews in Fisheries Science* 14:305–67.

Lima, S. L., and L. M. Dill. 1990. "Behavioral Decisions Made Under the Risk of Predation: A Review and Prospectus." *Canadian Journal of Zoology* 68:619–40.

Lindberg, D. R., J. A. Estes, and K. I. Warheit. 1998. "Human Influences on Trophic Cascades Along Rocky Shores." *Ecological Applications* 8:880–90.

Lindstrom, S. C., J. P. Houghton, and D. C. Lees. 1999. "Intertidal Macroalgal Community Structure in Southwestern Prince William Sound, Alaska." *Botanica Marina* 42:265–80.

Long, L. L., and C. J. Ralph. 1998. *Regulation and Observations of Human Disturbance Near Nesting Marbled Murrelets*. Arcata, CA: USDA Forest Service, Pacific Southwest Research Station, Redwood Science Laboratory.

Losos, E., J. Hayes, A. Phillips, D. Wilcove, and C. Alkire. 1995. "Taxpayer-Subsidized Resource Extraction Harms Species." *BioScience* 45:446–55.

Loughlin, T. R., and J. V. Tagart. 2006. *Compendium of Steller Sea Lion Related Research, 2000–2006*. Anchorage, AK: North Pacific Fishery Management Council.

Louv, Richard. 2005. *Last Child in the Woods: Saving Our Children From Nature-Deficit Disorder*. Chapel Hill, NC: Algonquin Books.

Lowry, L., and J. Bodkin. 2005. "Marine Mammals." In *The Gulf of Alaska: Biology and Oceanography*, edited by P. R. Mundy, 99–115. Fairbanks, AK: Alaska Sea Grant College Program.

Lubchenco, J., S. A. Navarrete, B. N. Tissot, and J. C. Castilla. 1993. "Possible Ecological Responses to Global Climate Change: Nearshore Benthic Biota of Northeastern Pacific Coastal Ecosystems." In *Earth System Responses to Global Change*, edited by H. A. Mooney, E. R. Fuentes, and B. I. Kronberg, 147–66. San Diego, CA: Academic Press.

Lucas, R. C. 1964. "Wilderness Perception and Use: The Example of the Boundary Waters Canoe Area." *Natural Resources Journal* 3:394–411.

MacCallum, B. 2001. *Status of the Harlequin Duck (*Histrionicus histrionicus*) in Alberta.* Alberta Wildlife Status Report 36. Edmonton, AB: Alberta Sustainable Resource Development and Alberta Conservation Association.

Madsen, J. 1985. "Impact of Disturbance on Field Utilization of Pink-Footed Geese in West Jutland, Denmark." *Biological Conservation* 33:53–63.

Mallory, M., and K. Metz. 1999. "Common Merganser (*Mergus merganser*)." In *The Birds of North America*, No. 442, edited by A. Poole and F. Gill, 1–27. Philadelphia, PA: Academy of Natural Sciences; Washington, DC: American Ornithologists' Union.

Manning, R. E. 2007. *Parks and Carrying Capacity: Commons Without Tragedy.* Washington, DC: Island Press.

Manning, R. E., and W. A. Valliere. 2001. "Coping in Outdoor Recreation: Causes and Consequences of Crowding and Conflict Among Community Residents." *Journal of Leisure Research* 33 (4): 410–26.

Mantua, N. J., S. R. Hare, Y. Zhang, J. M. Wallace, and R. C. Francis. 1997. "A Pacific Interdecadal Climate Oscillation with Impacts on Salmon Production." *Bulletin of the American Meteorological Society* 78:1069–79.

Marion, J. L. 1987. "Environmental Impact Management in the Boundary Waters Canoe Area Wilderness." *Northern Journal of Applied Forestry* 4:7–10.

———. 1991. *Developing a Natural Resource Inventory and Monitoring Program for Visitor Impacts on Recreation Sites: A Procedural Manual.* USDI National Park Service Natural Resources Report NPS/NRVT/NRR-91. Denver, CO: Natural Resource Publication Office.

———. 1995. "Capabilities and Management Utility of Recreation Impact Monitoring Programs." *Environmental Management* 19:763–71.

———. 1998. "Recreation Ecology Research Findings: Implications for Wilderness and Park Managers." In *Proceedings of the National Outdoor Ethics Conference, April 18–21, 1996, St. Louis, MO*, 188–96. Gathersburg, MD: Izaak Walton League of America, Gathersburg, MD.

Marion, J. L., and D. N. Cole. 1996. "Spatial and Temporal Variation in Soil and Vegetation Impacts on Campsites: Delaware Water Gap National Recreation Area." *Ecological Applications* 6:520–30.

Marion, J. L., and T. A. Farrell. 2002. "Management Practices That Concentrate Visitor Activities: Camping Impact Management at Isle Royale National Park, USA." *Journal of Environmental Management* 66:201–12.

Marion, J. L., and Y.-F. Leung. 1997. *An Assessment of Campsite Conditions in Great Smoky Mountains National Park.* Research/Resources Management Report. Gatlinburg, TN: USDI Park Service, Great Smoky Mountains National Park.

Mathews, E. A. 2000a. *Reactions of Steller Sea Lions (*Eumetopias jubatus*) to Vessels at a Haulout in Glacier Bay.* Gustavus, AK: USDI Park Service, Glacier Bay National Park and Reserve.

———. 2000b. *Measuring the Effects of Vessels on Harbor Seals* (Phoca vitulina Richardsi) *at North Marble Island, a Terrestrial Haulout in Glacier Bay National Park.* Gustavus, AK: USDI Park Service, Glacier Bay National Park and Reserve.

Mathisen, J. E. 1968. "Effects of Human Disturbance on Nesting of Bald Eagles." *Journal of Wildlife Management* 32:1–6.

McChesney, G. J., L. E. Eigner, T. B. Poitras, P. J. Kappes, N. M. Jones, D. N. Lontoh, P. J. Capitolo, R. T. Golightly, D. Le Fer, H. R. Carter, S. W. Kress, and M. W. Parker. 2007. *Restoration of Common Murre Colonies in Central California: Annual Report 2006.* Newark, CA: USDI Fish and Wildlife Service, San Francisco Bay National Wildlife Refuge Complex.

McCool, S. F., R. N. Clark, and G. H. Stankey. 2007. *An Assessment of Frameworks Useful for Public Land Recreation Planning.* General Technical Report PNW-GTR-705. Portland, OR: U.S. Forest Service, Pacific Northwest Research Station.

McCool, S. F., and D. N. Cole. 2001. "Thinking and Acting Regionally: Toward Better Decisions About Appropriate Conditions, Standards, and Restrictions on Recreation Use." *The George Wright Forum* 18 (3): 85–98.

McCool, S. F., and G. H. Stankey. 2004. "Indicators of Sustainability: Challenges and Opportunities at the Interface of Science and Policy." *Environmental Management* 33 (3): 294–305.

McGarigal, K., R. G. Anthony, and F. B. Isaacs. 1991. *Interactions of Humans and Bald Eagles on the Columbia River Estuary.* Wildlife Monographs 115. Bethesda, MD: Wildlife Society.

McIlroy, C. W. 1970. "Aspects of the Ecology and Hunter Harvest of Black Bear in Prince William Sound." MS thesis, University of Alaska.

McNamara, J. 2010. "Sea Kayaking in America: 1920s–30s." *Sea Kayaker Magazine* 139.

McRoy, C. P. 1988. "Natural and Anthropogenic Disturbances at the Ecosystem Level." In *Environmental Studies in Port Valdez, Alaska: A Basis for Management,* edited by D. G. Shaw and M. J. Hameedi, 329–44. Lecture Notes on Coastal and Estuarine Studies 24. New York: Springer-Verlag.

McShane, C., T. Hamer, H. Carter, G. Swartzman, V. Friesen, D. Ainley, R. Tressler, K. Nelson, A. Burger, L. Spear, T. Mohagen, R. Martin, L. Henkel, K. Prindle, C. Strong, and J. Keany. 2004. *Evaluation Report for the 5-Year Status Review of the Marbled Murrelet in Washington, Oregon, and California.* Portland, OR: USDI Fish and Wildlife Service.

Meffe, G. K., L. A. Nielsen, R. L. Knight, and D. A. Schenborn. 2002. *Ecosystem Management: Adaptive, Community Based Conservation.* Washington, DC: Island Press.

Melquist, W. E., and A. E. Dronkert. 1987. "River Otter." In *Wild Furbearer Management and Conservation in North America,* edited by M. Novak, J. A. Baker, M. E. Obbard and B. Malloch, 626–41. Toronto, ON: Ontario Ministry of Natural Resources and Ontario Trappers Association.

Melquist, W. E., and M. G. Hornocker. 1983. "Ecology of River Otters in West Central Idaho." *Wildlife Monographs* 83:3–60.

Melquist, W. E., P. J. Polechla, and D. Toweill. 2003. "River Otter." In *Wild Mammals of North America: Biology, Management, and Conservation*, edited by G. A. Feldhamer, B. C. Thompson, and J. A. Chapman, 708–34. Baltimore, MD: Johns Hopkins University Press.

Merilaita, S., and J. Lind. 2005. "Background-Matching and Disruptive Coloration, and the Evolution of Cryptic Coloration." *Proceedings of the Royal Society B* 272:665–70.

Merrell, T. R., Jr. 1988. "Fisheries Resources." In *Environmental Studies in Port Valdez, Alaska: A Basis for Management*, edited by D. G. Shaw and M. J. Hameedi, 203–24. Lecture Notes on Coastal and Estuarine Studies 24. New York: Springer-Verlag.

Microsoft. 2007. *Excel*. Redmond, WA: Microsoft.

Mikola, J., M. Miettinen, E. Lehikoinen, and K. Lehtila. 1994. "The Effects of Disturbance Caused by Boating on Survival and Behavior of Velvet Scoter *Melanitta fusca* Ducklings." *Biological Conservation* 67:119–24.

Misund, O. A., J. T. Øvredal, and M. T. Hafsteinsson. 1996. "Reactions of Herring Schools to the Sound Field of a Survey Vessel." *Aquatic Living Resources* 9:5–11.

Montopoli, G. J., and D. A. Anderson. 1991. "A Logistic Model for the Cumulative Effects of Human Intervention on Bald Eagle Habitat." *Journal of Wildlife Management* 55:290–93.

Monz, C. A. 1998. "Monitoring Recreation Resource Impacts in Two Coastal Areas of Western North America: An Initial Assessment." In *Personal, Societal and Ecological Values of Wilderness: Sixth World Wilderness Congress Proceedings on Research, Management, and Allocation*, vol. I, compiled by A. E Watson, G. H. Aplet, and J. C. Hendee, 117–22. USDA Forest Service Proceedings RMRS-P-4. Ogden, UT: Rocky Mountain Research Station.

Monz, C. A., D. N. Cole, Y.-F. Leung, and J. L. Marion. 2010. "Sustaining Visitor Use in Protected Areas: Future Opportunities in Recreation and Ecological Research." *Journal of Environmental Management* 45 (3): 551–62.

Monz, C. A., and P. Twardock. 2004. "Campsite Impacts in Prince William Sound, Alaska." In *The Environmental Impacts of Ecotourism*, edited by R. Buckley, 309–19. Wallingford, UK: CABI.

———. 2010. "A Classification of Campsites in Prince William Sound, Alaska, USA." *Journal of Environmental Management* 91 (7): 1566–72.

Morse, J. A., A. N. Powell, and M. D. Tetreau. 2006. "Productivity of Black Oystercatchers: Effects of Recreational Disturbance in a National Park." *Condor* 108:623–33.

Mowbray, E. E., D. Pursley, and J. A. Chapman. 1979. *The Status, Population Characteristics, and Harvest of the River Otter in Maryland*. Publications in Wildlife Ecology No. 2. Annapolis, MD: Maryland Wildlife Administration.

Mundy, P. R., and A. Hollowed. 2005. "Fish and Shellfish." In *The Gulf of Alaska: Biology and Oceanography*, edited by P. R. Mundy, 81–97. Fairbanks, AK: Alaska Sea Grant College Program.

Mundy, P. R., and P. Olsson. 2005. "Climate and Weather." In *The Gulf of Alaska: Biology and Oceanography*, edited by P. R. Mundy, 25–34. Fairbanks: Alaska Sea Grant College Program.

Murphy, K. A., L. H. Suring, and A. Iliff. 2004. *Western Prince William Sound Human Use and Wildlife Disturbance Model Assessment of Current Human Use Patterns. Exxon Valdez* Oil Spill Restoration Project Final Report. Anchorage, AK: USDA Forest Service, Chugach National Forest.

Murphy, S. M., and T. J. Mabee. 2000. "Status of Black Oystercatchers in Prince William Sound, Alaska, Nine Years After the *Exxon Valdez* Oil Spill." *Waterbirds* 23:204–13.

Naske, C.-M., and H. E. Slotnick. 1987. *Alaska: A History of the 49th State.* Norman: University of Oklahoma Press.

National Park Service. 1997. *VERP: The Visitor Experience and Resource Protection Framework—A Handbook for Planners and Managers.* Publication NPS D-1215. Denver, CO: NPS Denver Service Center.

Neary, J., B. Hunter, C. Prew, D. Sanders, and M. Emerick. 2003. *Region 10 Wilderness Campsite Definitions and Campsite Inventory Protocols as Agreed upon at the Ketchikan Wilderness Meeting, November 29, 2001 and as Modified in May/June 2003.* Ketchikan, AK: U.S. Forest Service.

Nelson, D. A. 1987. "Factors Influencing Colony Attendance by Pigeon Guillemots on Southeast Farallon Island, California." *Condor* 93:340–48.

Newbrey, J. L., M. A. Bozek, and N. D. Niemuth. 2005. "Effects of Lake Characteristics and Human Disturbance on the Presence of Piscivorous Birds in Northern Wisconsin, USA." *Waterbirds* 28:478–86.

Newsome D., S. A. Moore, and R. K. Dowling. 2001. *Natural Area Tourism: Ecology, Impacts, and Management.* Clevedon, UK: Channel View Books.

NMFS (National Marine Fisheries Service). 2008. *Recovery Plan for the Steller Sea Lion (*Eumetopias jubatus*).* Silver Spring, MD: National Marine Fisheries Service.

NOAA (National Oceanic and Atmospheric Administration). 2000. National Oceanic and Atmospheric Administration Environmental Sensitivity Index of Shoreline Types, Prince William Sound, Alaska. Silver Spring, MD: Office of Response and Restoration, National Oceanic and Atmospheric Administration.

NOAA (National Oceanic and Atmospheric Administration). 2010a. *A Final Supplemental Environmental Impact Statement for the* Exxon Valdez *Oil Spill Restoration Plan.* Seattle, WA: National Oceanic and Atmospheric Administration.

NOAA (National Oceanic and Atmospheric Administration). 2010b. *Endangered, Threatened, Proposed, Candidate, and Delisted Species in Alaska.* Anchorage, AK: National Marine Fisheries Service, National Oceanic and Atmospheric Administration. Accessed March 21, 2017. https://www.fws.gov/alaska/fisheries/endangered/pdf/consultation_guide/4_species_list.pdf.

North Pacific Fishery Management Council. 2002. *Fishery Management Plan for Groundfish of the Gulf of Alaska.* Anchorage, AK: North Pacific Fishery Management Council, Northwest Research Station.

Nysewander, D. R. 1977. "Reproductive Success of the Black Oystercatcher in Washington State." MS thesis, University of Washington.

Odum, W. E. 1982. "Environmental Degradation and the Tyranny of Small Decisions." *Bio-Science* 32 (9): 728–29.

Oring, L.W. 1964. "Behavior and Ecology of Certain Ducks During the Postbreeding Period." *Journal of Wildlife Management* 28:223–33.

Orth, R. J., T. J. B. Carruthers, W. C. Dennison, C. M. Duarte, J. W. Fourqurean, K. L. Heck, A. R. Hughes, G. A. Kendricks, W. J. Kenworthy, S. Olyarnik, F. T. Short, M. Waycott, and S. L. Williams. 2006. "A Global Crisis for Seagrass Ecosystems." *Bioscience* 56:987–96.

Ostrand, W. D., and Z. A. Gotthardt. 2005. "Habitat Selection Models for Pacific Sand Lance (*Ammodytes hexapterus*) in Prince William Sound, Alaska." *Northwestern Naturalist* 86:131–43.

Parrish, J. K. 1997. *Attendance and Reproductive Success of Tatoosh Island Common Murres.* Final Report 1996. Seattle: Zoology Department, University of Washington.

Patton, M. Q. 2002. *Qualitative Research and Evaluation Methods.* 3rd ed. Thousand Oaks, CA: Sage.

Pella, J., and J. Maselko. 2007. *Probability Sampling and Estimation of the Oil Remaining in 2001 from the* Exxon Valdez *Oil Spill in Prince William Sound, Alaska.* NOAA Technical Memorandum NMFS-AFSC-169. Washington, DC: U.S. Department of Commerce.

Pergams, O. R. W., and P. A. Zaradic. 2008. "Evidence for a Fundamental and Pervasive Shift Away from Nature-Based Recreation." *Proceedings of the National Academy of Sciences* 105 (7): 2295–300.

Peters, K. A., and D. L. Otis. 2006. "Wading Bird Response to Recreational Boat Traffic: Does Flushing Translate into Avoidance?" *Wildlife Society Bulletin* 34:1383–91.

Peterson C. H., S. D. Rice, J. W. Short, D. Esler, J. L. Bodkin, B. E. Ballachey, and D. B. Irons. 2003. "Long Term Ecosystem Response to the *Exxon Valdez* Oil Spill." *Science* 302:2082–86.

Pettebone, D., B. Meldrum, C. Leslie, S. Lawson, P. Newman, N. Reigner, and A. Gibson. 2013. "A Visitor Use Monitoring Approach on the Half Dome Cables to Reduce Crowding and Inform Park Planning Decisions in Yosemite National Park." *Landscape and Urban Planning* 118:1–9.

Picou, J. S., and C. G. Martin. 2007. *Long-Term Impacts of the* Exxon Valdez *Oil Spill: Patterns of Social Disruption and Psychological Stress Seventeen Years After the Disaster.* Final Report to the National Science Foundation. Mobile, AL: University of South Alabama.

Poe, A. J., M. I. Goldstein, B. A. Brown, and B. A. Andres. 2009. "Black Oystercatchers and Campsites in Western Prince William Sound, Alaska." *Waterbirds* 32:423–29.

Poe, C. B., and S. Greenwood. 2010. *Prince William Sound Human Use Hot Spots GIS Database and Spatial Analysis. Exxon Valdez* Oil Spill Criminal Restitution Project Final Report. Anchorage, AK: USDA Forest Service, Chugach National Forest.

Polechla, P. 1990. "Action Plan for North American Otters." In *Otters: An Action Plan for Their Conservation*, edited by P. Foster-Turley, S. Macdonald, and C. Mason, 74–79. Gland, Switzerland: International Union for Conservation of Nature, Species Survival Commission.

Propst, B. M., R. M. Schuster, and C. P. Dawson. 2008. "An Exploratory Analysis of Coping Schemes Used by Paddlers Who Camped in the St. Regis Canoe Area, New York." In *Proceedings of the 2008 Northeastern Recreation Research Symposium*, edited by D. Klenosky and C. Fisher, 139–46. Newtown Square, PA: U.S. Forest Service, Northern Research Station.

PWSAC (Prince William Sound Aquaculture Corporation). 2017. "Prince William Sound Aquaculture Corporation: Setting the World Standard for Responsible Aquaculture." Prince William Sound Aquaculture Corporation. Accessed February 15. http://pwsac.com/about/history.

Ramsey, F. L., and D. W. Schafer. 2002. *Statistical Sleuth: A Course in Methods of Data Analysis*. 2nd ed. Pacific Grove, CA: Duxbury.

Ream, C. H. 1976. "Loon Productivity, Human Disturbance, and Pesticide Residues in Northern Minnesota." *Wilson Bulletin* 88:427–32.

Reeburgh, W. S., and G. W. Kipphut. 1986. "Chemical Distributions and Signals in the Gulf of Alaska, Its Coastal Margins and Estuaries." In *The Gulf of Alaska: Physical Environment and Biological Resources*, edited by D. W. Hood and S. T. Zimmerman, 77–92. MMS Publication No. 86–0095. Anchorage, AK: National Oceanic and Atmospheric Administration.

Reid, S. E, and J. L. Marion. 2004. "Effectiveness of a Confinement Strategy for Reducing Campsite Impacts in Shenandoah National Park." *Environmental Conservation* 31:274–82.

Relph, E. 1976. *Place and Placelessness*. London: Routledge, Kegan, & Paul.

Richards, L., and J. M. Morse. 2007. *User's Guide to Qualitative Methods*. 2nd ed. Thousand Oaks, CA: Sage.

Ringelman, J. K. 1990. *Habitat Management for Molting Waterfowl*. Fish and Wildlife Leaflet 13.4.4. Washington, DC: U.S. Fish and Wildlife Service.

Rizzolo, D. J., D. Esler, D. D. Roby, R. L. Jarvis. 2005. "Do Wintering Harlequin Ducks Forage Nocturnally at High Latitudes?" *Condor* 107:173–77.

Robertson G. J., and R. I. Goudie. 1999. "Harlequin Duck (*Histrionicus histrionicus*)." In *The Birds of North America*, No. 466, edited by A. Poole and F. Gill, 1–30. Philadelphia, PA: Academy of Natural Sciences.

Rodgers, J. A., Jr., and S. T. Schwikert. 2003. "Buffer Distances to Protect Foraging and Loafing Waterbirds from Disturbance by Airboats in Florida." *Waterbirds* 26:437–43.

Rogers, D. E., and B. J. Rogers. 1986. "The Nearshore Fishes." In *The Gulf of Alaska: Physical Environment and Biological Resources*, edited by D. W. Hood and S. T. Zimmerman, 399–415. MMS Publication No. 86–0095. Anchorage, AK: National Oceanic and Atmospheric Administration.

Rojek, N. A., M. W. Parker, H. R. Carter, and G. J. McChesney. 2007. "Aircraft and Vessel Disturbances to Common Murres *Uria aalge* at Breeding Colonies in Central California, 1997–1999." *Marine Ornithology* 35:61–69.

Rostain, R. R., M. Ben-David, P. Groves, and J. A. Randall. 2004. "Why Do River Otters Scent-Mark? An Experimental Test of Several Hypotheses." *Animal Behaviour* 68:703–11.

Royer, T. C. 1981. "Baroclinic Transport in the Gulf of Alaska, Part II: A Fresh-Water Driven Coastal Current." *Journal of Marine Research* 39:251–66.

Ruggles, A. K. 1994. "Habitat Selection by Loons in Southcentral Alaska." *Hydrobiologia* 279/280:421–30.

Sales-Luis, T., D. Freitas, and M. Santos-Reis. 2009. "Key Landscape Factors for Eurasian Otter *Lutra lutra* Visiting Rates and Fish Loss in Estuarine Fish Farms." *European Journal of Wildlife Research* 55:345–55.

SAS Institute Inc. 2007. *JMP* (version 7.0.2). Cary, NC: SAS Institute Inc.

Sawicki, D. S., and D. R. Peterman. 2002. "Surveying the Extent of PPGIS Practice in the United States." In *Community Participation and Geographic Information Systems*, edited by W. J. Craig, T. M. Harris, and D. M. Weiner, 17–36. London: Taylor & Francis.

Schirato, G., and W. Parson. 2006. "Bald Eagle Management in Urbanizing Habitat of Puget Sound, Washington." *Northwestern Naturalist* 87:138–42.

Schwarz, A. L., and G. L. Greer. 1984. "Responses of Pacific Herring, *Clupea harengus pallasi*, to Some Underwater Sounds." *Canadian Journal of Fisheries and Aquatic Sciences* 41 (8): 1183–92.

Scognamillo, D. G. 2005. "Temporal and Spatial Harvest Patterns of River Otter in Louisiana and Its Potential Use as a Bioindicator Species of Water Quality." PhD diss., Louisiana State University and Agricultural and Mechanical College.

Scott, C., L. A. Brown, G. B. Jennings, and C. J. Utermohle. 2001. *Community Profile Database for Access 2000* (version 3.12). Juneau, AK: Alaska Department of Fish and Game, Division of Subsistence

Sea Otter Recovery Team. 2007. *Recovery Strategy for the Sea Otter (*Enhydra lutris*) in Canada.* Species at Risk Act Recovery Strategy Series. Vancouver, BC: Fisheries and Oceans Canada.

Shaw, D. G., and M. J. Hameedi, eds. 1988. *Environmental Studies in Port Valdez, Alaska: A Basis for Management.* Lecture Notes on Coastal and Estuarine Studies 24. New York: Springer-Verlag.

Shelby, B., and T. A. Heberlein. 1986. *Carrying Capacity in Recreation Settings.* Corvallis: Oregon State University Press.

Shelby, B., T. A. Heberlein, J. J. Vaske, and G. Alfano. 1983. "Expectations, Preferences, and Feeling Crowded in Recreation Activities." *Leisure Sciences* 6 (1): 1–14.

Shtatland, E. S., E. Cain, and M. B. Barton. 2001. "The Perils of Stepwise Logistic Regression and How to Escape Them Using Information Criteria and the Output Delivery System." In *Proceedings of the Twenty-Sixth Annual SAS Users Group International Conference*, 222–26. Cary, NC: SAS Institute.

Sieber, R. 2006. "Public Participation Geographic Information Systems: A Literature Review and Framework." *Annals of the Association of American Geographers* 96 (3): 491–507.

Sievanen, L., L. M. Campbell, and H. M. Leslie. 2012. "Challenges to Interdisciplinary Research in Ecosystem-Based Management." *Conservation Biology* 26:315–23.

Skaret, G., B. E. Axelsen, L. Nøttestad, A. Fernö, and A. Johannessen. 2005. "The Behaviour of Spawning Herring in Relation to a Survey Vessel." *ICES Journal of Marine Science* 62:1061–64.

Smelcher, J. E. 2006. *The Day That Cries Forever*. Anchorage, AK: Chenega Future.

Sokal, R. R., and F. J. Rohlf. 1995. *Biometry: The Principles and Practice of Statistics in Biological Research*. 3rd ed. New York: W. H. Freeman.

Speckman, S. G., J. F. Piatt, and A. M. Springer. 2004. "Small Boats Disturb Fish-Holding Marbled Murrelets." *Northwestern Naturalist* 85:32–34.

Spiegel, C. S. 2008. "Incubation Patterns, Parental Roles, and Nest Survival of Black Oystercatchers (*Haematopus bachmani*): Influences of Environmental Processes and Nest-Area Stimuli." MS thesis, Oregon State University. http://ir.library.oregonstate.edu/dspace/bitstream/1957/10239/1/Spiegel_Thesis.pdf.

Spies, R. B, ed. 2007. *Long-Term Ecological Change in the Northern Gulf of Alaska*. Amsterdam, Netherlands: Elsevier.

Springer, A. 2005. "Seabirds in the Gulf of Alaska." In *The Gulf of Alaska: Biology and Oceanography*, edited by P. R. Mundy, 69–79. Fairbanks, AK: Alaska Sea Grant College Program.

Springer, A. M., S. J. Iverson, and J. L. Bodkin. 2007. "Marine Mammal Populations." In *Long-Term Ecological Change in the Northern Gulf of Alaska*, edited by R. B. Spies, 352–78. Amsterdam, Netherlands: Elsevier.

Stalmaster, M. V., and J. L. Kaiser. 1998. "Effects of Recreational Activity on Wintering Bald Eagles." *Wildlife Monographs* 137. Bethesda, MD: Wildlife Society.

Stalmaster, M. V., and J. R. Newman. 1978. "Behavioral Responses of Wintering Bald Eagles to Human Activity." *Journal of Wildlife Management* 42:506–13.

Stankey, G. H. 1971. "The Perception of Wilderness Recreation Carrying Capacity: A Geographic Study in Natural Resources Management." PhD diss., Michigan State University.

———. 1973. *Visitors Perceptions of Wilderness Recreation Carrying Capacity*. USDA Forest Service Research Paper INT-142. Ogden, UT: Intermountain Research Station.

———. 1999. "The Recreation Opportunity Spectrum and Limits of Acceptable Change Planning Systems: A Review of Experiences and Lessons Learned." In *Ecosystem Management: Adaptive Strategies for Natural Resource Management in the Twenty-First Century*, edited by J. Aley, W. R. Burch, B. Conover, and D. Field, 173–88. Philadelphia, PA: Taylor & Francis.

Stankey, G. H., D. N. Cole, R. C. Lucas, M. E. Peterson, and S. S. Frissell. 1985. *The Limits of Acceptable Change (LAC) System for Wilderness Planning*. USDA Forest Service General Technical Report INT-176. Ogden, UT: Intermountain Forest Experiment Station.

State of Alaska. 2008. *Housing Unit Method Manual: Population Estimate Instructions and Reporting Forms*. Juneau, AK: Department of Commerce, Community, and Economic Development and Division of Community Advocacy.

Steidl, R. J., and R. G. Anthony. 1996. "Responses of Bald Eagles to Human Activity During the Summer in Interior Alaska." *Ecological Applications* 6:482–91.

———. 2000. "Experimental Effects of Human Activity on Breeding Bald Eagles." *Ecological Applications* 10:258–68.

Steidl, R. J., and B. F. Powell. 2006. "Assessing the Effects of Human Activities on Wildlife." *George Wright Forum* 23:50–58.

Stratton, L., and E. B. Chisum. 1986. *Resource Use Patterns in Western Prince William Sound: Chenega in the 1960s and Chenega Bay 1984–86*. Division of Subsistence Technical Paper No. 139. Juneau, AK: Alaska Department of Fish and Game.

Strong, P. I. V., and J. A. Bissonette. 1989. "Feeding and Chick-Rearing Areas of Common Loons." *Journal of Wildlife Management* 51:72–76.

Swimley, T. J., T. L. Serfass, R. P. Brooks, and W. M. Tzilkowski. 1998. "Predicting River Otter Latrine Sites in Pennsylvania." *Wildlife Society Bulletin* 26:836–45.

Szaniszlo, W. R. 2005. "California Sea Lion (*Zalophus californianus*) and Steller Sea Lion (*Eumetopias jubatus*) Interactions with Vessels in Pacific Rim National Park Reserve: Implications for Marine Mammal Viewing Management." MS thesis, University of Victoria.

Tabachnick, B. G., and L. S. Fidell. 2001. *Using Multivariate Statistics*. 4th ed. Needham Heights, MA: Allyn & Bacon.

Tallis, H. M., J. L. Ruesink, B. R. Dumbauld, S. Hacker, and L. M. Wisehart. 2009. "Oysters and Aquaculture Practices Affect Eelgrass Density and Productivity in a Pacific Northwest Estuary." *Journal of Shellfish Research* 28:251–61.

Tatitlek IRA Council. 2003. *Tatitlek Village Natural Resource Management Plan*. Tatitlek, AK: Indian Reorganization Act Council for Tatitlek Village.

Taylor, A. 2000. *The Strangest Town in Alaska: The History of Whittier, Alaska, and Portage Valley*. Seattle, WA: Kokogiak Media.

Tessler, D. F., J. A. Johnson, B. A. Andres, S. Thomas, and R. B. Lanctot. 2007. *Black Oystercatcher (*Haematopus bachmani*) Conservation Action Plan*. Manomet, MA: International Black Oystercatcher Working Group / Anchorage, AK: Alaska Department of Fish and Game / Anchorage, AK: U.S. Fish and Wildlife Service / Manomet, MA: Manomet Center for Conservation Sciences.

Testa, J. W., D. F. Holleman, R. T. Bowyer, and J. B. Faro. 1994. "Estimating Populations of Marine River Otters in Prince William Sound, Alaska, Using Radiotracer Implants." *Journal of Mammalogy* 75:1021–32.

Thayer, J. A., W. J. Sydeman, N. P. Fairman, and S. G. Allen. 1999. "Attendance and Effects of Disturbance on Coastal Common Murre Colonies on Point Reyes, California." *Waterbirds* 22:130–39.

Thomas, G. L., and R. E. Thorne. 2003. "Acoustic-Optical Assessment of Pacific Herring and Their Predator Assemblage in Prince William Sound, Alaska." *Aquatic Living Resources* 16:247–53.

Timken, R. L., and B. W. Anderson. 1969. "Food Habits of Common Mergansers in the Northcentral United States." *Journal of Wildlife Management* 33:87–91.

Titus, J. R., and L. W. VanDruff. 1981. *Response of the Common Loon to Recreational Pressure in the Boundary Waters Canoe Area, Northeastern Minnesota.* Wildlife Monographs 79. Bethesda, MD: Wildlife Society.

Transportation Research Board. 2000. *Highway Capacity Manual 2000.* Washington, DC: National Research Council.

Tseng, Y., G. T. Kyle, C. S. Shafer, A. R. Graefe, T. A. Bradle, and M. A. Schuett. 2009. "Exploring the Crowding-Satisfaction Relationship in Recreational Boating." *Environmental Management* 43:496–507.

Tuite, C. H., P. R. Hanson, and M. Owen. 1984. "Some Ecological Factors Affecting Winter Wildfowl Distribution on Inland Waters in England and Wales, and the Influence of Water-Based Recreation." *Journal of Applied Ecology* 21:41–62.

Twardock, P. 2004. *Kayaking and Camping in Prince William Sound.* Valdez, AK: Prince William Sound Books.

Twardock, P., and C. A. Monz. 2000. "Recreational Kayak Visitor Use, Distribution, and Economic Value in Western Prince William Sound, Alaska USA." In *Wilderness Science in a Time of Change Conference, Volume 4: Wilderness Visitors, Experiences, and Visitor Management*, compiled by D. N. Cole, S. F. McCool, W. T. Borrie, and J. O'Loughlan, 175–80. Proceedings RMRS-P-15-Vol-4. Fort Collins, CO: USDA Forest Service Rocky Mountain Research Station.

Twardock, P., Monz, C. A., Smith, M., and S. Colt. 2010. "Long-Term Changes in Resource Conditions on Backcountry Campsites in Prince William Sound, Alaska, USA." *Northwest Science* 84 (3): 223–32.

USDA Forest Service. 2002. *Chugach National Forest Revised Land and Resource Management Plan.* Alaska Region R10-MB-480c. Anchorage, AK: U.S. Forest Service.

———. 2005. *Chugach National Forest Revised Land and Resource Management Plan.* Alaska Region R10-MB-480c. Anchorage, AK: U.S. Forest Service.

———. 2008. *Chugach National Forest Revised Land and Resource Management Plan.* Alaska Region R10-MB-480c. Anchorage, AK: U.S. Forest Service.

———. 2010. *Connecting People With America's Great Outdoors: A Framework for Sustainable Recreation.* Washington, DC: U.S. Forest Service.

USDI Fish and Wildlife Service (USFWS). 2004. Beringian Seabird Colony Catalog Database (OBIS-SEAMAP ID 270). Anchorage, AK: Division of Migratory Bird Management, USDI Fish and Wildlife Service.

Vabø, R., K. Olsen, and I. Huse. 2002. "The Effect of Vessel Avoidance of Wintering, Norwegian Spring-Spawning Herring." *Fisheries Research* 58:59–77.

Vaske, J. J. 2008. *Survey Research and Analysis: Applications in Parks, Recreation and Human Dimensions.* State College, PA: Venture Publishing.

Vaughan, S. L., C. N. K. Mooers, and S. M. Gay III. 2001. "Physical Variability in Prince William Sound During the SEA Study (1994–1998)." *Fisheries Oceanography* 10 (S1): 58–80.

Vermeer, K. 1973. "Some Aspects of the Nesting Requirements of Common Loons in Alberta." *Wilson Bulletin* 85:429–35.

Viereck, L. A., C. T. Dyrness, A. R. Batten, and K. J. Wenzlick. 1992. *The Alaska Vegetation Classification*. USDA Forest Service General Technical Report PNW-GTR-286. Portland, OR: Pacific Northwest Research Station.

Villegas, M. J., A. Aron, and L. A. Ebensperger. 2007. "The Influence of Wave Exposure on the Foraging Activity of Marine Otter, *Lontra felina* (Molina, 1782) (Carnivora: Mustelidae) in Northern Chile." *Journal of Ethology* 25:281–86.

Walker, D. I., R. J. Lukatelich, G. Bastyna, and A. J. McComb. 1989. "Effect of Boat Moorings on Seagrass Beds near Perth, Western Australia." *Aquatic Botany* 36:69–77.

Walls, M., and J. Siikamaki. 2010. "Connecting Americans to the Great Outdoors." *Resources* 175:13–16.

Watson, J. W. 1993. "Responses of Nesting Bald Eagles to Helicopter Surveys." *Wildlife Society Bulletin* 21:171–78.

Weingartner, T. 2007. "The Physical Environment of the Gulf of Alaska." In *Long-Term Ecological Change in the Northern Gulf of Alaska*, edited by R. B. Spies, 12–47. Amsterdam, Netherlands: Elsevier.

White, D. D., T. E. Hall, and T. A. Farrell. 2001. "Influence of Ecological Impacts and Other Campsite Characteristics on Wilderness Visitors' Campsite Choices." *Journal of Park and Recreation Administration* 19 (2): 83–97.

Whittaker, D., B. Shelby, R. Manning, D. Cole, and G. Haas. 2010. *Capacity Reconsidered: Finding Consensus and Clarifying Differences*. Marienville, PA: National Association of Recreation Resource Planners.

Williams, D. R., M. E. Patterson, J. W. Roggenbuck, and A. E. Watson. 1992. "Beyond the Commodity Metaphor: Examining Emotional and Symbolic Attachments to Place." *Leisure Sciences* 14:29–46.

Wilson, B., and L. M. Dill. 2002. "Pacific Herring Respond to Simulated Odontocete Echolocation Sounds." *Canadian Journal of Fisheries and Aquatic Sciences* 59:542–53.

Wilson, J. G., and J. E. Overland. 1986. "Meteorology." In *The Gulf of Alaska: Physical Environment and Biological Resources*, edited by D. W. Hood and S. T. Zimmerman, 31–54. MMS Publication No. 86–0095. Anchorage, AK: National Oceanic and Atmospheric Administration.

Willson, M. F., and K. C. Halupka. 1995. "Anadromous Fish as Keystone Species in Vertebrate Communities." *Conservation Biology* 9:489–97.

Wolfe, M. L., and J. A. Chapman. 1987. "Principles of Furbearer Management." In *Wild Furbearer Management and Conservation in North America*, edited by M. Novak, J. A. Baker, M. E. Obbard, and B. Malloch, 101–12. Toronto, ON: Ontario Ministry of Natural Resources and Ontario Trapper Association.

Wolfe, P., H. R. Gimblett, R. Itami, and B. Garber-Yonts. 2008. "Monitoring and Simulating Recreation Use in Prince William Sound, Alaska." In *Monitoring, Simulation and Management of Visitor Landscapes*, edited by H. R. Gimblett, and H. Skov-Petersen, 349–69. Tucson: University of Arizona Press.

Wolfe, P. E., B. Garber-Yonts, and R. S. Rosenberger. 2006. "Monitoring and Analysis of Recreational Boat Use in Sensitive Wildlife Areas in Prince William Sound, Alaska: A Simulation Approach." In *ISSRM 2006: 12th International Symposium on Society and Resource Management: Social Sciences in Resource Management: Global Challenges and Local Responses: Book of Abstracts*, compiled by B. Beardmore, K. Englund, and W. Haider, 362. Burnaby, BC: School of Resource and Environmental Management, Simon Fraser University. http://www.adaptfish.rem.sfu.ca/Ben_Beardmore/pubs/ISSRM2006_Book_of_Abstracts.pdf.

Wood, C. C. 1987a. "Predation of Juvenile Pacific Salmon by the Common Merganser (*Mergus merganser*) on Eastern Vancouver Island. I: Predation During the Seaward Migration." *Canadian Journal of Fish and Aquatic Sciences* 44:941–49.

———. 1987b. Predation of Juvenile Pacific Salmon by the Common Merganser (*Mergus merganser*) on Eastern Vancouver Island. II: Predation of Stream-Resident Juvenile Salmon by Merganser Broods. *Canadian Journal of Fish and Aquatic Sciences* 44:950–59.

Wood, P. H., M. W. Collopy, and C. M. Sekerak. 1998. "Postfledging Nest Dependence Period for Bald Eagles in Florida." *Journal of Wildlife Management* 62:333–39.

Wood, P. H, T. C. Edwards Jr., and M. W. Collopy. 1989. "Characteristics of Bald Eagle Nesting Habitat in Florida." *Journal of Wildlife Management* 53:441–49.

Wooley, C., and J. Haggerty. 1993. "The Hidden History of Chugach Bay." In *Prince William Sound*, edited by P. Rennick. Anchorage, AK: Alaska Geographic.

Woolington, J. D. 1984. "Habitat Use and Movements of River Otters at Help Bay, Baranof Island, Alaska." MS thesis, University of Alaska.

Ydenberg, R. C., and L. M. Dill. 1986. "The Economics of Fleeing from Predators." *Advances in the Study of Behavior* 16:229–49.

Zar, J. H. 1999. *Biostatistical Analysis*. 4th ed. Upper Saddle River, NJ: Prentice Hall.

CONTRIBUTORS

Brad A. Andres started his professional career studying shorebirds in northern Alaska and has worked on questions of shorebird science, conservation, and management since that time. His PhD work focused on the persistent effects of oil spilled by the *Exxon Valdez* in Prince William Sound. His current position is with the U.S. Fish and Wildlife Service as national coordinator of the U.S. Shorebird Conservation Partnership. Current interests include the sustainability of shorebird hunting in the Western Hemisphere, development of flyway-scale approaches to shorebird conservation, and evaluation and monitoring of conservation success.

Chris Beck is a principal and founding partner of Agnew::Beck Consulting with more than 30 years of planning experience. He specializes in land use planning, public participation and facilitation, tourism and recreation, and economic development. He first came to Alaska in 1979, where he worked with the Alaska Department of Natural Resources (ADNR) and helped set up ADNR's regional land-planning process. In 1993, after getting master's degrees in city planning and landscape architecture from UC Berkeley, Chris returned to Alaska and started his own consulting business, specializing in planning for places where people feel particularly passionate about their communities and landscapes, places like Girdwood, Cordova, McCarthy, Prince William Sound, Talkeetna, and Bristol Bay. In 2002, Chris partnered with Thea Agnew to start Agnew::Beck Consulting, which has now grown to more than 20 people and works on projects from health care to physical planning, housing, fundraising, and communication. When not on the job, he tries to find time for the activities he frequently works on in the office: visiting towns that invite exploring on foot, as well as skiing, mountain biking, and kayaking Alaska's stunning coastline.

Nancy Bird worked at the Prince William Sound Science Center from its establishment in 1989 to 2011, serving as president from 2002 to 2011. She previously held various jobs in Cordova, including *Cordova Times* editor, seine fishing crew, college history instructor, and youth counselor.

Dale J. Blahna is a research social scientist with the U.S. Forest Service at the Pacific Northwest Research Station in Seattle, Washington. He has contributed extensively to social science research in outdoor recreation environments over his career. When outside of the office, he is almost always found in the outdoors: skiing, glacier hopping, kayaking Prince William Sound, camping, biking up mountains and down the city's trails, and enjoying the sun, snow, sleet, wind, and rain.

Harold Blehm is a retired public safety officer who has lived in Valdez for 28 years, during which time he has advocated for greater access to Chugach National Forest.

Sara Boario is the assistant regional director of external affairs for the U.S. Fish and Wildlife Service Alaska Region, where she has served since October 2014. She is responsible for strategic communications, government and tribal relations, and partnership efforts. Previously, she worked with the U.S. Forest Service, where she served as the public affairs and partnership staff officer on the Chugach National Forest, the most northern national forest in the country. Sara arrived on the Chugach as a presidential management fellow with the task of guiding community outreach and engagement efforts. In her tenure with the forest, she focused on collaborative stewardship, partnership initiatives, and integrated marketing communications, including managing the forest's effort to develop the Chugach Children's Forest and coleading the "Alaska's Forests: More Than a Place to Visit, It's Where We Live" public education campaign. Prior to joining the Forest Service, Sara served as chief of staff for Alaska state senator Georgianna Lincoln, representing the largest geographical state senate district in the country, encompassing more than 100 rural communities and Alaska Native villages. She holds a BA in history from Lewis and Clark College and a master's of public administration with a dual focus in nonprofit management and community development, from the Hatfield School of Government at Portland State University. Sara was raised in communities across Alaska, where she grew up fishing, hunting, and enjoying the outdoors.

Bridget A. Brown is a GIS and technology program leader for HDR Inc. She has experience leading information management and GIS services integral to the successful delivery of engineering and environmental projects in Alaska, Washington, and the Middle East. Prior to her employment at HDR, she worked as a wildlife biologist, focusing on spatial and temporal relationships of recreation and wildlife in Southcentral Alaska.

Courtney Brown worked in Alaska for seven years for the USDA Forest Service in recreation planning and special use management, as well as conducting social science research at a landscape level. Courtney currently works for U.S. Fish and Wildlife Service refuges in Hawaii, focusing on connecting people to nature, engaging partners in conservation collaboration, and providing opportunities for urban residents to discover, appreciate, and participate in caring for our wild wetlands. She has a BS degree in journalism and geography from the University of Colorado and a master's degree in human dimensions and recreation resource management from Utah State University. Her research and master's thesis focused on visitor management strategies in the era of ecosystem management, a regional approach to management of Alaska's agency-managed bear viewing areas. When not at work, Courtney enjoys spending time with her family on Maui, adventuring in the rainforests and playing in the ocean.

Greg Brown is an associate professor at the University of Queensland and former director of the Environmental Studies Program at Central Washington University. His passion for unique landscapes and cultural diversity has resulted in an adventurous career path around the world. He has held positions as program director for the master's program in environmental studies at Green Mountain College in Vermont; as program director for the Biodiversity, Environmental, and Park Management Program at the University of South Australia in Adelaide; and as associate professor and chair of environmental science at Alaska Pacific University in Anchorage, Alaska. He has also held teaching and research positions in Idaho, Illinois, California, and Arizona. Greg has published in the areas of land use planning, natural resource policy, the human dimensions of ecosystem management, parks and protected areas management, and socioeconomic assessment of rural communities. His current research involves developing methods to expand public involvement in environmental planning by having individuals map spatial measures of landscape values and special places—a type of public participation geographic information systems.

Milo Burcham is a wildlife biologist coordinating the subsistence program for the Chugach National Forest in Cordova, Alaska. Milo has worked for the Chugach Forest for over 15 years. His background has been largely working with ungulates, in particular moose and elk. He earned bachelor's and master's degrees from the University of Montana. Milo is now cooperating with the Alaska Department of Fish and Game, studying the black bear population in Prince William Sound. He is an avid wildlife photographer and enjoys travelling and photographing wildlife in remote parts of the world. His photographs have appeared in *National Geographic* and many other publications, and he has worked with BBC while filming the *Planet Earth* series.

Kristin Carpenter is the executive director of the Copper River Watershed Project, a community-based nonprofit that works to ensure the long-term sustainability of the Copper River watershed's salmon-based economy and culture. She has lived in Cordova for 17 years, working on fish habitat restoration, tourism planning, and other regional issues throughout the Copper River watershed.

Ted Cooney was raised and educated in Helena, Montana, attended the University of Montana-Missoula, and received a doctorate degree in marine science from the University of Washington. He was an outstanding fisheries oceanographer at the University of Alaska Fairbanks. Sadly, Ted passed away on Wednesday, February 13, 2013, but his body of work lives on in his many publications—including his chapter in this book.

Patience Andersen Faulkner is an Eyak Tribal Member who is on the Native Village of Eyak Council and represents the Cordova District Fishermen's United on Prince William Sound Regional Citizens' Advisory Council. She is a legal technician and paralegal, teaches Alaska Native crafts and culture in Cordova and Nuuciq ("Nuchek") Spirit Camp, and is a guest cultural instructor with the Louisiana Bayou tribes impacted by the BP oil spill. She enjoys the labor-intensive activities of basket-making, drum making, sea otter fur sewing, herbal plant processing, and weaving. She can also use an assistant to learn as well as help around, so come on by!

The work in this volume is from **Maryann Smith Fidel**'s thesis for an interdisciplinary master's degree in environmental and social science. She spent six summer seasons in Prince William Sound as a guide and a Kayak Ranger for the USDA Forest Service. The inspiration she gets from spending time out in nature with family and friends has driven her research interests, which focus on the human relationship to the natural world, through individual attitudes and policy. She has spent the past six years working in partnership with Alaska Native communities on collaborative research that explores the effects of climate change to traditional ways of life. She is especially interested in the process of coproducing knowledge and data products that address decision-making needs in environmental policy. She currently lives in her hometown of Anchorage with her husband and husky, and works as an environmental scientist for the Yukon River Inter-Tribal Watershed Council.

Jessica B. Fraver was a graduate student in the School of Natural Resources and Environment at the University of Arizona, and her chapter on river otters was completed as her graduate thesis. Jessica received her master of science with a major in wildlife and fisheries science. Since graduation, Jessica has been working for the Arizona Land Trust. While in Alaska, Jessica was on the field team working in Prince William Sound, collecting human use and river otter data.

Jennifer Gessert has been an Alaskan wilderness guide since 2000 and has also been a sea kayak ranger for both the Tongass National Forest and Chugach National Forest. She currently lives with her partner and son in McCarthy and Anchorage, Alaska.

Randy Gimblett is a professor in the School of Natural Resources and Environment at the University of Arizona. He has been engaged in research work in studying human-landscape interactions and their associated conflicts and public policies related to protection of special environments and environmental experiences for over three decades. He has expertise in the assessment of recreation planning and management, environmental planning and impact assessment, agent-based modeling, visitor use management, and modeling human-landscape systems across space and time at multiple scales. He works closely with communities and stakeholder groups. When not writing, he spends his time camping, hiking, trail running, mountain biking, and kayaking in the beautiful landscapes he loves.

Michael I. Goldstein has a PhD in wildlife ecology. As the planner for the Alaska region of the U.S. Forest Service, Mike loves working at the interface of science and management. Mike also serves as coeditor-in-chief of Elsevier's *Encyclopedia of the Anthropocene* and section editor for "Ecology and Conservation" in Elsevier's online major reference work *Earth Systems and Environmental Sciences.*

Samantha Greenwood has been a resident of Prince William Sound for the last 25 years. She has worked in both federal and local governments as biologist, GIS analyst, and city planner. She considers the Prince William Sound area to be an amazing place to live and work.

Lynn Highland grew up in Nebraska, about as far from an ocean as one can get, but always loved being around what passes for bodies of water in that part of the world. He graduated from Georgia Tech then moved to Alaska in 1978, when oil and money were flowing equally rapidly. He began boating in Prince William Sound in 1979 in a wooden boat that provided a lot of education. He spent many years with only paper charts, a radio, and a depth finder. His first long-range navigation set was a very big deal. Along the way, he acquired a 100-ton captain's license. He has introduced four generations of Highlands to the Sound and will continue boating there as long as he possibly can.

Marybeth Holleman is author of *The Heart of the Sound*, coauthor of *Among Wolves*, and coeditor of *Crosscurrents North*, among others. A Pushcart Prize nominee, her essays, poems, and articles have appeared in such venues as *Orion*, *Sierra*, *Christian Science Monitor*, *North American Review*, *The Future of Nature*, and National Public Radio. Raised in the southern Appalachians, she transplanted to Alaska's Chugach Mountains 30 years ago, after falling head over heels for Prince William Sound. She runs the blog *Art and Nature* at http://artandnatureand.blogspot.com, and her online home is at http://www.marybethholleman.com.

Shay Howlin is a biometrician with Western EcoSystems Technology Inc., a company that provides state-of-the-art statistical consulting in the design, conduct, and analysis of ecological field data. She is a graduate of Oregon State University with a MS in statistics and of the Pennsylvania State University with a BS in wildlife and fisheries science. Shay has worked in wildlife science for over 20 years, starting as an avian field biologist, and has specialized in statistical consulting for over 15 years. Shay has consulted on research projects on a variety of wildlife species and their habitats in Alaska and beyond. She is a coauthor of *Synthesis of Nearshore Recovery Following the 1989* Exxon Valdez *Oil Spill: Sea Otter Liver Pathology and Survival in Western Prince William Sound, 2001–2008*, Exxon Valdez *Oil Spill Restoration Project Final Report*.

Tanya Iden applies her understanding of community development and planning in projects ranging from revitalizing neighborhoods to helping agencies and communities be more efficient, effective, and mission driven. Tanya's training and experience complements her natural skill as a community-minded facilitator, developing a harmonious rapport with diverse project teams and stakeholders. Born and raised in Anchorage, Alaska, Tanya enjoys helping people and communities articulate shared strengths and values and find solutions that maintain Alaska's environmental, cultural, and economic integrity for future generations.

Robert M. Itami is director of GeoDimensions Pty Ltd., in Victoria, Australia. He has international experience in Australia, the United States, and Canada working with government to integrate simulation and spatial modeling into complex planning and management problems in national parks and wilderness areas. He has expertise in computer simulation modeling with over 10 years of experience in vessel traffic simulation. He also consults in outdoor recreation management, especially in the integration of community values and analytical planning techniques.

Lisa Jaeger has been an instructor for the National Outdoor Leadership School since 1987, captain of the Matanuska Susitna Borough Water Rescue Team, engineer and rescue technician for the West Lakes Fire Department, and adjunct faculty at Alaska Pacific University. She is married to Terry Boyle.

Laura A. Kennedy was a graduate student in the School of Natural Resources and Environment at the University of Arizona, and her chapter on harlequin ducks was completed as her graduate thesis. Laura received her master of science with a major in wildlife and fisheries science. While in Alaska, Laura was on the field team working in Prince William Sound collecting human use and harlequin duck data and kayaking in one of the world's most beautiful landscapes.

Spencer Lace has lived in Illinois, Ohio, Arizona, and Utah. He earned a BA in environmental studies from Prescott College and an MS in renewable natural resources from the University of Arizona. In addition to working as a research assistant at Arizona, Spencer has worked for a simulation software startup, nonprofit organizations, and city governments. He currently lives and works in Park City, Utah, where he is the GIS administrator for Park City Municipal Corporation. Park City has a rich silver mining history and rapidly growing housing and tourism markets. Spencer helps this organization expand geospatial technology use in planning, sustainability, transportation, and water utility departments to support a high quality of life for residents, a vibrant economy, and environmental health.

Nancy Lethcoe was born in Seattle in 1940, received her doctorate from the University of Wisconsin, came to Alaska to teach at Alaska Methodist University, and then taught at Stanford University before starting Alaska's first sailing charter business in Whittier in 1974, with her husband, Jim. Nancy and Jim spent winters researching and writing books on the Prince William Sound region.

Currently living in Cordova, **Kate McLaughlin** has been a resident of Prince William Sound since 1998, working as an independent environmental consultant. A certified master class hummingbird bander, Kate runs the Alaska Hummingbird Project Inc., operating the northernmost hummingbird banding stations in the world.

Rosa H. Meehan is a program manager at the Alaska Ocean Observing System and assists with data source identification, integration, and product development. Her emphasis is on further integration of biological resource information that addresses stakeholder needs, including resource and coastal land management issues. Previously, she worked for the U.S. Fish and Wildlife Service, most recently as the Alaska marine mammal program manager.

Christopher Monz, PhD, is professor of recreation resources management in the Department of Environment and Society at Utah State University. His research specialty is recreation ecology, with current research interests focusing on the integration of biophysical science, social science, and park planning approaches. He has conducted over 20 years of research in national parks and other protected areas worldwide. Chris has served as principal investigator or coinvestigator on over four million dollars in funded research from the National Science Foundation, USDA Forest Service, National Park Service, U.S. Fish and Wildlife Service, and the Paul Sarbanes Transit in Parks Program. He has worked extensively in the United States, including Alaska, and internationally in five countries, most recently as a 2014 Fulbright scholar in Norway. He is the coauthor of the textbook *Wildland Recreation: Ecology and Management*, published in 2015.

Karen A. Murphy arrived in Alaska in 1984 to volunteer for the Chugach National Forest just after finishing a bachelor's degree in wildlife biology from Colorado State University—before later completing a master's degree in environmental management at Duke University. In her 32 years in Alaska, she has worked on wetland studies, on the *Exxon Valdez* oil spill restoration plan, and as a district biologist for the Chugach National Forest. She moved to the U.S. Fish & Wildlife Service, where she worked on comprehensive conservation plans for refuges in Alaska, then as the Alaska regional fire ecologist. Her work in fire ecology started her into a climate change career, working with University of Alaska scientists to project changes in fire regimes and patterns under different climate scenarios. In 2010, she joined the fledgling network of landscape conservation cooperatives (LCCs) as the coordinator for the western Alaska LCC. She loves working with the amazing people of western Alaska and many other talented partners from around the state.

Lisa Oakley was in the right place at the right time to work collaboratively with the Chugach National Forest and partners on the creation of Sound Stories in her role as projects director with Alaska Geographic. She's a nonprofit management professional who has focused her work largely on conservation organizations in Alaska in media projects, fundraising, and communications.

Aaron J. Poe has worked in Alaska for 18 years in wildlife biology and landscape planning, as well as partnership development and community engagement. His research and management efforts have largely focused on understanding the relationships between human use and wildlife resources. In his current position, he is focused on building research partnerships to address landscape-level environmental stressors in the Aleutian and Bering Sea Islands region. He has bachelor's degrees in fisheries and wildlife management and geography, specializing in GIS and remote sensing, from Utah State University and a master's degree in natural resource management from the University of Arizona. Away from work, he enjoys spending time in the outdoors with his wife and two young children and being a vocal advocate for youth and for public schools.

Chandra B. Poe has lived, played, and worked in Alaska for 15 years, making memories with her husband and two children in Alaska's magical places, including Prince William Sound. After 10 years working in wetlands mapping and GIS analysis, she is now a business analyst working for Resource Data, Inc., in Anchorage.

Karin Preston was a GIS analyst on the Chugach National Forest during the work reported in chapter 11 of this book. She has a BS degree from Charter Oak State College with a concentration in geography.

Jeremy Robida arrived in Valdez, Alaska, for a summer kayaking guide job in 2006, and like many other Alaskan transplants, the job ended and he stayed. Since 2006, he's been with the Prince William Sound Regional Citizens' Advisory Council, a nonprofit representing those affected by the *Exxon Valdez* oil spill in 1989. The council serves as a conduit between industry, the regulatory community, and citizens, ultimately working to encourage understanding and communication among parties and to provide a citizen perspective and a voice at the table. Jeremy works specifically with monitoring spill response equipment, spill drill exercises, and the general response side of the house. Before arriving in Alaska, he earned his BA in cultural anthropology from the University of Wisconsin-Madison and spent eight years working with wine and cheese at Whole Foods Market after college. His hobbies include backcountry skiing, cycling, photography, and DJing.

Clare M. Ryan is a natural resource policy specialist who looks at the ways that scientific information can be integrated into policy and management decisions. She studies the processes by which policies are made and implemented, ways to foster collaboration in management, and how to address conflicts that can arise when multiple stakeholders participate in decision-making. Before returning to academia, she spent several years as a scientist and manager at both the state and federal levels. As a reflection of her wide range of expertise, she was the director of the Program on the Environment from July 2011 to July 2016 and holds appointments in the School of Environmental and Forest Sciences, Daniel J. Evans School of Public Policy and Governance, School of Marine and Environmental Affairs, and School of Law.

Gerry Sanger is a retired biologist who specialized in Alaskan marine birds, ending his career with post–*Exxon Valdez* oil spill bird surveys in Prince William Sound. For the past several years, he has been captain and guide on small boat nature tours in the Sound.

Born in Bridgeport, Connecticut, nature writer and author **Bill Sherwonit** has called Alaska home since 1982. He has contributed essays and articles to a wide variety of newspapers, magazines, journals, and anthologies and is the author of more than a dozen books. His most recent books include *Animal Stories: Encounters with Alaska's Wildlife*; *Changing Paths: Travels and Meditations in Alaska's Arctic Wilderness*; and *Living with Wildness: An Alaskan Odyssey*. Sherwonit's work primarily focuses on Alaska's wildlife and wildlands, but he's passionate about wild nature in all its varied forms, including the nature of his adopted hometown, Anchorage, and the spirited wildness we carry within us.

Lowell H. Suring joined Northern Ecologic LLC in 2009 after serving as a wildlife ecologist with the USDA Forest Service at several locations in Minnesota, Alaska, and Idaho since 1980. He also held permanent positions with the U.S. Fish and Wildlife Service in New Mexico and with the New York State Department of Environmental Conservation. Lowell is a certified wildlife biologist and is active in the Wildlife Society at state chapter, section, and international levels. His professional interests include the development and implementation of analytical techniques and tools that may be used to evaluate the capability of habitats to support wildlife and the effects of land management activities on habitat capability. He has had extensive experience in development and implementation of techniques for broad-scale habitat assessment, landscape planning to maintain species of conservation concern, development and implementation of habitat capability models, assessment of habitat use patterns of species of concern, integration of wildlife science into natural resource planning and management, and management and recovery of threatened and endangered species.

Paul Twardock is the author of *Kayaking and Camping in Prince William Sound*. He is a professor of outdoor studies at Alaska Pacific University in Anchorage, Alaska, and studies human use and impacts in Prince William Sound. He has spent many days—perhaps even years—paddling and exploring his way around Prince William Sound.

As a field biologist, **Sarah Warnock** followed shorebirds from their breeding grounds in Alaska to wintering areas in California and beyond for more than 10 years. She is also a science educator, working to develop and implement programs that bring kids and science together in the out of doors.

Sadie Youngstrom is a dedicated and engaging biology field research professional with expertise in marine wildlife and experience that includes extensive data collection and analysis spanning tropical and cold/arctic environments, and salt- and freshwater ecosystems. She is passionate about studying animal behavior and promoting animal welfare. Sadie has contributed to various organizations and causes, including but not limited to humpback whale, Steller's sea lion, harbor and Hawaiian monk seal population analysis, seabird monitoring, salmonid habitat and abundance, vessel operation, public training, and education. She has held a U.S. Coast Guard captain's license since 2002 and currently is endorsed as a master up to 100 tons in near coastal waters. Sadie earned her BS in marine biology and minored in environmental science from Alaska Pacific University, Anchorage, Alaska, in 2008 (her thesis was titled "Monitoring Human Disturbance on Molting Harlequin Ducks [*Histrionicus histrionicus*] and Common Mergansers [*Mergus merganser*] in Prince William Sound, Alaska: Effects of Motorized Vessels and Sea Kayaks"). In her off time, Sadie explores the wonders that planet Earth has to offer.

INDEX

Note: Page numbers in *italic* represent illustrations.